D0966330

Correspondence Analysis
in the
Social Sciences

Correspondence Analysis in the Social Sciences

Recent Developments and Applications

edited by

Michael Greenacre
University of South Africa
Pretoria, South Africa

and

Jörg Blasius
University of Cologne
Germany

ACADEMIC PRESS

San Diego London Boston New York Sydney Tokyo Toronto

ACADEMIC PRESS
a division of Harcourt Brace & Company
525 B Street, Suite 1900, San Diego, CA 92101-4495, USA

Academic Press Limited
24-28 Oval Road, London NW1 7DX, UK

International Standard Book Number: 0-12-104570-6

PRINTED IN THE UNITED STATES OF AMERICA
99 00 01 02 03 IBT 11 10 9 8 7 6 5 4 3 2

Contents

Preface

In recent years a new technique has become more and more attractive in social science statistics: correspondence analysis. Correspondence analysis is a multivariate method for exploring cross-tabular data by converting such tables into graphical displays, called 'maps', and related numerical statistics. Since cross-tabulations are so often produced in the course of social science research, correspondence analysis is valuable in understanding the information contained in these tables.

In the last few years the number of theoretical publications as well as published empirical applications of correspondence analysis have increased enormously. As is often the case with the introduction of new technology, a large gap exists between the theory and practice of the method. Many useful aspects of correspondence analysis are described in the statistical literature but these may be inaccessible to social scientists. On the other hand, computer software to perform correspondence analysis is freely available so that it is increasingly possible to apply the method. There is a continuing need for a transfer of the statistical theory to applications in the social sciences. This book will assist this transfer process on the one hand by presenting different formal aspects of the method with a minimum of technical details, and on the other hand by providing a wide range of applications to different kinds of data in the social sciences. The aim of the examples is twofold: first to provide the statistician with substantive material which can be used for teaching and for the transfer to empirical applications, and second to give social science researchers sufficient illustrations of the method for adapting correspondence analysis to their problems.

When we say that correspondence analysis is an exploratory technique we mean that it is primarily intended to reveal features in the data rather than to confirm or reject hypotheses about the underlying processes which generate the data. In order to explore data we need to make as few assumptions about the data as possible. In correspondence analysis there are no assumptions about the underlying distribution of the data, but there are a few 'structures' imposed on the data which are inherent in the method and which could be viewed as assumptions. Nevertheless, the technique is so flexible that it can be applied to almost any cross-tabulation as well as a number of other types of data where these structures can be justified. Thus, the usual process in statistical data analysis of 'checking the assumptions' is replaced in the present context by a substantive justification of the graphical elements which form the basis of the correspondence analysis maps.

This way of describing data via visualization includes a particular kind of thinking which is typical of most French social scientists and for many statisticians in France. This philosophy is associated with a famous quotation of Benzécri, one of the inventors of the method: 'The model must follow the data, and not the other way around'. This was Benzécri's radical way of expressing the importance of the data and of the features in the data 'revealing themselves' rather than limiting one's vision of the data to restrictive and subjectively defined statistical models.

Jean-Paul Benzécri was the one who, from the early 1960s, popularized the graphical interpretation of correspondence analysis, almost exclusively in France, leading to the publication of his voluminous treatise on *'Analyse des Données'* in 1973. The algebraic formulation of the method is recognized to date back to a paper in 1935 by Hirschfeld (later known as Hartley) on the 'correlation' between the rows and columns of a contingency table. R. A. Fisher and Louis Guttman independently published articles in the early 1940s which essentially redefined correspondence analysis, with Guttman's work being the closest in spirit to the subsequent work of Benzécri. Chikio Hayashi also developed these ideas in Japan as an improvement of scaling techniques to achieve what he called the 'quantification of qualitative data'. Nevertheless, outside France it was only in the middle of the 1980s when correspondence analysis became known by social scientists in the English-speaking world, especially market researchers. This growth of interest in correspondence analysis was assisted by the publication of the first books in English by Greenacre in 1984 and by Lebart, Morineau and Warwick in the same year.

The first published applications of correspondence analysis in the social sciences can be traced back to the 1970s in France where the method had become the most applied multivariate technique. Even a journal, *Les Cahiers de l'Analyse des Données*, edited and mostly written by Benzécri, was dedicated for the most part to correspondence analysis and Benzécri's philosophy of data analysis. One of the most famous applications in the social sciences is given by the French sociologist Pierre Bourdieu who used this method in his book *La Distinction*, published in 1979. This work as well as other widely read books and articles by Bourdieu – some of which are translated into many languages – and the inclusion of the method in the major statistical packages such as BMDP in 1988, SAS in 1990 and SPSS in 1990, have led to an increasing number of applications appearing in Europe and in the United States.

Correspondence analysis can be subdivided into two approaches: simple correspondence analysis (in the following abbreviated as CA), and multiple correspondence analysis (MCA). Input data to CA is usually a matrix of raw frequencies of one or more contingency tables. In this context there is usually a specific *variable to be described*, for example we could consider the variable 'preferred leisure activities' with categories such as 'playing chess', 'watching television' and 'reading'. The number of categories can be quite high in such a case, say 15 to 20 activities. On the other hand there may be one or more *describing variables* which might have an association with the 'preferred leisure activities' of the respondents, variables such as 'education', 'age group', 'income group' or even a variable such as 'preferred leisure activities of the partner', with the same categories 'playing chess', 'watching television', etc. Each of the describing variables is cross-tabulated with the variable to be described in order to investigate the importance of this association. Usually, a chi-squared statistic (or some other measure of association) is computed to measure the statistical significance of this association. Such tests lead to the following type of result: that education, for example, is significantly associated with preferred leisure activities. The nature of this association is usually investigated informally by comparing the relative frequencies of each education group across the leisure activities.

Correspondence analysis can be used to investigate both the magnitude and the substantive nature of the association between the row and column categories of the cross-tabulation. Correspondence analysis visually depicts the education groups, for example the rows of the cross-tabulation, as a scatter of points in the map, as well as the leisure activities, the columns. More specific conclusions about the association between categories of the two variables can now be made, for example that 'highly educated groups have a strong association with playing chess'. Several cross-tabulations with respect to the same variable to be described can be analysed jointly by concatenating, or 'stacking', them before applying CA. This allows several relationships to be visualized simultaneously in the same map.

In the multiple case, there are several categorical variables which all have the same status, for example they are all questions on a particular theme such as political interests, either all variables to be described or all describing variables. This is a situation reminiscent of principal components analysis of measurement data, where we are interested in exploring the structure of association amongst a set of variables by identifying underlying dimensions. Multiple correspondence analysis is the application of the CA algorithm to the matrix of intervariable associations which is constructed from all the pairwise cross-tabulations of the variables. This matrix, often referred to as the Burt matrix, is the analogue of the covariance matrix. In this sense the only difference between CA and MCA appears to be the distinction in CA between the variables to be described and describing variables, in which case the cross-tabulations between these two types of variable are analyzed, whereas in MCA there is no such distinction and all cross-tabulations among the single group of variables are analysed simultaneously. Since there is usually a distinction amongst the variables of the type described above, it seems that there are fewer situations in the social sciences where MCA would be applicable.

There is, however, an additional aspect to MCA which distinguishes it from simple CA: MCA can analyse respondent-level data, that is a matrix in which the rows are the respondents (or some objects of interest such as trade unions) and the columns the categories of response. In such a matrix, usually called an indicator matrix, each row is a description of a particular respondent so that it is possible to interpret the positions of single respondents in the display and compare them with one another. The rows of the indicator matrix can be thought of as the variable to be described and the columns as the describing variables. A sociological application of MCA is given by Bourdieu (1984) in which he describes professors by variables such as 'membership in faculty', 'number of children' and 'occupation of father'. It makes sense in such an application to interpret the spatial position in the display of each professor relative to the describing variables. All these persons are well-known to the (French) community (one of the subjects is Bourdieu himself, others are well-known persons such as Fourastié and Lévi-Strauss). Another application of MCA would be to describe politicians of a parliament by their attitudes as measured via their ballots (with alternatives 'yes' and 'no') on various issues. Or one could distinguish important groups in society, for example the trade unions, by doing a content analysis collected from newspapers to compare their public statements toward issues such as nuclear power ('for' or 'against') and several strikes (to 'support' or 'prevent' them).

The first part of the book runs under the heading 'General introduction' and introduces the reader to the field of simple correspondence analysis. In the first chapter Greenacre gives an introduction to the geometric interpretation and uses small contingency tables to illustrate the three basic concepts of 'profile', 'mass' and 'chi-squared distance'. These are the graphical structures which essentially define correspondence analysis. By using as few statistical details as possible, he explains how one can obtain a graphical display of the rows and columns of a table by using simple geometric ideas. In doing so he demonstrates the very important difference between two types of maps available in CA: symmetric and asymmetric maps. Both are discussed and the reader will receive many specific instructions on how to interpret these maps.

In the second chapter, Blasius gives examples of CA in the social sciences. In addition to explaining the graphical displays he focuses on the numerical output and the interpretation of the so-called 'contributions'. He starts off with a light-hearted example on 'cultural competences' in a community such as the ability to 'play chess' or 'dance to pop music'. By using a combined variable of 'sex' and 'age' as the describing variable, he obtains the two-dimensional solution which was expected almost by definition. In an example such as this where the results are so clear and unambiguous, it is easier to explain and interpret the various numerical coefficients associated with CA output. A further aim of this article is to show what happens when further describing variables are added which have no relationship to 'cultural competence'. In an additional example Blasius describes inter- and intra-regional mobility, using data from the Frankfurter Nationalversammlung which was the first democratic parliament in Germany.

In Chapter 3, Blasius and Greenacre give a step-by-step explanation of the computations involved in performing a CA. While most readers are prepared to accept the results of whatever computer program they prefer to perform CA, this chapter is valuable for two reasons. First, there might be readers who have programmed or would like to program their own algorithms who can refer to this chapter for the computational steps and to check their own calculations. Second, by laying out all the calculations for a specific example and showing numerically the values of all the profiles, masses, distances and coordinates, this chapter serves to demystify the technique as well as the algebra which is shown along with the numerical computations.

Chapter 4, written jointly by van der Heijden, Mooijaart and Takane, focuses on the modelling of contingency tables. These authors have 'some difficulty in calling correspondence analysis model-free' and introduce another definition of a model: 'a model is a nonlinear projection of the data on a (...) parameter space'. This nonlinear projection can be optimized by criteria such as (generalized) least squares or maximum likelihood. In addition to the maximum likelihood solution of CA, they show the close relations between CA and latent class analysis, log-linear models, log-bilinear models and ideal point discriminant analysis. These closely related techniques are illustrated using data from elections in the Netherlands where the input data are the frequencies of a contingency table cross-tabulating 'choice of political party' by 'type of city'.

Chapter 5, written by Böckenholt and Takane, starts with the idea that constraints will be needed to search for meaningful patterns. As an example of linear constraints they introduce the equal spacing of categories on the same dimensions, which leads to an easily interpretable solution of the results. Secondly, Böckenholt and Takane partial out the effects of a concomitant variable from the CA solution, and finally they discuss what will happen when partialing out the effects of a subset of variables and then relating the residual information to the association structure in the table.

Chapter 6 is the last paper in this part and documents the French perspective of correspondence analysis as seen by members of the Bulletin of Sociological

Methodology (Karl van Meter, Philippe Cibois, Lise Mounier, and Marie-Ange Schiltz). Starting with the work and investigation of Benzécri the authors distinguish three periods of 'French Data Analysis' (*Analyse des Données*). Background information and literature are provided to clarify the way of thinking of many French statisticians and in particular the majority of French social scientists. The authors conclude the paper with a comparison of the pros and cons of 'French Data Analysis' against the traditional modeling approach.

The second part of the book deals with the analysis of multivariate categorical data. It starts with Greenacre's introduction of joint and multiple correspondence analysis (Chapter 7). As a first step, he considers the CA of a concatenated set of two-way tables which have a common variable, usually the variable to be described, and shows that this leads to a map which is the 'average' map of the CAs of the individual tables. This idea is then used to generalize CA to the multiple case, by jointly analyzing all pairwise cross-tabulations of a set of variables. The advantages of this approach, called 'joint correspondence analysis', are discussed and compared with multiple correspondence analysis in a context of a data set from a socioeconomic survey on living conditions.

Chapter 8 by Lebart shows the complementary use of correspondence analysis and cluster analysis. Lebart starts with a small artificial symmetric matrix where there is an exact relationship between the principal axes of CA and the nodes of a cluster analysis. He shows that the two clusters which are merged last by the cluster analysis determine the strongest contrast in CA, loading on opposite sides of the first principal axis. The example is chosen specifically to demonstrate the advantages of clustering over CA and to give credence to the idea that both methods can be used in tandem to explore real data. The joint use of both techniques is illustrated on a data set from the 'National Survey about the Aspirations and Living Conditions of the French'. Clusters of respondents can be depicted on a CA map as well as the positions of supplementary (or 'illustrative') socioeconomic categories, leading to a very rich description of the survey data.

Chapter 9 by Heiser and Meulman deals with homogeneity analysis, the approach to multiple correspondence analysis which is included in the CATEGORIES module of SPSS under the name HOMALS. The authors use the term 'homogeneity analysis' as a collective name for a group of exploratory techniques for nonlinear multivariate analysis. Their chapter starts by emphasizing the role of homogeneity analysis in the study of reliability of measurement. A set of categorical variables is called homogeneous if they have the same 'center', that is they are all the same up to a certain class of information-preserving transformations. Loss of homogeneity is defined as a variance which measures deviations of optimally transformed variables from their center. Although the basic ideas are different from correspondence analysis, the solutions of homogeneity analysis are the same as those of multiple correspondence analysis. Heiser and Meulman also discuss how to analyze metric variables using the same approach. Various techniques are illustrated in an example of a Dutch survey of opinions on abortion and sexual freedom.

The final chapter in this part, given by Rovan, describes different kinds of visualizations in more than two dimensions. Starting with an example of a Burt matrix, he presents the two-dimensional map and, using the numerical diagnostics, he shows that some of the variables are represented poorly. In theory a three-dimensional map would improve the situation, but such maps are not simple to interpret. As an alternative, Rovan proposes using Andrews' curves which allow the visualizing of points in higher-dimensional spaces and illustrates their application to the same example. Advantages and disadvantages of these approaches are discussed. As an empirical example he uses the same data set as Greenacre in Chapter 7.

The third part of the book contains three articles on the analysis of longitudinal data, applying correspondence analysis to event history data, panel data and trend data respectively. The first paper of this part (Chapter 11) by Martens compares correspondence analysis and cluster analysis on event history data on job mobility. In traditional event history analysis one has a categorical variable, for example 'family status', and changes in the categories of this variable (e.g., from 'single' to 'married') are described over time by a number of variables such as 'occupational status' and 'income' which are observed with 'family status' at the same time points. Martens analyzes job mobility over 10 years with an interval of one year between observations by coding the data in indicator matrix form suitable for an MCA. As variables he uses the number of employees in the companies and different departments within the companies.

Chapter 12 by Thiessen, Rohlinger and Blasius, deals with the analysis of panel data. As an empirical example they use the division of labour in young family households among 13 tasks such as 'laundry', 'preparing main meals' and 'driving the car'. The object of their study is to show how these household tasks have changed over the first four years after marriage. This is an example of a three-wave panel and each couple in the study was asked which tasks were his job, her job or whether they did it jointly or alternately. Using the first time point as reference frame to determine the dimensions of the CA map, the data at the other time points were projected onto the map as supplementary points. In addition, for the second and third time points the sample is subdivided into two groups, according to whether the wife keeps her job or not. The use of numerical contributions to inertia is again illustrated here, both for the active and supplementary points.

The third paper in this section, Chapter 13 by Müller-Schneider, deals with the analysis of trend data. Changes in the readership of different newspapers and participation in several leisure activities between the years 1953 and 1987 are studied. The problem with these data is that, besides the social changes there are also structural changes in the whole population which should have no influence: for example, the increasing educational level. Müller-Schneider proposes applying correspondence analysis to what he calls 'historic profiles'. This study illustrates how data can be coded to allow comparison of different studies over time. For describing the changes in readership and leisure activities, he also uses supplementary points and then discusses the issue of which time point should be used as a reference frame for establishing the CA map.

The last part of the book is reserved for further applications of correspondence analysis in the social sciences. Here we selected applications which give an idea of the versatility of the method and its interpretation. In Chapter 14 Giegler and Klein use CA for analyzing textual data from personal advertisements in different newspapers. The paper starts by describing how the data are prepared using a content analysis of the advertisements. A number of categories are established which serve as descriptors of the text, which in our description above constitute the variable to be described. In addition there are three categorical variables which serve as the describing variables. In this case the three variables are not concatenated but are considered in all possible combinations, making it possible to visualize all the interaction patterns. The data consist of the frequencies of the content categories for each of these combinations. Using CA leads to a visual description which shows differences between the newspapers, how males differ from females in their wording of the advertisements for each of the newspapers, and how the advertisements differ when people describe themselves, their desired partner or desired relationship.

The next application, Chapter 15 by Wuggenig and Mnich, deals with the description of lifestyles. While Bourdieu developed a questionnaire for analyzing different kinds of lifestyles, Wuggenig and Mnich interpret lifestyles by using a new survey technique called 'photo-questioning'. The method is based on the idea that research subjects should play an active role in the research. Respondents were asked to take eight photographs in their homes, four of objects or parts of rooms which they liked and four which they disliked. A total of 730 objects appearing in the photographs were eventually analyzed, and these were classified into 37 groups, for example 'pictures', 'musical instruments', 'beds' and 'toilets'. By using CA, typical life styles of different 'class fractions' (i.e. groups defined according to occupation), age groups and genders can be described.

The final paper, Chapter 16 by Snelders and Stokmans, describes a study of several telephones according to attributes such as 'big', 'horizontal receiver', 'kitsch' and 'elegant'. These attributes are suggested by the respondents during a process called 'natural grouping' of the telephones. A matrix is then constructed from the frequencies of each attribute associated with each product and analyzed by CA to obtain a 'perceptual map' of the telephones and attributes. In the second part of their paper, Snelders and Stokmans show how CA can be used to display results of a conjoint analysis in which the importances of each attribute are quantified in a decision for or against the purchase of a telephone. They define four homogeneous groups of consumers and then construct a three-way matrix of groups by attributes by importance rankings, which is then converted into a two-way matrix by using all combinations of groups and attributes as the rows of the table. Correspondence analysis is then used to visualize the gradient of preferences in the group/attribute combinations.

We hope that the given selection of papers provides all readers of this book with ideas which are useful for their own research. We have tried to maintain a balance between statistical details and social science applications. On the one hand, the reader interested mainly in the statistical details will be able to grasp the important issues, and then refer to the additional references for further details. On the other hand, for the reader more orientated to applications, the statistical details should be enough, and further details would probably be confusing rather than convincing. However, there has to be a balance in the amount of social science theory too. In this respect we tried to find a good median between statistical details and social science theory. We had much fun in searching for this median and we hope that you will like it.

Acknowledgements

The idea of producing a book on correspondence analysis specifically aimed at social scientists was conceived after the conference 'Recent Developments and Applications in Correspondence Analysis' in 1991 held at the *Zentralarchiv für empirische Sozialforschung* of the University of Cologne. This was the first international conference devoted exclusively to correspondence analysis, and included a short course on correspondence analysis presented by Michael Greenacre and two days of invited and contributed papers on applications of correspondence analysis in the social sciences. We would especially like to thank Erwin K. Scheuch and Ekkehard Mochmann from the *Zentralarchiv* for their generous support of this conference and for inviting Michael Greenacre to Cologne on several occasions.

This book has been compiled from two main sources. Different aspects of the methodology of correspondence analysis are presented by various researchers who have been involved with the method for many years. To complement the methodological section, the editors have invited a range of application papers from the participants of the Cologne conference. The book has been edited to give a self-contained and balanced perspective of the theory and practice of correspondence analysis in the social sciences.

The 'father' of this project is Walter Kristof, of the Institute for Sociology, University of Hamburg. Walter's enthusiasm for correspondence analysis has been the latent factor in inspiring his former student Jörg Blasius' interest in the method, in bringing Michael Greenacre to Hamburg and Cologne and in setting up the very successful conference mentioned above. Without Walter, this book would not have come to fruition. To the contributors who have repeatedly revised their manuscripts so patiently we thank them for their hard work and hope that the final product is pleasing to all.

Many other people have contributed directly or indirectly to this book. We would like to thank: Friederika Priemer of the *Zentralarchiv* for retyping and proof-reading the papers as well as preparing the book's bibliography; François Steffens, head of Department of Statistics of the University of South Africa, who made it possible to invite Jörg Blasius to South Africa; Dan Bradu of the University of South Africa for his detailed commentary on earlier parts of the manuscript; and our students Jacqueline Berg, Udo Dillman, Gabriele

Franzmann and Wolfgang Schäfer for their assistance during the conference and in preparing the book. And final thanks go to Academic Press who have supported this project and have accommodated our needs as researchers to bring you this book.

Michael Greenacre and Jörg Blasius
October 1993

About the Authors

Jörg Blasius works at the Zentralarchiv für empirische Sozialforschung, University of Cologne. His primary research interests are multivariate explorative analyses, urban research, life styles, and methods of empirical social research.

Address: Zentralarchiv für empirische Sozialforschung, Universität zu Köln, Bachemer Str. 40, 50931 Köln, Germany.

The BMS (Bulletin de Méthodologie Sociologique or Bulletin of Sociological Methodology) is a quarterly scientific journal founded in 1983 and publishes in English and French. The Board of Editors are Karl M. van Meter, Marie-Ange Schiltz, Philippe Cibois and Lise Mounier. Van Meter, head of the journal, is a member of the Laboratory of Informatics for Human Sciences (LISH) of the French National Center for Scientific Research (CNRS). Schiltz is a member of the Center for Application of Social Mathematics (CAMS) of the CNRS. Cibois is a member of the René Descartes-Paris V University. Mounier is a member of the Laboratory of Mathematical and Statistical Analysis and Secondary Analysis (LASMAS) of the CNRS.

Address: French National Center of Scientific Research, LISH-CNRS, 54 boulevard Raspail, 75006 Paris, France.

Ulf Böckenholt is Associate Professor at the Department of Psychology, University of Illinois at Urbana-Champaign. His primary research interests are in the analysis of multivariate categorical data and mathematical models of judgment and choice.

Address: Department of Psychology, University of Illinois at Urbana-Champaign, 603 E. Daniel St., Champaign, IL 61820, USA.

Helmut Giegler is associate professor at the Institute for Sociology at the University in Erlangen-Nürnberg. His main research fields are quantitative and qualitative social research, social structure analysis, life styles, leisure research, and sociology of the mass media.

Address: Institut für Soziologie, Universität Erlangen-Nürnberg, Kochstr. 4, 91054 Erlangen, Germany.

Michael Greenacre is professor of Statistics at the University of South Africa, Pretoria, where he runs a statistical consultation service. His primary research interest is in multivariate data analysis, particularly the application of correspondence analysis to social science and environmental data. He has published two books on correspondence analysis.

Address: Department of Statistics, University of South Africa, P.O. Box 392, Pretoria 0001, South Africa.

Willem Heiser is professor of Data Theory in the Faculty of Social Sciences, Leiden University. His research areas are multidimensional scaling and unfolding, nonlinear multivariate analysis, and classification. Recent work includes robust/resistant methods for multivariate analysis, analysis of asymmetric association tables, and constrained latent class analysis.

Address: Department of Data Theory, University of Leiden, P.O.Box 9555, 2300 RB Leiden, The Netherlands.

Harald Klein works at the Institute for Sociology at the Friedrich Schiller University in Jena. His main research fields are empirical methods, especially computer-aided content analysis, the impact of mass media, sociology of media, and mass communication research.

Address: Institut für Soziologie, Friedrich-Schiller Universität, Otto-Schott-Str. 41, 07740 Jena, Germany.

Ludovic Lebart is Head of Research at the Centre National de la Recherche Scientifique in Paris. His research interests include the statistical analysis of qualitative and textual data, with emphasis on large data sets, algorithms and methods, problems of validation, survey data processing and survey methodology. He has co-authored several books on statistical data analysis.

Address: Ecole Nationale Supérieure des Télécommunications, 46 rue Barrault, 75013 Paris, France.

Bernd Martens works in the Department of Sociology at the University of Tübingen. He has a particular interest in sociological data analysis and his recent work is connected with sociology of technology and environmental issues, including monitoring of biotechnological trends and social problems of waste management.

Address: Institut für Soziologie, Universität Tübingen, Wilhelmstr. 36, 72074 Tübingen, Germany.

Jacqueline Meulman is associate professor of Data Theory in the Faculty of Social Sciences, Leiden University. Her research areas are nonlinear multi-variate analysis and multidimensional scaling, and in particular the integration of these two classes of multidimensional data analysis techniques.

Address: Department of Data Theory, University of Leiden, P.O.B. 9555, 2300 RB Leiden, The Netherlands.

Peter Mnich is a research assistant for Methods of Empirical Research at the University of Lüneburg. His research interests include empirical sociology of politics as well as cohort and network analysis.
Address: Universität Lüneburg, 21332 Lüneburg, Germany.

Ab Mooijaart is Associate Professor in Psychometric Research Methods at the Department of Psychology of the University of Leiden. His main research interests are structural equation modelling and modelling of contingency tables.
Address: Department of Psychometrics and Research Methods, University of Leiden, P.O. Box 9555, 2300 RA Leiden, The Netherlands.

Thomas Müller-Schneider works at the University of Bamberg. His main research interests are methods of empirical social research, social structure analysis, and sociology of culture.
Address: Otto Friedrich Universität Bamberg, Fakultät Sozial- und Wirtschaftwissenschaften Kapuzinerstr. 16, 96047 Bamberg, Germany.

Harald Rohlinger works at the Zentralarchiv für empirische Sozialforschung, University of Cologne. His main interest focuses on historical social research and on instruments for exploratory data analysis, especially smallest space analysis and correspondence analysis.
Address: Zentralarchiv für empirische Sozialforschung, Universität zu Köln, Bachemer Str. 40, 50931 Köln, Germany.

Jože Rovan is assistant professor of Statistics at the Faculty of Economics, University of Ljubljana. He is primarily interested in multivariate analysis, particularly in the dimension-reduction techniques and the graphical representation of their solutions. His present research work is mainly devoted to applications of correspondence analysis in the economic sciences and the comparison of correspondence analysis and cluster analysis solutions.
Address: Ekonomska fakulteta, Univerza v Ljubljana, Kardeljeva ploščad 17, 61109 Ljubljana, Slovenia.

Dirk Snelders is a Ph.D. student at the Department of New Product Development, School of Industrial Design Engineering, Delft University of Technology. He has carried out research on the lay understanding of money, the effect of price-cuts on brand image, and the effect of payment on work motivation. His current research interest is in the role of emotion in consumer judgment and the resulting subjectivity in consumer response.

Address: Department of New Product Development, School of Industrial Design Engineering, Delft University of Technology, Jaffalaan 9, 2628 BX Delft, The Netherlands.

Mia Stokmans is assistant professor at the Department of Language and Literature at Tilburg University. Her current research interests concern the validity of conjoint analysis in relation to information search pattern during consumer decision-making processes.

Address: Department of Language and Literature, Tilburg University, P.O.B. 90153, 5000 LE Tilburg, The Netherlands.

Yoshio Takane is professor of Psychology at McGill University in Montreal. His recent interests include constrained principal component analysis/ correspondence analysis, latent variable models, statistical expert systems, and statistical bases of connectionist models.

Address: Department of Psychology, McGill University, Stewart Biological Sciences Building, 1205 Dr Penfield Avenue, Montreal QC, Canada H3A 1B1.

Victor Thiessen is professor of Sociology and Social Anthropology at Dalhousie University, Halifax. His recent book *Arguing with Numbers* (1993) focuses on sources of methodological artefacts in social research. His main interest is in interpersonal dynamics within families and he is currently conducting research on adolescents' images of work in the context of their parents' work experience.

Address: Department of Sociology and Social Anthropology, Dalhousie University, Halifax, Nova Scotia, Canada.

Peter G. M. van der Heijden is professor of Statistics for the Faculty of Social Sciences at Utrecht University. His main research interests are exploratory and modeling approaches to the analysis of categorical data, and the analysis of compositional data.

Address: Department of Methodology and Statistics, Utrecht University, 3508 TC Utrecht, The Netherlands.

Ulf Wuggenig is associate professor of Cultural Sociology and the Sociology of Arts at the University of Lüneburg. His research areas include general sociology, sociology of inequality and life styles, and the study of visual arts and media from a sociological point of view.

Address: Universität Lüneburg, 21332 Lüneburg, Germany.

Part 1

General Introduction

1

Correspondence Analysis and its Interpretation

Michael Greenacre

1.1 INTRODUCTION

The primary goal of correspondence analysis is to transform a table of numerical information into a graphical display, facilitating the interpretation of this information. This goal is shared by such familiar graphical techniques as histograms, box-plots, star diagrams and various types of scattergrams. The aim of all these methods is to communicate numerical information by expressing it in a different form. These techniques are all exploratory in the sense that they describe, rather than analyze, the data.

In social science research results are very often summarized as a simple frequency count of a set of response options, or categories. Graphical displays of such a result are often found in the mass media in the form of bar-charts or pie-charts. Such displays represent the data exactly and can vary only according to the graphic artist's creativity and stylistic use of graphical tools such as colour and texture. Another common result found in the social sciences is a two-way cross-tabulation, usually referred to as a *contingency table*. This table also contains frequency counts but is one level of complexity greater in that it breaks down the counts of a set of responses according to another set of categories. This other set of categories often constitute a biographical or demographical variable of interest, such as age group or nationality (these are called 'describing variables' in the Preface). Such a two-way table of data can be displayed as a series of bar-charts, one for each age group, for example, or as a three-dimensional histogram. Correspondence analysis, however, approaches the display of such tables in a different and unique way.

Figures 1.1–1.3 show three different cross-tabulations as well as their displays by correspondence analysis, which are called 'maps'. The first data set is a cross-tabulation of the responses to the question 'Are civil rights people pushing too fast?' according to the age groups of the respondents. There are four response options 'too fast', 'about right', 'too slow' and 'don't know', and seven age groups. The map in Figure 1.1 shows the youngest age group on the left, then a big jump to the next set of three age groups, a smaller jump to the fifth age group and then to the sixth, and finally a large jump to the oldest age group. The age groups thus line up in order, but there are different intervals separating one from the next. Looking at the response options now, we see that the answers 'about right' and 'too slow' are on the left, 'too fast' is in the middle, and 'don't know' is on the right. In a nutshell, without having explained how correspondence analysis functions, we can interpret this map as indicating the association of the age groups with the response categories: on the one hand there is a gradient of response which is correlated with age, with the younger ages being associated with the response that 'civil rights people are pushing too slow or about right' and the older ages expressing the more conservative opinion of 'too fast'; in addition to this pattern, it appears that the 'don't know' responses are predominant in the oldest age group.

The next example in Figure 1.2 is taken from two surveys in consecutive years on preferences of TV programs. In this case the response options are the types of TV show: 'Variety', 'Comedy', 'Situation comedy', 'Drama' and 'Western'. In each survey the responses are cross-tabulated against sex and the two tables are concatenated by stacking in order to produce a single table. Looking at the map in Figure 1.2 we can come to the following conclusions about the data in the stacked table. The horizontal dimension in the map separates the females on the left from the males on the right, showing for example that males have a preference (compared to women) for 'Westerns'. The vertical dimension separates out the successive surveys, with the first survey in the bottom half of the display and the second survey on top, and we have used dotted lines to link the points from year to year. Since the program types 'Situation comedy' and 'Comedy' are on top, it looks like the trend for both sexes has been towards comedy programs. It seems as if females have moved relatively strongly away from 'Drama' programs and males away from 'Westerns'.

The third example in Figure 1.3 is concerned with the responses to the question 'How is the UN doing in solving the problems it has had to face?' and the response options are on a four-point scale of 'very good job', 'good job', 'poor job', 'very poor job', as well as the option 'don't know'. The responses are now cross-tabulated by nationality of the respondent, either from (West) Germany, Britain, France, Japan or the United States. In this case the table is in the form of percentages, where each row adds up to 100%.

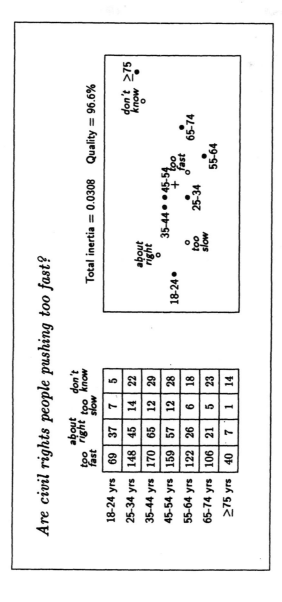

FIGURE 1.1   Example 1: A 7 × 4 table of frequencies from a social attitudes survey together with the two-dimensional symmetric CA map and the amount of inertia accounted for by the map.

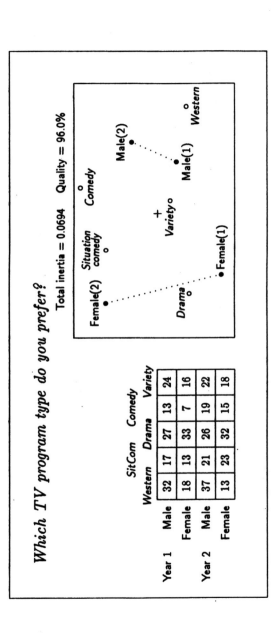

FIGURE 1.2   Example 2: Two stacked 5 × 5 tables of frequencies from two television surveys in consecutive years, together with the two-dimensional symmetric CA map and the amount of inertia accounted for by the map.

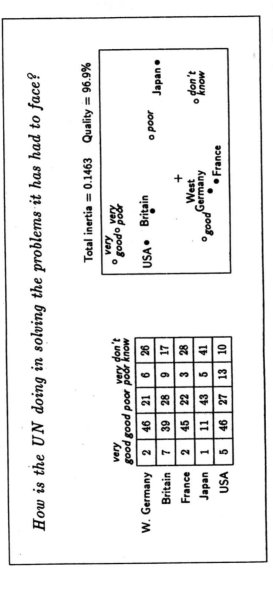

FIGURE 1.3   Example 3: A 5 × 5 table of percentages (rows sum to 100%) from a series of surveys in five different countries, together with the two-dimensional symmetric CA map and the amount of inertia accounted for by the map.

Since correspondence analysis actually displays the relative frequencies in either the rows or columns (or both), it follows that the method can handle data which are already in percentage form. The map in Figure 1.3 shows that there is a dichotomy between Japan on the right and the other countries, with the USA on the left being the most opposed to Japan in the sense that it is the furthest from Japan. Britain lies close to the USA on the left while France and Germany lie in the middle but towards the bottom of the map. Associated with the USA and, to a lesser extent, Britain, are the extreme categories 'very good job' and 'very poor job', indicating a relative predominance of polarized views in these countries. Japan, on the other hand, should have a relatively high number of responses that the UN is doing a 'poor job', as well as 'don't know' responses.

It should be clear from these three introductory examples that the maps are quite informative of the data content. It would require some time to arrive at such interpretations by simply studying the data in their original form or in the form of appropriately derived percentages. On the other hand, our interpretation of the maps is intentionally somewhat vague and we use phrases such as 'tends towards', 'relatively high', 'on the right' and 'associated with'. When we say that the USA has relatively high numbers of both 'very poor job' and 'very good job' responses, in actual fact the percentages in question are 5% and 13% in these categories, the 5% being the second highest (Britain has 7%) and the 13% being the highest (next highest is Britain with 9%). So the map displays the data in each row (or column) compared to the corresponding data for all other rows (or columns). In other words, there is a particular range of differences between rows, for example, and the map gives a picture of the relative highs and lows within this range.

We shall now systematically go through the various concepts involved in correspondence analysis and explain how the maps are constructed. This chapter follows the approach of Greenacre (1993) and readers can refer to that publication for a more detailed explanation. In this chapter we concentrate on simple correspondence analysis, denoted by CA, i.e. correspondence analysis of two-way contingency tables. In Part 2 of the book we turn our attention to more general forms of the method involving multi-way tables.

1.2   CORRESPONDENCE ANALYSIS CONCEPTS

The concepts of CA are geometric ones rather than statistical ones. The only statistical concept to which CA is linked is the Pearson chi-square (χ^2) statistic, which is regularly computed for cross-tabulations in order to assess the significance of the association between the row and column variables. Later we show how the χ^2 statistic can be interpreted in terms of the geometric concepts of CA. There are three primary concepts in CA, namely profiles, masses and chi-squared distances, and we shall introduce these in turn using the examples of Figures 1.1–1.3 as illustrations.

1.2.1 Profiles

When inspecting a cross-tabulation it makes little sense to compare the actual frequencies in each cell. Each row and each column has a different number of respondents, called the *base* of respondents. For comparison we need to reduce either the rows or columns to the same base. It is customary to reduce the rows or columns to the common base of 100% by computing percentages relative to the row or column totals. There is almost always a natural tendency for one to look at the table either as a set of rows or as a set of columns. This decision usually manifests itself in the way we choose to compute percentages, either row-wise or column-wise. As a convention, we usually use the columns for the categories of the variable to be described and the rows for the categories of one or more describing variables, in which case it is usually the row profiles which are of interest. For example, in Figure 1.1 we would be more likely interested in computing percentages of the response options in each age group, i.e. the row percentages. Thus, 69 of the 118 respondents in the 18–24 year age group, or 58.5%, said that civil rights people are pushing 'too fast'; 40 out of the 62 respondents in the category 75 years and over, or 64.5%, gave the same response. These percentages allow direct comparisons to be made between the age groups. Such a set of percentages, calculated for a row or a column of frequencies, is called a *profile*. Thus the profile of the first age group across the response options is the set of percentages [58.5% 31.4% 5.9% 4.2%], or the set of proportions [0.585 0.314 0.059 0.042]. In Table 1.1 the set of row profiles of the data in

TABLE 1.1
Row profiles of the table in Figure 1.1, in percentage form.

Age group (years)	Too fast	About right	Too slow	Don't know	Total
18–24	58.5	31.4	5.9	4.2	100.0
25–34	64.6	19.7	6.1	9.6	100.0
35–44	61.6	23.6	4.3	10.5	100.0
45–54	62.1	22.3	4.7	10.9	100.0
55–64	70.9	15.1	3.5	10.5	100.0
65–74	68.4	13.5	3.2	14.8	100.0
>75	64.5	11.3	1.6	22.6	100.0
Average profile	64.2	20.3	4.5	11.0	100.0

Figure 1.1 is given. Notice the last row which is called the *average profile*. This is the profile of the column totals of the table, which are 814, 258, 57 and 139, respectively, with a grand total equal to the sample size of 1268. Thus 64.2% of the whole sample gave the response 'too fast', so that the response in the oldest age group is close to average, while the response in the youngest group is below average (and in fact the lowest). When interpreting the data of Figure 1.1 we would usually inspect the percentages in Table 1.1 to draw comparisons between the age groups. This is in effect what CA does, but in a graphical format.

The profiles are examples of mathematical 'vectors'. Vectors have a geometric interpretation since they define points in a multidimensional space. For example, the elements of the first row profile, 0.585, 0.314, 0.059 and 0.042, can be used as coordinates to situate the vectors in a four-dimensional space. Each profile is thus condensed into a unique point in this space. In most applications, this space is of dimensionality greater than three, so it is very difficult to imagine the positions of each profile. But we shall show the dimensionality of the profiles can be reduced so that we can visualize the profiles in a more accessible format, for example in a two-dimensional plane.

1.2.2 Masses

The second fundamental concept in CA is that of a *mass* associated with each profile. Each profile is made up of a certain number of cases or respondents, for example there are 118 respondents in the youngest age group in Figure 1.1 who constitute the profile [0.585 0.314 0.059 0.042]. This profile is assigned a weight proportional to the 118 respondents. Similarly, the profile of the oldest age group [0.645 0.113 0.016 0.226] is assigned a weight proportional to the 62 respondents in this group, approximately half the weight of the youngest age group. To be precise, the marginal frequencies of 118, 229, 276, etc., are divided by the grand total (sample size) of the table, 1268, to obtain quantities 0.093, 0.181, 0.218, etc., which are called *row masses* and which are used to weight each row profile differently in the analysis. The object of this weighting system is to allow each respondent to contribute equally to its corresponding profile point.

The average profile (of the rows in this case) in the last row of Table 1.1, which is the profile of the column totals of the data in Figure 1.1, turns out to be the weighted average of the individual row profiles, where the weights are the corresponding masses. In other words, the average profile can be thought of as lying in an average, or central, position in the cloud of profile points, but tends to lie more towards the profiles which have higher mass. In this particular example, the average profile could also be calculated as a weighted average of the seven row profiles as follows: 0.093 multiplied by the first row profile, plus 0.181 times the second row profile, plus 0.218 times the third profile, etc.

1.2.3 Distances

Up to now we have talked about the profile vectors in the rows of Table 1.1 as being points situated in a four-dimensional space. For example, the profile of the youngest age group is a point with four coordinates, 0.585 on the first dimension of this space, 0.314 on the second, 0.059 on the third and 0.042 on the fourth. With some imagination it is possible to have an intuitive idea of the position of a profile in this space and of distances between profile points. We need to fix this notion more clearly because it is the concept of distance between two points in space which will allow us to view the profiles ultimately as a map. The physical concept of distance to which we are accustomed is called *Euclidean distance*, also known as *Pythagorean distance*, and can be thought of as the 'straight-line' distance between points in a physical space. The Euclidean distance between the profiles of the youngest and oldest age groups in Table 1.1 would be equal to the square root of the sum of squared differences in the profile values:

$$\sqrt{(0.585 - 0.645)^2 + (0.314 - 0.113)^2 + (0.059 - 0.016)^2 + (0.042 - 0.226)^2} = 0.282$$

In CA, however, we use a variation of Euclidean distance called a *weighted* Euclidean distance to measure, and thus to depict, distances between profile points. Here the weighting refers to differential weighting of the dimensions of the space, not to the weighting of the profiles themselves as discussed in section 1.2.2. In practice this weighting has the effect that response options which occur less frequently are made to contribute more highly to the inter-profile distance, while those that occur more frequently are made to contribute less. The specific way this is done is to divide each of the squared differences in the distance calculation by the corresponding element of the average profile, so that in the above example the distance is:

$$\sqrt{\frac{(0.585 - 0.645)^2}{0.642} + \frac{(0.314 - 0.113)^2}{0.203} + \frac{(0.059 - 0.016)^2}{0.045} + \frac{(0.042 - 0.226)^2}{0.110}} = 0.743$$

Because of the analogy with the chi-squared concept of calculating squared differences between proportions relative to their expected, or mean, values, this distance is known as the *chi-square distance.*

There are many different ways to justify the choice of chi-square distance. From a theoretical point of view, the division of each squared term by the expected frequency is variance-standardizing and compensates for the larger variance in high frequencies and the smaller variance in low frequencies. In practice this means that if no such standardization is performed the differences between larger proportions will tend to be large and thus dominate the distance calculation, while the small differences between the smaller proportions tend to be swamped. The weighting factors in the chi-square distance function thus tend to equalize the roles of the response options in measuring distances between the profiles. We should also point out that all the appealing properties of CA stem from the choice of chi-square distance function, for example the decomposition of the chi-square statistic and the joint display of rows and columns discussed later in this chapter, so for this reason the chi-square distance is the natural one for the problem.

1.2.4 Inertia

The term *inertia* (or more specifically 'moment of inertia') is borrowed from mechanics. Every physical object has a centre of gravity, or centroid. Every particle constituting that object has a certain mass (r) and a certain distance (d) from the centroid. The (moment of) inertia of the object is defined as the sum of the quantities rd^2 for the whole object. In our case we have a set of profile points with masses adding up to 1, these points have a centroid (the average profile) and there is a measure of distance (the chi-square distance) between profiles. The inertia of this set of points, or cloud of points, can thus be computed. Each profile point contributes a certain amount towards this inertia. For example, the first age group profile in Table 1.1 contributes an amount equal to:

$$0.093 \times \left(\frac{(0.585 - 0.642)^2}{0.642} + \frac{(0.314 - 0.203)^2}{0.203} + \frac{(0.059 - 0.045)^2}{0.045} \right.$$

$$\left. + \frac{(0.042 - 0.110)^2}{0.110} \right) = 0.0103$$

which is the mass (0.093) times the square of the chi-square distance between the first row profile and the average row profile. This quantity is the inertia of the first row and adding up these quantities for all the rows gives the *total inertia* (see Chapter 3, sections 3.2.14 and 3.2.22). The inertia has a geometric interpretation as a measure of the dispersion of the profiles in multidimensional space. The higher the inertia, the more spread out they are (see Chapter 2).

An alternative way of defining the inertia is in terms of all weighted squared distances between pairs of profiles. For example, in section 1.2.3 we saw that the chi-square distance between the youngest and oldest age groups had the value 0.743. Let us denote this quantity by d_{17}, where the subscript denotes the first and seventh rows. These age groups have masses 0.0931 and 0.0489, respectively, denoted by r_1 and r_7, respectively. Multiplying these two masses by the squared distance between the first and seventh profiles gives an amount of $r_1 r_7 \, d_{17}^2 = 0.0931 \times 0.0489 \times (0.743)^2 = 0.00252$. A similar calculation can be performed for all pairs of profile points, of which there are $(7 \times 6)/2 = 21$

in total. Adding up these quantities also gives the total inertia of the seven profiles. This is another illustration of how the total inertia measures the dispersion of the profiles in multidimensional space, since the higher the profile-to-profile distances are, the higher are the contributions of the weighted squared distances to the total inertia.

Figure 1.4 illustrates four different situations with increasing levels of inertia. In each case the contingency table has five rows and three columns and we are interested in the display of the five row profiles. In the first example, the row profiles are very similar and this is reflected by a bunch of profile points in the middle of the display and a low inertia of 0.0076. As the profiles become more dissimilar in the second and third examples, so the profile points

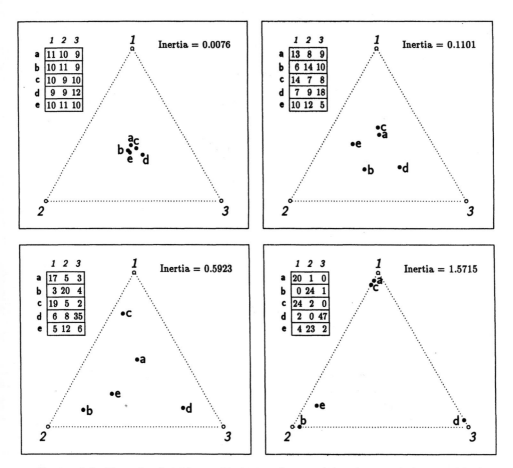

FIGURE 1.4   Four 5 × 3 tables, with increasing total inertias, and the graphical displays of the row profiles and column vertices.

are more dispersed and the inertia values increase to 0.1101 and 0.5923, respectively. When the profiles are very polarized, that is they are almost entirely concentrated in one category as in the last example of Figure 1.4, the differences between the profiles approach a maximum. Here row 'a' is almost entirely associated with column '1', 'b' with column '2', 'c' with '1', 'd' with '3' and 'e' with '2'. The total inertia in this case has a value of 1.5715.

1.2.5 Geometry of profiles and vertices

In each of the maps in Figure 1.4, a triangle has been drawn which links three points labelled '1', '2' and '3'. These special profile points, called *vertex points*, or *vertices*, represent fictitious row profiles of the form [1 0 0], [0 1 0] and [0 0 1], respectively, in which a profile would be totally concentrated into one category. In the last example, profiles 'b' and 'e' lie close to the vertex '2' because their profiles are very high in the second category, e.g. the profile of row 'b' is [0 0.96 0.04]. In these simple examples of profiles with three elements, the triangles delimit the space in which the profile points lie. It is impossible for a profile point to lie outside this triangular space. The greatest dispersion attainable by the profiles is when they lie exactly at the vertices of the triangular profile space. In such a situation the inertia would attain its maximum value of 2 in this particular example. In general the total inertia has a maximum equal to 1 less than the number of columns – or 1 less than the number of rows – whichever is smaller.

The vertices have a special significance in the interpretation of CA maps, since the closer a profile point (a row point, say) comes to one of the vertices (representing a column), the more the corresponding row and column are associated. This is often referred to as the *barycentric* property of CA, the word 'barycentre' being a synonym for 'weighted average'. The position of each profile point in the profile space can be obtained by taking a weighted average of the vertex points, where the weights are the elements of the profile. For example, consider the position of the point 'e' in the four displays of Figure 1.4. In the first display 'e' has frequencies [10 11 10], hence a profile of [0.323 0.355 0.323] (e.g. 0.323 = 10/31). Using these profile values as weights assigned to the three respective vertices, the point 'e' finds a position in the centre of the triangle but just slightly towards vertex '2' because of the slightly higher second profile value. In the other three displays, the profile of 'e' becomes more and more concentrated in category '2', with profiles [0.370 0.444 0.185], [0.217 0.522 0.261], and [0.138 0.793 0.069], respectively. In the last display, 'e' is situated at the weighted average of the vertices with weight 0.138 on vertex '1', 0.793 on '2' and 0.069 on '3'. This explains why 'e' lies very close to vertex '2' in the last display.

The profile positions and the profile space are easy to see in Figure 1.4 because the row profiles consist of three elements only. Analyzing larger tables is based on the same concepts of a set of profile points with masses in a space which is structured by the chi-square distance function, but where the points are situated in a higher-dimensional space which cannot be easily visualized. Vertices, or fictitious unit profiles which have a single element of one and the remaining elements zero, still define the outer limits of the space of the profiles within the multidimensional space. Such imaginary spaces of points in hyperspace are of little use to the practical data analyst, however, and we need a way of displaying larger tables in a simple low-dimensional format such as a plane. This leads us to another important aim of CA, namely to reduce the dimensionality of a data matrix.

1.3 REDUCTION OF DIMENSIONALITY

Although it requires a stretch of the imagination to picture profile points in a space of dimensionality greater than three, it is not difficult to consider the possibility that the points lie close to a line or plane. This is analogous to the situation in regression analysis where a straight line, or a plane in some higher-dimensional space, is fitted to a set of points in order to explain the values of a particular response variable of interest. In the regression situation the fit is measured by the square of the so-called 'multiple correlation coefficient', denoted by R^2. In view of the large amount of 'white noise' in social science data a value of R^2 above 0.5 would generally be regarded as very satisfactory. This value is an omnibus measure of quality of fit for the whole data set, and does not indicate which response values are explained relatively well or poorly.

In CA we have an analogous situation in that we are looking for a low-dimensional space, usually a plane, which reflects as accurately as possible the chi-square distances between the profiles. In CA this turns out to be equivalent to looking for the plane which is in some sense 'closest' to all the points. The particular definition of this closeness is in terms of weighted least-squares, which we shall now explain. In section 1.2.4 we described how the inertia of the ith row profile was calculated as the mass r_i times the squared distance d_i^2 from the profile to the average profile, the distance being the chi-square distance. Now for any given plane the distance from the profile point to the plane can be computed as the smallest chi-square distance between the profile and the plane. The point in the plane which is closest to the profile is called the *projection* of the profile onto the plane. The distance from the profile to its projection is denoted by e_i and the distance in the plane from the centroid to the projection by $\hat{d}_i$, as illustrated in Figure 1.5. The centroid, projection and profile points form a right-angled triangle, in which the well-known theorem of Pythagoras applies:

$$d_i^2 = \hat{d}_i^2 + e_i^2$$

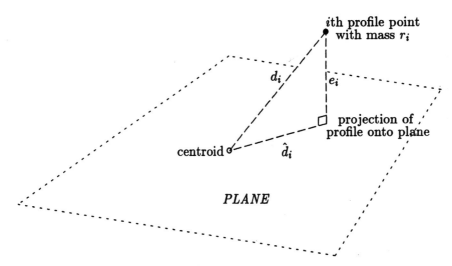

FIGURE 1.5   Projection of a profile point onto a plane, showing the centroid (in the plane), and the distances in the right-angled triangle formed by the centroid, profile and projection.

The total inertia $\Sigma_i r_i d_i^2$ is thus decomposed into two components:

$$\sum_i r_i d_i^2 = \sum r_i \hat{d}_i^2 + \sum_i r_i e_i^2 \tag{1.1}$$

which can be stated verbally as:

total inertia = inertia in plane + residual inertia

In CA the closeness of the profile points to the plane is measured by the weighted sum of squared distances from the points to the plane, i.e. the residual inertia $\Sigma_i r_i e_i^2$, and the objective of the analysis is to find the plane which minimizes this quantity. Alternatively, (1.1) shows that the minimization of the residual inertia is equivalent to the maximization of $\Sigma r_i \hat{d}_i^2$, the inertia in the plane (or, in general, in the low-dimensional subspace chosen). The residual inertia is a measure of how much variance, or more specifically inertia, has been lost in reducing the profiles to a two-dimensional format, and the analysis finds the plane for which this loss is a minimum. Or equivalently, it is the plane which retains the maximum inertia which is possible in a two-dimensional representation.

These best-fitting planes form the basis of the displays in Figures 1.1 to 1.3. In these maps we have an approximation of the true positions of the profile points, as they are observed from the optimal viewing plane (in the sense just described) in the profile space. These maps give no indication of the closeness of the points to the plane, but the quality of display can be summarized by an overall measure on a percentage scale. As in regression analysis this measure quantifies the amount of variance, or inertia, accounted for by the map. Thus in Figure 1.1, for example, we see that 96.6% of the inertia of the row (or column) profiles is explained by the map, in other words only 3.4% of the inertia lies in the remaining third dimension which is not portrayed in the map. These two percentages are, respectively, the inertia in the plane and the residual inertia in (1.1), expressed as percentages of the total inertia.

1.4   JOINT DISPLAY OF ROWS AND COLUMNS

Up to now we have been discussing one set of profile points, their associated masses, their interpoint distances and a way to reduce their dimensionality to obtain approximate maps of their positions. In Figure 1.1 our discussion applied to the rows of the table, the age groups, and we have explained how the positions of the age group points (depicted by solid circles in the map) are obtained. Yet the map also includes points for the columns (depicted by empty circles). Likewise in Figure 1.2, our interest would be in the profiles of the columns across the categories of TV program type, which are positioned by solid circles in the map; and in Figure 1.3 the data are already in the form of column profiles, depicted by solid circles in the map, while the other set of points (the rows in Figures 1.2 and 1.3) are also found in the map, depicted by empty circles. We need to explain how this joint display comes about.

As we discussed in section 1.2.1, our attention is usually concentrated on the profiles of the rows or of the columns of a table, depending on the research question. It is a remarkable property of CA that the analysis of the row profiles and the column profiles of the same table have interrelated results which show certain strong similarities. As an example, consider the table in Figure 1.1. The seven row profiles of this table have a total inertia of 0.0308. The four column profiles also turn out to have an inertia of 0.0308, hence the measures of variation of the two sets of points in their respective spaces are identical. Even though the row and column profiles have different numbers of elements, it turns out that their dimensionalities are also identical. In this particular example of a 7×4 table the dimensionality is equal to 3 (see Chapter 3, section 3.2.15). When it comes to reducing the dimensionality, e.g. projecting the row profiles onto their best-fitting plane on the one hand and projecting the column profiles onto their best-fitting plane on the other, we also find that exactly the same amount of inertia is explained, 0.0298 in each case, or 96.6% of the total inertia.

These similarities between the problems of the row profiles and the column profiles are still not sufficient to explain why both sets of points are shown to co-exist in the maps of Figures 1.1–1.3. For example, why is the point 'about right' on the left side of the map in Figure 1.1, on the same side as the age group '18–24 years'? There is a fundamental relationship between the two sets of points which permits us to overlay the row map and the column map and to make inferences from the joint map. To understand this relationship we should return to Figure 1.4 and recall the notion of an extreme point, or vertex, of the profile space. In Figure 1.4 the extreme vertices of the profile space were easy to display because the row profiles were exactly two-dimensional. To imagine the vertices in the higher-dimensional spaces under-lying Figures 1.1–1.3 is more difficult. As an example consider the data in Figure 1.1. We can imagine four imaginary row profiles, each one entirely concentrated into one of the four response categories. For example, the first of these fictitious profiles is [1 0 0 0], in other words an imaginary group of respondents that have only given the response 'too fast'. Similarly, the other profiles are [0 1 0 0], [0 0 1 0] and [0 0 0 1]. Each of these profiles exists in the space of the row profiles, and define the vertices of a four-pointed tetrahedron which encloses all the row profiles. The map of the seven row profiles in Figure 1.1 shows the positions of these seven profiles almost exactly, but we can also project the four extreme vertex profiles onto this planar map. This is what is shown in Figure 1.6, which we call the 'asymmetric map' of the data in Figure 1.1, as opposed to the 'symmetric map' in Figure 1.1 itself.

Why are the positions of the vertex points important? The main reason is that they do in fact co-exist in the same space as the seven row profiles. Thus the positions of the seven row profiles can be referred directly to the vertex points for purposes of interpretation. Thus in Figure 1.6 we can say that because the age group '18–24 years' lies closest to the vertex points 'too slow' and 'about right', it should have the highest percentage of these responses. From the data we can see that this is true, namely 31.4% + 5.9% = 37.3% of this age group give these responses, whereas the percentage is less than 30% for the other age groups (Table 1.1).

There is an alternative way of interpreting the asymmetric map which is useful, as described in section 1.2.5. Each profile point in Figure 1.6 is the weighted average of the four vertex points, where the weights are the profile values themselves. Hence, if we put a weight of 0.585 on the vertex 'too fast' (see data in Table 1.1), 0.314 on 'about right', 0.059 on 'too slow' and 0.042 on 'don't know', then the average of these positions is the profile position '18–24 years'. In this sense the profile is attracted to the vertex points for which it has a relatively high profile value.

Because of the straightforward joint interpretation of Figure 1.6, why then do we not always use this asymmetric form of the map in correspondence analysis? The main reason is that in most applications the profile points occupy a very small region of the entire profile space, as in the first example in Figure 1.4. In other words the differences between profiles are often quite

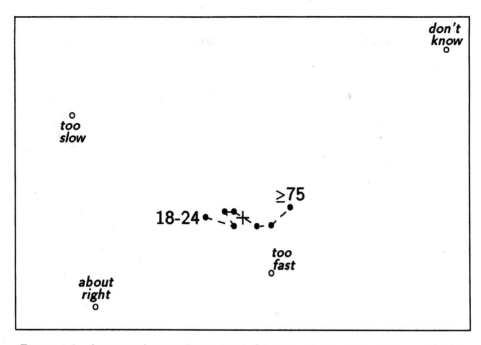

FIGURE 1.6   Asymmetric two-dimensional CA map of the data of Example 1 in Figure 1.1, showing the projections of the row profiles (age groups) and column vertices (response categories).

small and the profiles tend to bunch up in the middle of the asymmetric display. In Figure 1.6, for example, we did not put all the row profile labels on the plot because the points were too close together in the middle of the map. This problem is exacerbated when there are many profiles to be displayed. Figure 1.7 shows the asymmetric map for the data in Figure 1.3, where the total inertia is greater (0.1463), of which 0.1417, or 96.9%, is shown in the map. Now the profile points (rows) tend more away from the centre of the map towards the vertex points (columns) and the asymmetric map is more attractive as a result, with labelling of all the profiles easily possible. In cases of high inertia, therefore, the asymmetric map is a practical possibility. On the other hand, in many situations where there is no clear distinction between the row and column variables both sets of profiles are of equal interest and the symmetric map would be preferred on substantive grounds.

Finally, let us compare the asymmetric map of Figure 1.6, for example, with its symmetric equivalent in Figure 1.1. The positions of the row profiles (represented by solid dots) are identical in both cases. The relative positions of the column profiles (solid circles) in Figure 1.1 look remarkably similar to

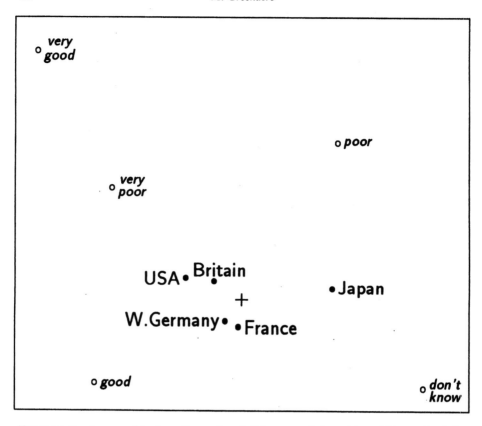

FIGURE 1.7 Asymmetric two-dimensional CA map of the data of Example 3 in Figure 1.3, showing the projections of the row profiles (countries) and column vertices (response categories).

those of the column vertices (empty circles) in Figure 1.6. This is the crux of the row–column duality in correspondence analysis. It turns out that the positions of the column profiles are simply a scalar contraction of those of the column vertices, where the scalar contraction is different on each of the principal axes of the map. In practical terms this means that there is a particular scaling factor relating these two sets of points on the horizontal axis and another scaling factor relationship between them on the vertical axis. It will be shown in Chapter 3 that these scaling factors are just the square roots of the principal inertias on each axis. As a specific example, the first principal axis in Figure 1.1 (and Figure 1.6) accounts for an inertia of 0.0260, of which the square root is 0.161. Along the horizontal axis, therefore, the position of each of the column profile points in Figure 1.1 is about one-sixth (0.161) of that of the corresponding vertex point in Figure 1.6. Along the vertical axis, the square root of the second principal inertia (0.0038) is 0.062. Thus the contraction of profiles relative to their vertices is even more severe along this axis — the coordinate positions of the column profiles are one-sixteenth (0.062) of the positions of the vertices along the vertical axis.

1.5 PRACTICAL CONSIDERATIONS

In practice, the symmetric map as illustrated in Figures 1.1 to 1.3 is the map which is the most popular. Let us summarize the properties of this map:

(a) It is comprised of the optimal displays of the row profiles and of the column profiles, even though these two sets of points — strictly speaking — occupy different spaces.

(b) The row and column points are equally spread out in the map, both in the horizontal and vertical directions (or, in general, along each principal axis).

(c) The rationale for the joint representation is essentially the underlying asymmetric map, with either the rows or columns imagined as vertex points of the profile space. There is a constant scaling factor between the vertex positions, as projected onto an optimal map, and their corresponding profile positions projected onto the optimal map of the same (low) dimensionality.

(d) There is no direct row-to-column distance interpretation, but there is certainly a joint interpretation of the row and column points with respect to the principal axes of the map. Left-to-right and up-and-down oppositions in the map are interpreted in the same way for both rows and columns and the correspondence between their deviations outward along these axes can be directly interpreted as association.

(e) The magnitude of the row-to-column association in an absolute sense cannot be directly observed in a symmetric map, but can only be judged from the numerical magnitudes of the principal inertias themselves.

Finally, let us summarize the interpretation of Figures 1.7 and 1.3. In Figure 1.7, only the row profiles (countries) are mapped onto their best-fitting plane, and 96.9% of the total inertia is displayed. This is therefore an excellent display of the positions of the country profiles across the response options. There are five column vertices which have been projected onto the plane from their outlying positions. The positions of the vertices are reference points for the interpretation of the profiles. For each of these reference points, the positions of the five countries can be compared, showing, for example, that the USA and Britain are relatively high on the responses on the left side of the map, while Japan is relatively high on the responses on the right side. Similarly, the subtler contrast between the countries on the lower side, France and West Germany, is due to a contrast in the relative frequencies between the responses 'good' and 'don't know' in the lower part and the remaining responses. Thus the only difference in interpretation between Figures 1.3 and 1.7 is that the row vertex points in Figure 1.7 are fixed reference points in the space of the profiles. The interpretation in Figure 1.3 still relies on comparing which rows and which columns lie in the main quadrants of the map, but we do not interpret the actual distances between a row point and a column point.

Several of these aspects of interpretation will become clear in the subsequent chapters, especially in the practical applications. More details about the geometric interpretation of correspondence analysis are given by Greenacre (1993).

2

Correspondence Analysis in Social Science Research

Jörg Blasius

2.1 INTRODUCTION

The aim of this chapter is to give some examples on how to use and how to interpret correspondence analysis solutions in the social sciences. In this chapter we refer to simple correspondence analysis (CA) only – multiple correspondence analysis (MCA) will be introduced in Part 2.

Correspondence analysis is an exploratory method. One of the strongest criticisms of CA is the absence of hypothesis testing and thus the charge that it is 'theoryless'. On the one hand, as mentioned in the preface, there are possibilities of 'hypothesis testing' by setting linear constraints (Chapter 5) or by using a modelling approach to CA (Chapter 4). Nevertheless, these approaches are very new and, till today, discussed in statistical literature only. On the other hand, we agree with the critics that there should be a theory, which includes assumptions on how the variables to be analyzed are connected to one another.

Usually, theories considered in CA do not refer to causal relationships, they are based on social science theories such as the 'assumptions' of Bourdieu's *La Distinction* (1979). These kinds of theories cannot be tested in the way proposed by Popper (1966); they can be described (confirmed) only. This means one can find something like clusters of categories which are connected in an equal manner with other clusters of categories, in agreement with prior assumptions. Correspondence analysis should be used for describing the data, for getting ideas about the structure of variables, and for developing hypotheses. In a second step (on a different sample) these new assumptions (or new hypotheses) can be tested (by a confirmatory method) or reinforced (by an exploratory method such as CA). Furthermore, if one does not find a significant structure by using an inferential statistical method such as loglinear modelling, one can use CA for vizualizing the residuals, which could provide one with new ideas about the structure of the data (Chapter 4).

2.2 USING CORRESPONDENCE ANALYSIS IN THE SOCIAL SCIENCES

As described in Chapter 1 correspondence analysis has three main concepts: chi-squared distances, profiles, and masses. These concepts are the basis for understanding the method, and they always have to be considered when interpreting correspondence analysis solutions.

With respect to these three concepts, Pearson's chi-square (i.e. the sum of squared differences of expected and empirical values, divided by the expected ones), calculated on the matrix of input data, and then divided by the total N, is a measure of the amount of variation in the data, called 'total inertia'. Chi-square, denoted by χ^2, can be used as an overall measure for deciding if there is enough variation in the data to warrant further analyses. However, even when the variation in the data is small and not worth interpreting substantively, the chi-square would be significant if the sample size was large enough. Conversely, if the variation is high the chi-square could be insignificant if the sample size was small.

Furthermore, a detailed study of the terms which compose the chi-square provides information about which cells of the table have frequencies above or below the average. In effect, one can see which rows and which columns contribute to what extent to the total variation in the data. The total inertia (its maximum value is the minimum {number of rows, number of columns} − 1) can be used as a crude measure for the variation in the data, too. If the value is low there is little or no variation in the data, which one would call white noise. A measure for comparing chi-squares independent of the sample size and the number of cells is the well-known Cramer's V:

$$V = \sqrt{\frac{\chi^2}{n(\min\{I, J\} - 1)}}$$

where I = number of rows and J = number of columns.

If one excludes the square root and the divisor min $\{I, J\} - 1$, one obtains the total inertia, also known as Pearson's mean square contingency coefficient. This coefficient is not standardized to a range from 0 to 1 like Cramer's V, but because of the known minimum of the number of rows and the number of columns, total inertia can be used as a crude measure of the amount of variation in the data.

Correspondence analysis deals with profiles. This means one does not interpret the raw frequencies which are used as input data, one always interprets their values relative to the marginals. Instead of interpretations such as 'most of …' or 'none of …', descriptions such as '… above the average' or '… relatively seldom' have to be used. In effect, comparing rows or columns means comparing their profiles.

As Chapter 1 shows, the marginals of rows and columns are also considered in the analysis. In terms of correspondence analysis these frequencies relative to the total *n* are called 'masses', they are used as 'weighting factors' (Chapters 1 and 3). Usually the marginals of contingency tables are different, for example, the number of persons having a primary education level is different from the number of persons having a secondary education level. These differences will be considered in the analysis.

As the first chapter of this book shows, there is a strong distinction between the presentation of the rows and the presentation of the columns. To avoid misinterpretations both row- and column-displays can be given and interpreted separately, or the asymmetric scaling is used in a joint map to enable a valid simultaneous (but not symmetric) interpretation of rows and columns. In the asymmetric plot random variation becomes visible when all profiles are located near the centroid and far from the vertices. In the symmetric plot one can see if there is enough variation in the data to warrant analysis by referring to the scale of the map, usually indicated by an interval of 0.1: a wide interval would indicate low association (for the differences of symmetric and asymmetric scaling see Chapters 1 and 3, or, in more detail, Greenacre 1993).

In order to give an example of the use of CA we refer to Bourdieu's *La Distinction* (1979), his well-known assumptions about different life-styles in different class fractions. Bourdieu develops the characteristics of life-styles (e.g. the patterns of preferred clothes, food, beverages, and taste) to describe 'horizontal' inequalities between different class fractions (measured by professional positions). For this reason, he introduces a 'space of professional positions' which includes descriptions of the different occupational positions, e.g. the percentages of artists having different kinds of interior in their apartments (with categories such as 'modern', 'clean', and 'full of fantasy') and serving different kinds of dinner for friends such as 'improvised', 'fine and exquisite', and 'exotic'. Furthermore, he describes a 'space of life-styles' which includes descriptions of the life-styles, e.g. the percentages of persons having a life-style such as 'planning dinner for friends: fine and exquisite', cross-tabulated by the 'professional position'. By using CA, Bourdieu links both spaces to one, the 'social space' (Bourdieu 1985). In the 'social space' both the 'space of social positions' and the 'space of life-styles' are displayed and interpreted simultaneously.

Thinking in terms of CA, Bourdieu starts with a multi-response table as input data. The columns are represented by the professional positions, the rows by the categories of life-style. Within the analysis the row profiles refer to the 'space of life-styles', whereas the column profiles refer to the 'space of professional positions'. Notice that both spaces are based on the same table of raw frequencies, only the profiles of rows and columns are different. In this study the masses are the weights of the life-styles and of the social positions. The greater the number of persons belonging to a life-style or to a social position, the greater the weight. If one excluded the weights, for example by using the profiles of the life-styles as input data instead of raw frequencies, every life-style would have the same weight, independent of how often the life-style was counted over all social positions.

In applying CA usually one has one (or more) variable(s) to be described and one or more describing variables (see Preface). In the given example the professional position is the variable to be described, and the different 'life-styles', measured by variables such as 'preferred shops for buying interior' and 'kinds of planning dinner for friends', are the describing variables.

After this short introduction into the theoretical possibilities of applying CA we give some hints for interpreting CA solutions. One of the most interesting questions in applying a multivariate method pertains to the stability of the results, in other words what will happen if there are changes in the variables to be included in the model. To demonstrate the stability of CA solutions we use an entertaining example involving variables on different topics of 'cultural competence'. These variables are intended more to illustrate the graphical display and the numerical results than to have substantive meaning. The second example is driven by social science theory: analyzing mobility data.

2.3 THE STABILITY OF SOLUTIONS

2.3.1 First example: Analyzing variables on 'cultural competence'

In this section it will be shown that CA is a very robust technique in the presence of additional variables. As an illustration we use data from a representative survey on various topics in empirical social research (ALLBUS) in Germany in 1986[1], focusing on a subset of questions concerning different aspects of activities with respect to a combined variable of sex and age (men up to 39 years, men between 40 and 59 years, men 60 years and older; and women in the same age categories). The respondents were asked if they were able to dance to pop music, to waltz, to play chess and to shorten trousers, with possible responses 'yes', 'just a bit' or 'no'. In this application, there is

---

[1] The data from the survey 'Allgemeine Bevölkerungsumfrage der Sozialwissenschaften — ALLBUS 1986' are available at the Zentralarchiv für empirische Sozialforschung, study number S1500.

no need for an explicit theory – for example, dancing to pop music should be age-affected and shortening trousers should be sex-affected.

Input data are the raw frequencies of four contingency tables corresponding to the four above-mentioned 'cultural competences', each with three rows for the three response categories. The four tables are concatenated by maintaining the given column structure, in the given case by the six categories of the combined variable 'sex × age'. The matrix of input data is given in Table 2.1.

For a first general inspection of the data, chi-square is commonly used. Because the input data is the concatenation of four 3×6 contingency tables the chi-square includes four values with $(3-1) \times (6-1) = 10$ degrees of freedom. Each of the four chi-squares divided by the total of the respective contingency tables is a total inertia, reflecting the variation in the tables (Table 2.2). Finally, for each contingency table Cramer's V as well as the sum of chi-squared distances can be computed. The sum of chi-squared distances can be seen as a measure of spread where the chi-squared distances of all pairs of categories within the single variables are added up.

As Table 2.2 shows the highest variation belongs to the variable 'shortening trousers' which can be confirmed by the greatest total inertia (= 0.500), the greatest distance (the sum of chi-squared distances between all pairs of categories belonging to the same variable, $D = 3.172$) and the highest Cramer's V (= 0.500). The lowest variation belongs to the variable 'waltzing' (total inertia = 0.120, Cramer's V = 0.245, $D = 1.736$). Whereas the categories of 'shortening trousers' should be relatively far away from one another, the three categories of 'waltzing' should be relatively close together in the map.

The sum over the four chi-squares is 3428.1 with an average total inertia of 0.282 (1.127/4). The chi-square computed for the full input data (Table 2.1)

TABLE 2.1
Example 1: input data.

A	M_T39Y	M_40T59Y	M_60Y+	W_T39Y	W_40T59Y	W_60Y+
WALTZYES	208	298	208	426	425	323
WALTZJAB	178	103	64	158	69	52
WALTZNO	235	84	52	109	27	54
CHESSYES	287	146	114	87	47	23
CHESSJAB	143	68	44	118	54	27
CHESSNO	195	269	167	485	419	375
TROUSYES	167	109	91	637	500	406
TROUSJAB	132	76	62	32	12	12
TROUSNO	308	291	165	23	10	11
POPDAYES	386	145	33	535	214	67
POPDAJAB	138	105	33	92	88	32
POPDANO	95	234	249	58	208	312

TABLE 2.2

Example 1: sample size, chi-squares, Cramer's V, sum of chi-squared distances (HD), and inertias.

	Chi-square	Inertia	Sum of χ^2-distances	Cramer's V	Sample size
Waltzing	368.14	0.120	1.736	0.245	3073
Playing chess	554.18	0.181	2.076	0.301	3068
Shortening trousers	1520.56	0.500	3.172	0.500	3044
Dancing to popmusic	985.18	0.326	2.626	0.404	3024
Sum	3428.06	1.127	—	—	—

is 3423.5, the total inertia is 0.280. These values are slightly different because of missing data (see the different sample sizes of the tables). If one assumes no missing values the chi-squares and the total inertias would be the same. It follows that the chi-square of a concatenated table is the sum over all chi-squares of the contingency tables included in the model, and the total inertia of a concatenated table is the average total inertia of all contingency tables included in the model (Chapter 7). Chi-square as well as total inertia indicate that there is more than white noise in the data, hence the variation can be interpreted substantively.

In addition to the chi-square (or the total inertia) one can use their components, the squared deviations of expected and empirical values (divided by the expected values) in each cell for detecting variables contributing white noise only. This information can be used for deciding whether certain variable(s) need to be excluded. The matrix CELLIN (Table 2.3) contains the values of these squared deviations. For example, the highest variation of the data belongs to the men below 39 years of age (first column) and to the ability and disability of shortening trousers (seventh and ninth rows).

As will be shown in Chapter 3, the average total inertia (= 0.280) is equal to the sum of eigenvalues corresponding to the principal axes of the vector configuration. The first eigenvalue is 0.1792, the proportion of explained inertia is 63.9%, the second eigenvalue is 0.0904 (32.2% explained inertia), the others are negligible (the third eigenvalue is 0.0077 and explains 2.7% of the total variation, the fourth is 0.0017, and the last is 0.0014).

As in principal components analysis (PCA) there is no clear rule to determine how many dimensions should be interpreted. In both techniques one criterion is a scree test where one can see if there is a jump in the descending order of the eigenvalues. Another criterion is related to the eigenvalue-criteria in PCA where dimensions explaining less variance than one variable are dropped out of the analysis. Transforming this criterion in terms of CA one

TABLE 2.3
Example 1: Chi-square components.

CELLIN	M_T39Y	M_40T59Y	M_60Y+	W_T39Y	W_40T59Y	W_60Y+
WALTZYES	79.45	0.00	0.48	0.00	34.02	14.22
WALTZJAB	21.12	0.20	0.04	2.03	12.89	13.81
WALTZNO	129.78	0.24	0.81	2.50	48.91	7.30
CHESSYES	146.40	10.91	21.73	32.71	44.01	57.10
CHESSJAB	28.38	0.19	0.28	2.30	6.91	20.57
CHESSNO	95.05	3.53	5.62	6.56	27.65	45.65
TROUSYES	124.84	123.01	59.85	97.54	95.19	75.01
TROUSJAB	65.98	11.68	22.53	23.59	33.95	24.42
TROUSNO	127.46	209.26	75.73	139.55	117.92	91.19
POPDAYES	40.66	24.40	86.42	159.45	1.76	80.92
POPDAJAB	15.55	10.13	6.49	3.04	0.32	18.83
POPDANO	82.62	14.50	134.16	158.20	0.70	143.30

has to exclude factors which have an amount of explained inertia less than average (average = 100/number of dimensions). In the given example the number of dimensions is five (minimum of the number of rows and columns, minus one), therefore the average explained variance is 20%. Another criterion is to retain as many factors as can be interpreted substantively. Nevertheless, as in PCA, all criteria are somewhat arbitrary and it is the task of the social science researcher to determine the number of relevant axes. In the given example the decision is easy: under each of the proposed criteria two factors should be kept for interpretation. With two out of five possible factors 96.1% of the total variation can be explained; the solution is two-dimensional.

Figure 2.1 shows a symmetric display of rows and columns. Although one is not allowed to interpret the distances between rows and columns (see Greenacre and Hastie 1987, Greenacre 1989, 1993, as well as Chapters 1 and 3), the graphical presentation gives a good first illustration of how the variables are connected to one another in multidimensional space. As computed in Table 2.2 the sum of distances between the rows can be confirmed by measurement with a ruler from Figure 2.1 (small discrepancies would be due to the small variation in the sample sizes of the four tables, as indicated earlier, and the fact that 96.1% of the total variation is explained by the first two dimensions). As in Table 2.2, the greatest sum of (squared) distances belongs to the three categories of 'shortening trousers', the smallest one to 'waltzing'.

The first axis (horizontal line) explains 63.9% of the total inertia. By projecting the six column variables on the first axis it can be seen that this factor is dominated by the contrast between 'female' and 'male' respondents: the three categories of men are located on the right-hand side, the three

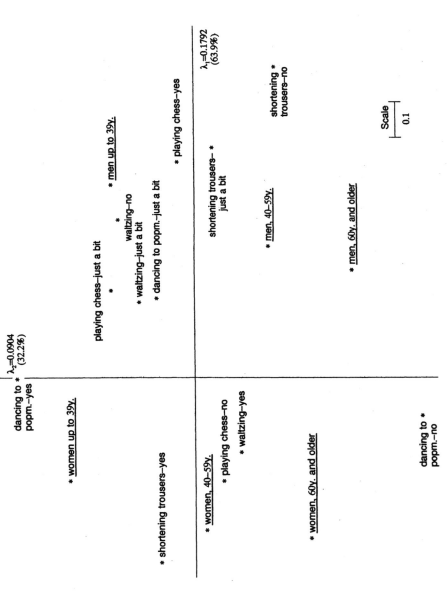

FIGURE 2.1   Example 1: graphical presentation.

categories of women on the left-hand side. As in PCA the axes can be labelled: due to the reported contrast, we call the first axis a 'sex dimension'.

The second axis explains 32.2% of the total inertia. By projecting the six column variables on the second axis it can be seen that the categories 'men up to 39 years' and 'women up to 39 years' are located on the positive part, whereas the categories 'men, 60 years and older' and 'women, 60 years and older' are located on the negative part. With respect to these projections we call the second axis an 'age dimension'.

By matching the solutions of the columns (the six categories of the variable combining sex and age) and the twelve rows (playing chess, waltzing, shortening trousers, dancing to pop music with three categories each) to the first axis, it can be seen that especially the ability to shorten trousers and − to a lesser degree (see the closer location to the origin of the configuration) − the ability to waltz and the inability to play chess are located on the left-hand side. These correspond to the three categories of women. It can be summarized that women relatively often have the ability to shorten trousers and to waltz, and that they relatively seldom have the ability to play chess. The categories on the right-hand side of axis one are 'shortening trousers−no', 'playing chess−yes', and 'shortening trousers−just a bit' as well as − again to a lesser degree − the categories 'playing chess−just a bit', 'waltzing−just a bit', and 'dancing to pop music−just a bit'. With respect to these results, men have below average ability to shorten trousers but relatively often the ability to play chess. Please note again that we are always interpreting profiles and never absolute values. This means, when we are talking about high values we are talking about high values compared to the average.

By focusing on the second axis it can be seen that the variable 'dancing to pop music−yes' is located at the top, whereas 'dancing to pop music−no' is located at the bottom. Matching the solution of the row presentation with the solution of the column presentation it can be deduced that young persons (men and women) relatively often have the ability to dance to pop music while older persons (men and women) relatively seldom have this ability.

In the given example more than 96% of the variation can be explained by the first two dimensions. Therefore the error in the two-dimensional display is less than 4% which is, compared with other studies using multivariate methods, a very low value. For introducing the coefficients available in CA, the numerical output is given in Table 2.4.

In Table 2.4, columns 4 and 7 (LOC1, LOC2) give information on each variable's position on the first two axes, i.e. their coordinates in the two-dimensional vector subspace. If the map was error-free, the cosines between each row-vector and the principal axes could be calculated directly from the map, by using a ruler and a protractor. The angle cosines can be treated as correlations of the variable categories with the axes, and are interpreted similarly to factor loadings in PCA. In CA the squared factor loadings are

TABLE 2.4
Example 1: numerical solution.

GENSTAT	MASS	SQCOR	INR	LOC1	QCOR1	INR1	LOC2	QCOR2	INR2
M_T39Y	0.202	0.981	0.280	0.562	0.816	0.357	0.252	0.165	0.143
M_40T59Y	0.158	0.893	0.119	0.375	0.665	0.124	−0.220	0.228	0.084
M_60Y+	0.105	0.967	0.121	0.313	0.304	0.058	−0.463	0.663	0.249
W_T39Y	0.226	0.988	0.183	−0.293	0.377	0.108	0.373	0.611	0.347
W_40T59Y	0.170	0.956	0.124	−0.441	0.952	0.184	−0.031	0.005	0.002
W_60Y+	0.139	0.950	0.173	−0.467	0.624	0.169	−0.338	0.326	0.175
WALTZYES	0.155	0.902	0.037	−0.212	0.661	0.039	−0.128	0.241	0.028
WALTZJAB	0.051	0.955	0.015	0.219	0.595	0.014	0.170	0.360	0.016
WALTZNO	0.046	0.798	0.055	0.467	0.645	0.056	0.227	0.153	0.026
CHESSYES	0.058	0.968	0.091	0.654	0.962	0.137	0.051	0.006	0.002
CHESSJAB	0.037	0.977	0.017	0.253	0.496	0.013	0.249	0.480	0.026
CHESSNO	0.156	0.975	0.054	−0.296	0.909	0.076	−0.080	0.066	0.011
TROUSYES	0.156	0.992	0.168	−0.536	0.954	0.251	0.106	0.037	0.020
TROUSJAB	0.027	0.979	0.053	0.738	0.975	0.081	−0.047	0.004	0.001
TROUSNO	0.066	0.976	0.222	0.929	0.916	0.318	−0.239	0.060	0.042
POPDAYES	0.113	0.990	0.115	−0.001	0.000	0.000	0.531	0.990	0.353
POPDAJAB	0.040	0.580	0.016	0.225	0.456	0.011	0.118	0.124	0.006
POPDANO	0.095	0.983	0.156	−0.072	0.011	0.003	−0.670	0.972	0.470

listed (see the columns QCOR1 and QCOR2 in Table 2.4) instead of the simple factor loadings which are given by PCA. Because the squared correlations are always positive, the signs of the location parameter (LOC1 and LOC2) have to be used to ascertain the directions of the factor loadings. The squared correlations over the axes to be taken into account (in the given example two) are summed up in column 2 (SQCOR). The values in this column can be interpreted similarly to the communalities in PCA – they show to what degree the variable categories are explained by the relevant axes.

In general, together with the location parameter, the squared correlations indicate on which axis certain variable categories load jointly. By referring to these coefficients it is possible to interpret the combined column categories of 'sex × age' and the row categories of cultural abilities, simultaneously. Thus, an indirect comparison of rows and columns is available.

As we showed when interpreting the graphical solution, the positive part of the first axis is dominated by the three categories of men (M_T39Y, M_40T59Y, M_60Y +), the negative part by the three categories of women (W_T39Y, W_40T59Y, W_60Y +). Highly associated with these column-variables are the different categories of the row-variables 'ability to shorten trousers', 'playing chess', and 'waltzing'. By considering the signs in column LOC1 it can be seen that men are less skilled with respect to 'shortening trousers' and 'waltzing' than women. Thereby the values for the three categories of 'waltzing' are relatively low, which allows the conclusion that 'shortening trousers' distinguishes men and women to a higher extent than 'waltzing'. Conversely, men have a higher ability to 'play chess' – whereby the middle category 'playing chess–just a bit' is correlated relatively poorly with the first axis (QCOR1 = 0.496).

The first column (MASS) of Table 2.4 lists the relative mass of each variable category, separately standardized for rows and columns to one (these two vectors are the average profiles of the rows and the columns). Multiplying the masses by the squared distances to the origin of the vector configuration (the LOC parameters) and expressing these raw inertias relative to the principal inertias (the eigenvalues) of the dimensions, one obtains the contributions of inertia on each axis and on the total model (Chapter 3). The contributions of inertia show to what extent each variable category determines the geometric orientation of the axes (INR1, INR2) as well as the total model (INR). The interpretation of the contributions of inertia is opposite to that of the squared factor loadings. In the latter case the values indicate to what extent each variable category is determined by the single axis, in the former case the values indicate to what extent the geometric orientation of each axis (as well as the total model) is determined by the single variable categories. The contributions of inertia are standardized to one, again separately for rows and columns.

At the row level, it can be seen that the total model receives its geometric orientation mainly from the inability to shorten trousers (TROUSNO,

INR = 0.222), from the ability to shorten trousers (TROUSYES, INR = 0.168), as well as from the inability and the ability to dance to pop music (POPDAYES and POPDANO, INR = 0.115 and 0.156). The groups which have the two above-named abilities 'just a bit' have a small amount only. By focusing on the separate axes it can be seen that the geometric orientation of the first axis is determined to 56.9% by the two extreme categories of 'shortening trousers' (the two values in the column INR1 have to be added), the second axis is determined to 82.3% by the two extreme categories of 'dancing to pop music'.

At the column level the variable categories 'men up to 39 years' (INR1 = 0.357), 'women between 40 and 59' (W_40T59Y, INR1 = 0.184), and 'women, 60 years and older' (W_60Y +, INR1 = 0.169) have the highest influence on the geometric orientation of the first axis. Therefore, the main difference with respect to the analyzed 'cultural competences' – and especially 'shortening trousers – yes, no' (INR1 = 0.251 and 0.318) and to a lesser degree 'playing chess–yes' (CHESS–YES, INR1 = 0.137) – is between the men up to

TABLE 2.5
Example 2: input data.

A	M_T39Y	M_40T59Y	M_60Y+	W_T39Y	W_40T59Y	W_60Y+
WALTZYES	208	298	208	426	425	323
WALTZJAB	178	103	64	158	69	52
WALTZNO	235	84	52	109	27	54
CHESSYES	287	146	114	87	47	23
CHESSJAB	143	68	44	118	54	27
CHESSNO	195	269	167	485	419	375
TROUSYES	167	109	91	637	500	406
TROUSJAB	132	76	62	32	12	12
TROUSNO	308	291	165	23	10	11
POPDAYES	386	145	33	535	214	67
POPDAJAB	138	105	33	92	88	32
POPDANO	95	234	249	58	208	312
LAMPYES	565	436	264	195	127	54
LAMPJAB	32	24	22	70	30	32
LAMPNO	28	27	42	412	356	334
PHOTOYES	516	412	237	532	354	190
PHOTOJAB	93	52	44	21	38	130
PHOTONO	18	22	44	21	38	130
TAXYES	259	280	201	240	199	113
TAXJAB	177	109	47	211	105	74
TAXNO	176	92	75	213	201	227
WATCHYES	533	345	173	374	173	92
WATCHJAB	48	42	29	92	43	32
WATCHNO	41	90	107	204	282	283

39 years versus the women between 40 and 59 and the women of 60 years and older. With respect to the second axis the main difference refers to 'men, 60 years and older' (M_60Y +, INR2 = 0.249) versus 'women up to 39 years' (W_T39Y, INR2 = 0.347). This distinction corresponds with 'dancing to pop music', the ability versus the inability (INR2 = 0.353 and 0.470), while the category 'just a bit' has no influence on the geometric orientation of the second axis (POPDAJAB, INR2 = 0.006).

In general, when interpreting the contributions of inertia one has to take into account that the masses of the variables are included as multipliers. Therefore, a high contribution of inertia can be due to a high mass only, whereas a low contribution of inertia does not allow the conclusion that the variable category is poorly correlated with the axis. Finally, when interpreting the numerical results of CA all information such as the contributions of inertia, the (squared) correlations (including their signs) and the masses have to be considered simultaneously. This complete information for comparing rows and columns should lead to a careful interpretation of the numerical output of CA.

2.3.2 Adding further variables

Before introducing a substantive example, we show that CA produces very robust solutions. For this purpose four further variables are included in the model by maintaining the column structure of Table 2.1. Techniques like PCA, for instance, are very sensitive to tasks such as this. In the given example the following cultural abilities were added, again with three categories each: filling out tax forms, adjusting a quartz watch, photographing, and fixing a lamp. The matrix of input data is given in Table 2.5, the graphical display of the extended model in Figure 2.2, the numerical solution in Table 2.6.

By comparing Figures 2.1 and 2.2 one can see that the locations of the first used row and column categories have been changed only slightly. As in the first analysis, the horizontal axis (68.7% explained variance) is again determined by 'sex' and the second axis (24.4% explained variance) by 'age', therefore the same labels are used for describing the axes. In both models, the ability – as well as the inability – to shorten trousers and the ability/inability to dance to pop music are connected in an equivalent manner with the same parts of the same axes. The additional row-variables do not distort the associations between the twelve rows and the six columns of the previous analysis. Referring to the differences in the explained variances between both models, the first axis (68.7% versus 63.9%) reflects a small increase in the relative importance of this dimension. Compared with the 'cultural abilities' of the first analysis it can be followed that the four additional row-variables are affected to a higher degree by sex than by age.

By focusing on the additional 'cultural competences' one can see that the categories 'photographing–no' and 'photographing–just a bit' are correlated

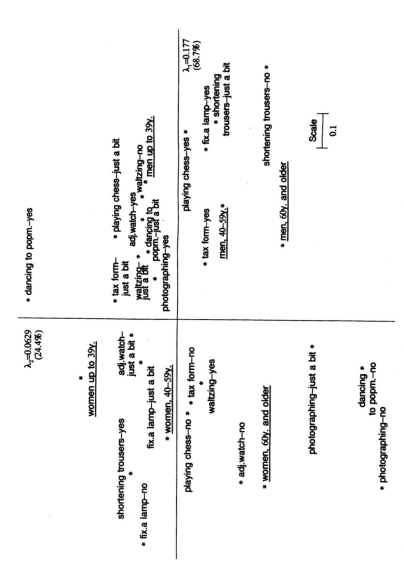

FIGURE 2.2 Example 2: graphical presentation.

TABLE 2.6
Example 2: numerical solution.

GENSTAT	MASS	SQCOR	INR	LOC1	QCOR1	INR1	LOC2	QCOR2	INR2
M_T39Y	0.205	0.931	0.253	0.529	0.882	0.324	0.125	0.049	0.051
M_40T59Y	0.160	0.896	0.127	0.398	0.776	0.143	-0.156	0.120	0.062
M_60Y+	0.106	0.924	0.092	0.257	0.298	0.040	-0.373	0.625	0.235
W_T39Y	0.221	0.980	0.149	-0.218	0.273	0.059	0.350	0.706	0.431
W_40T59Y	0.167	0.905	0.125	-0.418	0.900	0.164	0.032	0.005	0.003
W_60Y+	0.140	0.935	0.255	-0.584	0.727	0.270	-0.312	0.208	0.218
WALTZYES	0.078	0.771	0.022	-0.222	0.688	0.022	-0.077	0.083	0.007
WALTZJAB	0.026	0.919	0.008	0.221	0.628	0.007	0.150	0.291	0.009
WALTZNO	0.023	0.658	0.029	0.443	0.602	0.026	0.135	0.056	0.007
CHESSYES	0.029	0.935	0.048	0.632	0.933	0.066	-0.026	0.002	0.000
CHESSJAB	0.019	0.935	0.009	0.262	0.541	0.007	0.223	0.394	0.015
CHESSNO	0.079	0.933	0.031	-0.306	0.924	0.042	-0.030	0.009	0.001
TROUSYES	0.079	0.999	0.097	-0.534	0.906	0.127	0.171	0.093	0.037
TROUSJAB	0.014	0.958	0.028	0.707	0.927	0.038	-0.130	0.031	0.004
TROUSNO	0.033	0.988	0.119	0.898	0.881	0.152	-0.313	0.107	0.052
POPDAYES	0.057	0.990	0.065	0.044	0.006	0.001	0.539	0.983	0.264
POPDAJAB	0.020	0.614	0.008	0.233	0.501	0.006	0.111	0.113	0.004
POPDANO	0.048	0.987	0.084	-0.153	0.052	0.006	-0.649	0.935	0.321
LAMPYES	0.068	0.995	0.097	0.598	0.969	0.137	-0.098	0.026	0.010
LAMPJAB	0.009	0.530	0.003	-0.162	0.299	0.001	0.142	0.231	0.003
LAMPNO	0.050	0.998	0.122	-0.787	0.977	0.173	0.115	0.021	0.010
PHOTOYES	0.093	0.828	0.011	0.131	0.567	0.009	0.089	0.261	0.012
PHOTOJAB	0.016	0.511	0.028	-0.095	0.020	0.001	-0.474	0.491	0.056
PHOTONO	0.011	0.811	0.047	-0.581	0.317	0.022	-0.725	0.494	0.095
TAXYES	0.054	0.608	0.014	0.182	0.474	0.010	-0.097	0.134	0.008
TAXJAB	0.030	0.933	0.007	0.053	0.047	0.000	0.230	0.886	0.025
TAXNO	0.041	0.726	0.017	-0.275	0.710	0.017	-0.042	0.017	0.001
WATCHYES	0.070	0.992	0.040	0.360	0.874	0.051	0.132	0.117	0.019
WATCHJAB	0.012	0.532	0.003	-0.059	0.057	0.000	0.170	0.474	0.005
WATCHNO	0.042	0.986	0.062	-0.572	0.853	0.077	-0.226	0.133	0.034

with the age dimension to a relatively high degree. Both categories are positively associated with the two categories 'men, 60 years and older' and 'women, 60 years and older'. Focusing on the first axis too, 'photographing– no' is positively correlated with 'women, 60 years and older', and negatively with 'men, 60 years and older'. It can be concluded that 'photographing–no' is sex *and* age-affected (QCOR1 = 0.317, QCOR2 = 0.494); this variable category describes 'women' as well as 'elderly people'. Furthermore, 'photographing–yes' as well as some 'middle categories' on cultural abilities are sex *and* age-affected, too, e.g. 'playing chess' and 'waltzing' (see the squared correlations of the rows '... –just a bit'), but they do not have an important impact on the determination of the geometric orientation of the axes (see the low INR1 and INR2 values).

Additional sex-specific attributes are 'fixing a lamp' ('yes' and 'no') and – to a lesser degree (see the closer locations to the origin of the vector configur- ation and the lower contributions of inertia) – 'adjusting a quartz watch' ('yes' and 'no'). Both abilities are more typical for men than for women. Finally, the three categories of 'filling out a tax form' are poorly associated with both axes: their locations are close to the origin of the vector configuration and their contributions of inertia are low.

Four of the six columns load highly on the first dimension, the other two ('men, 60 years and older' and 'women up to 39 years') are more associated with the second one (QCOR2 = 0.625 and 0.706). Whereas the four groups can be distinguished by sex-affected variables, the older men and the younger women differ by the age-affected 'ability to dance to pop music' in particular.

CA solutions should be interpreted by a substantive theory (substantive assumptions). But in this example there should be no need for explaining why women have above average ability to shorten trousers and why men relatively often have the ability to fix lamps. Instead of a substantive interpretation some more details of reading the numerical print-out are given. These details are not really necessary when reporting the results but they help to understand the method.

The category 'shortening trousers–just a bit' is highly correlated with the positive part of the first axis (QCOR1 = 0.927) while the contribution of inertia is relatively low (INR1 = 0.038). In this case, the low inertia relies on the low mass of this category (MASS = 0.014). (Please note again that the contributions of inertia of the axes are calculated by multiplying the squared distances to the origin of the vector configuration by the masses of the variable categories.) This means that the category 'shortening trousers–just a bit' is typical for men, in a similar manner as the category 'shortening trousers–no' (see also the close locations for both categories). The difference is only that there are less men who asserted 'shortening trousers–just a bit' than 'shortening trousers–yes', but their profiles are quite similar.

Another example for a high squared correlation and a relatively low inertia is 'adjusting a quartz watch–yes' (QCOR1 = 0.874, INR1 = 0.051). In this case the mass is relatively high (MASS = 0.070), but (with respect to the first axis) the category is relatively close to the origin of the vector configuration (LOC1 = 0.360). As a result, 'adjusting a quartz watch–yes' has a relatively small contribution to the geometric orientation of the first axis. In other words, this 'cultural competence' can be attached as typical for men, but there are other row-categories considered in the analysis which mirror sex-specific abilities to a much higher extent.

By comparing the solutions of examples one and two we can see that no substantive changes can be reported in the display or in the numerical output. The solutions of the first analysis remain stable by including additional variables – variables which are highly correlated with the previous ones and which belong to the same types of latent dimensions, either to the 'sex-dimension' or to the 'age-dimension'. Compared with the first example, the second set of variables is more <u>affected by sex than by age and the first</u> <u>dimension becomes more important.</u>

2.3.3 Generating an artificial third dimension

The next methodological question is what happens if a variable is added which is neither correlated with age nor with sex. As an illustration a variable will be considered which distinguishes, on the one hand, men up to 39 years and women older than 59 years, and on the other hand, men 60 years and older and women up to 39 years. Even careful inspection of the original survey data did not provide such a variable which define a third dimension. Therefore a dummy variable had to be constructed with this characteristic (Table 2.7). This variable, also consisting of three categories, was added to the first data set (Table 2.1). The graphical solution of the third analysis is given in Figure 2.3 and the numerical solution in Table 2.8.

Comparing Figures 2.1 and 2.3 shows that there are no major changes in the locations of the row and column variables. The three additional categories of the dummy variable are close to the origin of the two-dimensional subspace. The main difference between the analyses is the explained variance of the axes:

TABLE 2.7
Example 3: the dummy variable.

A	M_T39Y	M_40T59Y	M_60Y+	W_T39Y	W_40T59Y	W_60Y+
DUMMY1	418 (2/3)	162 (1/3)	36 (1/9)	77 (1/9)	174 (1/3)	285 (2/3)
DUMMY2	70 (1/9)	162 (1/3)	73 (2/9)	154 (2/9)	174 (1/3)	95 (2/9)
DUMMY3	139 (2/9)	162 (1/3)	218 (2/3)	463 (2/3)	174 (1/3)	47 (1/9)

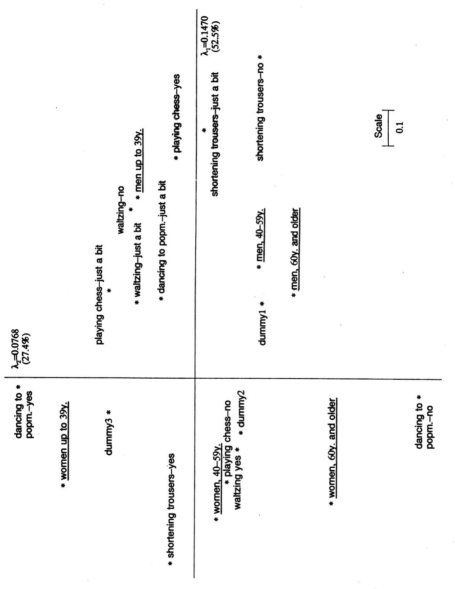

FIGURE 2.3 Example 3: graphical presentation.

TABLE 2.8
Example 3: numerical solution.

GENSTAT	MASS	SQCOR	INR	LOC1	QCOR1	INR1	LOC2	QCOR2	INR2	LOC3	QCOR3	INR3
M_T39Y	0.203	0.997	0.277	0.544	0.771	0.407	0.174	0.079	0.080	−0.238	0.147	0.246
M_40T59Y	0.158	0.868	0.101	0.320	0.571	0.110	−0.179	0.178	0.066	0.146	0.119	0.072
M_60Y+	0.105	0.932	0.126	0.234	0.164	0.039	−0.283	0.239	0.110	0.421	0.529	0.401
W_T39Y	0.226	0.995	0.208	−0.304	0.357	0.142	0.391	0.591	0.449	0.110	0.047	0.059
W_40T59Y	0.170	0.932	0.106	−0.397	0.904	0.182	−0.059	0.020	0.008	−0.040	0.009	0.006
W_60Y+	0.139	0.980	0.181	−0.356	0.346	0.120	−0.399	0.436	0.288	−0.269	0.198	0.216
WALTZYES	0.123	0.953	0.030	−0.215	0.681	0.039	−0.122	0.218	0.024	0.061	0.054	0.010
WALTZJAB	0.041	0.972	0.012	0.212	0.564	0.013	0.181	0.408	0.017	0.003	0.000	0.000
WALTZNO	0.037	0.934	0.044	0.482	0.690	0.058	0.196	0.114	0.018	−0.209	0.130	0.034
CHESSYES	0.046	0.973	0.073	0.653	0.961	0.133	0.069	0.011	0.003	0.017	0.001	0.000
CHESSJAB	0.030	0.988	0.014	0.247	0.472	0.012	0.257	0.513	0.026	−0.020	0.003	0.000
CHESSNO	0.125	0.979	0.043	−0.295	0.899	0.074	−0.088	0.080	0.013	−0.002	0.000	0.000
TROUSYES	0.125	0.995	0.135	−0.532	0.937	0.240	0.077	0.020	0.010	−0.108	0.039	0.031
TROUSJAB	0.021	0.981	0.042	0.735	0.969	0.078	−0.014	0.000	0.000	0.080	0.012	0.003
TROUSNO	0.053	0.982	0.178	0.918	0.896	0.303	−0.184	0.036	0.023	0.218	0.050	0.054
POPDAYES	0.090	0.995	0.092	−0.014	0.001	0.000	0.527	0.971	0.326	−0.082	0.023	0.013
POPDAJAB	0.032	0.573	0.013	0.225	0.455	0.011	0.102	0.094	0.004	−0.053	0.025	0.002
POPDANO	0.076	0.973	0.124	−0.060	0.008	0.002	−0.656	0.933	0.423	0.123	0.033	0.024
DUMMY1	0.075	0.998	0.102	0.212	0.118	0.023	−0.190	0.095	0.035	−0.546	0.785	0.482
DUMMY2	0.048	0.529	0.019	−0.153	0.208	0.008	−0.127	0.143	0.010	0.142	0.179	0.021
DUMMY3	0.079	0.950	0.079	−0.104	0.038	0.006	0.256	0.231	0.067	0.439	0.680	0.325

for the first axis it decreases from 63.9% to 52.5% and for the second axis from 32.2% to 27.4%, whereas the third dimension (reported in Table 2.8 only) now explains 16.6% of the total variation. The geometric orientation of this 'new' dimension – which can be labelled 'dummy-dimension' – is determined to more than 80% by the categories 'dummy1' and 'dummy3' (INR3 = 0.482 and 0.325). No further row category has a major influence of this axis. Among the columns the 'constructed' positive association between 'men up to 39 years' and 'women, 60 years and older' as well as between 'men, 60 years and older' and 'women up to 39 years' is confirmed (see the LOC3 and QCOR3 parameters in Table 2.8).

Compared with the first example, the lower correlations of some of the column variables with the first two axes have to be traced back to their (high) correlations with the third axis. Therefore, by comparing the numerical outputs referring to the first two axes only, one obtains slightly different solutions. Yet interpreting all three dimensions simultaneously as well as the graphical displays of both analyses shows that the interpretation of the first two axes is still the same: the substantive meanings of the axes did not change.

2.4   ANALYZING MOBILITY DATA

The analysis of mobility tables has a long history in social science research. One of the most famous works is given by Blau and Duncan (1967) who described occupational mobility in the United States by using multidimensional scaling (MDS). In subsequent years this technique was often used to show similarities on different kinds of mobility. Multidimensional scaling is an exploratory technique which allows similarities (dissimilarities) between objects or between variables or between objects and variables to be displayed as distances on a two-dimensional map. Because distances are very seldom measured in the social sciences, the data has to be initially transformed into any kind of similarity (dissimilarity) which can be interpreted as distances. For instance, in the case of mobility tables the 'dissimilarity index' is very often used (Blau and Duncan 1967, Best 1990). CA can be considered as a special case of MDS, where the dissimilarities are quantified using the chi-square distance and where each object (or variable) is weighted proportional to the mass.

2.4.1   The study of Best

In the 1980s Best collected a data set of 809 members of 'Frankfurter Nationalversammlung' which was the first democratic government in Germany (the parliament was in session from May 1848 to May 1849). One of Best's (1990) main points of interest were the analyses of job mobility between the main jobs of the parliamentarians in May 1848 and their fathers' occupations, as well as between the jobs of the parliamentarians when entering the labour market and their main jobs in May 1848. According to functional differences, Best classified the different jobs of the parliamentarians into ten main categories: education (including research), administration, self-employment (in particular journalists, artists, writers), justice, lawyer, military, clergy, farmer, business, and petit bourgeoisie. With respect to the main functions of Parsons' system theory (Parsons 1971), Best expected four clusters for describing the inter-generation and the intra-generation mobility. Within the clusters there should be high mobility whereas between the clusters the mobility should be low.

With respect to Best's application of Parsons' theory to the members of 'Frankfurter Nationalversammlung', 'business' and 'petit bourgeoisie' rely on a so-called 'economic subsystem' which is labelled 'adaption' (Parsons 1971). The second cluster should include persons working in the educational system as well as self-employees such as artists and journalists (these jobs can be classified as educational ones, too), this cluster is labelled 'pattern maintenance'. Considering that in the middle of the last century the clergy had a major influence in educating the population, persons working for the church belong to the second cluster, too. The third cluster includes persons working in the field of 'social control', e.g. lawyers and members of the administration and justice. This cluster is labelled 'integration'. The last cluster − labelled 'goal attainment' − includes farmers and military. Please note that farmers were not poor, hard working people and that military were not simple soldiers and as a common rule one can say: the first son got the fields, the second went to the military (see, for example, Best 1990). In general, all members of 'Frankfurter Nationalversammlung' had enough time and enough money to engage in politics.

Since the highest mobility can be expected within the single job categories, CA will be dominated by the main diagonals of the mobility tables. Furthermore, if the assumptions are true, mobility between job categories of different clusters should be relatively low and thus demonstrable on higher dimensions only. In the following, we consider the mobility within the job categories; if one is more interested in the analysis of the off-diagonal elements, one can use a variation of CA which treats the diagonal as missing elements (see, for example, Greenacre 1984, section 8.6).

In studying the intra-generation mobility of the parliamentarians Best used phi-coefficients as input information for the MDS. For describing the inter-generation mobility he used the dissimilarity index. In both analyses he was able to confirm his postulates, but he did not provide the reader with a direct comparison of both kinds of mobility. In the following application of CA to 'Frankfurter Nationalversammlung' data we use the raw frequencies of the cross-tabulations of 'main job in May 1848' by 'job of the father' as well as

TABLE 2.9
Analyzing mobility tables, input data.

A	mjustice	madmin	meduc	mmilit	mclergy	mbusin	mfarmer	mlawyer	mselfemp	mpetbour
fjustice	22	15	12	0	0	1	4	15	1	0
fadmin	27	34	20	4	4	2	8	24	5	0
feduc	6	16	9	2	7	1	0	6	8	0
fmilit	8	8	1	3	0	0	1	3	2	0
fclergy	8	6	13	0	3	1	4	6	5	0
fbusin	15	9	13	2	3	31	1	7	9	2
ffarmer	11	12	6	3	1	4	35	5	5	0
flawyer	4	10	2	0	0	1	2	10	2	0
fselfemp	5	6	8	0	2	3	0	8	8	0
fpetbour	13	14	28	1	5	12	5	20	19	7
sjustice	117	83	19	0	1	6	17	76	6	0
sadmin	11	37	6	0	0	1	7	7	6	0
seduc	4	3	67	1	4	0	3	5	13	0
smilit	0	5	1	16	0	2	8	1	0	0
sclergy	0	1	11	0	26	0	0	1	0	0
sbusin	1	0	0	0	0	14	0	0	0	1
sfarmer	0	4	0	0	0	0	19	1	0	0
slawyer	8	2	1	0	0	0	0	22	1	0
sselfemp	0	7	20	0	4	4	1	0	37	0
spetbour	0	1	1	1	0	19	0	0	3	6

by 'job when entering the labour market' as input information, concatenating both tables row-wise. In contrast to MDS, when applying CA there is no need for transforming the data into a similarity measure.

2.4.2 CA on mobility data

From the theoretical point of view, when 'Frankfurter Nationalversammlung' starts its session (May 1848) the main occupation is the most important. Therefore, this variable is used as reference variable by comparing the occupations when entering the labour market and the occupations of the fathers. The matrix of input data is given in Table 2.9.

Inspection of the inertia in each cell (not shown here) shows that the highest squared deviations between the expected and the empirical values (divided by the expected ones) are in the main diagonals of the two concatenated contingency tables. Furthermore, it can be seen that the squared deviations in the table 'main occupation in May 1848 by occupation when entering the labour market' are higher than in the table 'main occupation in May 1848 by occupation of the father'. This can be traced back to the higher variation referring to occupational stability than occupational heritage (total inertia = 2.653 versus 0.553). Yet the total variation in the data is decomposed under the condition that the highest amount of inertia is explained by the first axis of the vector-subspace (Chapters 1 and 3), the second highest amount of inertia by the second axis and so on. It can be assumed that the first axes are mainly explained by the main diagonals. The graphical solution of the CA of the mobility tables is given in Figure 2.4.

2.4.3 Interpreting the CA solution

Careful inspection of Figure 2.4 − focusing on the angles from the vector end-points to the origin of the vector subspace − shows that the four expected clusters can be separated. On the left above the first axis there are the nine categories of Parsons' main function 'pattern maintenance' (−clergy, −educ, −self-emp; the first character marks the variables: m = main occupation in May 1848, s = starting occupation when entering the labour market, f = occupation of the father). Furthermore on the left, but below the second axis, there are the six categories of Parsons 'economic subsystem' (−petbour, −busin). Not so well separated in the map, but close to one another within the clusters, are the nine categories of 'integration' (−admin, −justice, −lawyer) as well as the six categories of 'goal attainment' (−milit, −farmer). With respect to this two-dimensional solution Parsons' main functions seem to be confirmed.

The first axis explains 24.7% of the total variation in the data. The (visualized) main difference is between the categories 'adaption' and 'pattern

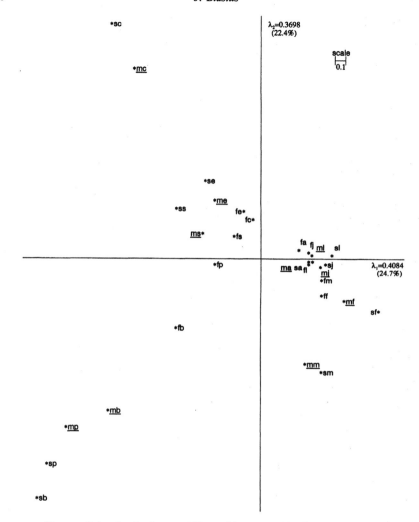

FIGURE 2.4   Analyzing mobility tables, graphical presentation.
The first letter refers to the variable: m = main occupation in May 1848, s = occupation when entering the labour market, f = occupation of the father.
The second letter refers to the occupation: c = clergy, e = education, s = self employment (pattern maintenance); j = justice, a = administration, l = lawyer (integration); b = business, p = petit bourgeoisie (adaption); f = farmer, m = military (goal attainment).

maintenance' versus 'goal attainment' and 'integration'. The second axis explains 22.4% of the total variation, the (visualized) major difference on this axis is between the categories 'pattern maintenance' versus 'goal attainment' and 'adaption'. Furthermore, by inspecting the two-dimensional solution some variation within the clusters can be reported: while the variation between the categories of Parsons' cluster 'pattern maintenance' is relatively high – the distances between the row categories as well as the distances between the column categories are relatively high – the variation within the cluster 'integration' is relatively low. Note that the reported two-dimensional map explains less than 50% of the total inertia. Therefore, further dimensions have to be considered. This can be done by using additional maps, e.g. onto the plane of axes 3 and 4, or by interpreting the numerical solutions in higher dimensions. We choose the latter (see Table 2.10) because of the added advantage of having the contributions of inertia as well as the squared correlations.

By inspecting Table 2.10 one can see that the overall structure is less clear than in the earlier examples. As in PCA, by interpreting the numerical printout one should start with the choice of a 'critical value' for considering the squared correlations, for example 0.25. This means, 25% of the variation of each variable category – in terms of PCA a factor loading greater than 0.50 – has to be explained by one axis. By using this admittedly arbitrary cutpoint there is a distinction between the column categories of 'integration' versus the categories of 'adaption', and – to a lesser degree (below the 'critical value' of 0.25) – the categories of 'pattern maintenance' which are reflected on the first axis.

At the column level, the two categories of 'adaption' determine the geometric orientation of the first axis to 40.6%, whereby the relatively low contribution of inertia of 'mpetbour' (compared with 'mbusin') is based on the low mass of this category (MASS = 0.011). For both categories, the squared correlations with the first axis as well as the distances on the first axis to the origin of the vector-subspace are similar; the mobility between the members of both occupations is relatively high (see also Table 2.9).

Although the category 'mclergy' is only lowly correlated with the first axis (QCOR1 = 0.193) and although its mass is relatively low (MASS = 0.042), it has a high influence on its geometric orientation (INR1 = 0.132). This depends on the long distance on this axis to the origin of the vector configuration. According to the first axis, there are two possible interpretations: first, there are movements from the categories '−busin' or '−petbour' above the average, and second, there are common movements with the categories of 'adaption' to the categories of 'integration' below the average.

At the row level, the same categories as at the column level are located on the same parts of the first axis (only the squared correlation of 'slawyer' is below the 'critical value' of 0.25). Furthermore, the same row and column categories have a high influence on the geometric orientation of the axes. These common high squared correlations (and their common locations on the same parts of the axis) and these common high contributions of inertia can be traced back to relatively high movements within the clusters of 'adaption' and 'integration'. In contrast, the mobilities between occupations of both clusters

TABLE 2.10

Analyzing mobility data, numerical solution.

GENSTAT	MASS	SQCOR	INR	LOC1	QCOR1	INR1	LOC2	QCOR2	INR2	LOC3	QCOR3	INR3	LOC4	QCOR4	INR4
mrjustice	0.180	0.838	0.053	0.445	0.411	0.087	-0.064	0.009	0.002	-0.398	0.329	0.096	-0.207	0.089	0.037
madmin	0.189	0.502	0.041	0.396	0.433	0.073	-0.025	0.002	0.000	-0.132	0.048	0.011	-0.083	0.019	0.006
meduc	0.165	0.762	0.093	-0.410	0.180	0.068	0.536	0.308	0.128	0.066	0.005	0.002	0.502	0.270	0.200
mmilit	0.023	0.641	0.142	0.357	0.012	0.007	-0.966	0.091	0.058	2.342	0.534	0.422	-0.195	0.004	0.004
mclergy	0.042	0.985	0.170	-1.141	0.193	0.132	1.703	0.429	0.326	0.691	0.071	0.067	-1.407	0.293	0.395
mbusin	0.071	0.970	0.174	-1.366	0.459	0.323	-1.392	0.476	0.370	-0.201	0.010	0.010	-0.317	0.025	0.034
mfarmer	0.080	0.633	0.136	0.707	0.177	0.097	-0.378	0.051	0.031	1.059	0.397	0.301	0.148	0.008	0.008
mlawyer	0.150	0.646	0.052	0.408	0.289	0.061	0.019	0.001	0.000	-0.415	0.299	0.087	-0.180	0.057	0.024
mselfemp	0.090	0.689	0.082	-0.555	0.205	0.068	0.243	0.039	0.014	-0.008	0.000	0.000	0.816	0.444	0.288
mpetbour	0.011	0.666	0.056	-1.749	0.365	0.083	-1.539	0.283	0.071	-0.308	0.011	0.004	-0.236	0.007	0.003
fjustice	0.048	0.825	0.009	0.399	0.495	0.019	0.053	0.009	0.000	-0.319	0.316	0.017	-0.044	0.006	0.000
fadmin	0.089	0.749	0.007	0.295	0.628	0.019	0.089	0.058	0.002	-0.053	0.021	0.001	-0.077	0.043	0.003
feduc	0.038	0.544	0.010	-0.151	0.052	0.002	0.439	0.436	0.020	0.095	0.020	0.001	-0.126	0.036	0.003
fmilit	0.018	0.408	0.009	0.494	0.309	0.011	-0.184	0.043	0.002	0.187	0.044	0.002	-0.098	0.012	0.001
fclergy	0.032	0.806	0.004	-0.057	0.017	0.000	0.352	0.647	0.011	0.018	0.002	0.000	0.163	0.139	0.004
fbusin	0.064	0.890	0.045	-0.767	0.506	0.092	-0.631	0.343	0.069	-0.166	0.024	0.006	-0.143	0.018	0.006
ffarmer	0.057	0.605	0.058	0.490	0.141	0.033	-0.331	0.064	0.017	0.812	0.389	0.126	0.129	0.010	0.005
flawyer	0.021	0.539	0.006	0.401	0.339	0.008	-0.048	0.005	0.000	-0.298	0.187	0.006	-0.060	0.008	0.000
fselfemp	0.028	0.661	0.005	-0.244	0.199	0.004	0.211	0.150	0.003	-0.223	0.167	0.005	0.209	0.146	0.006
fpetbour	0.086	0.616	0.018	-0.416	0.489	0.036	-0.040	0.004	0.000	-0.105	0.031	0.003	0.179	0.091	0.013
sjustice	0.225	0.960	0.067	0.518	0.544	0.148	-0.048	0.005	0.001	-0.398	0.321	0.120	-0.210	0.090	0.048
sadmin	0.052	0.283	0.022	0.422	0.259	0.023	-0.019	0.001	0.000	-0.113	0.019	0.002	0.059	0.005	0.001
seduc	0.069	0.721	0.084	-0.497	0.124	0.042	0.716	0.257	0.096	0.157	0.012	0.006	0.809	0.328	0.218
smilit	0.023	0.734	0.141	0.503	0.025	0.014	-1.038	0.106	0.067	2.476	0.601	0.472	-0.177	0.003	0.003
sclergy	0.027	0.991	0.167	-1.339	0.176	0.119	2.115	0.439	0.327	0.854	0.072	0.066	-1.761	0.304	0.403
sbusin	0.011	0.911	0.069	-1.998	0.391	0.108	-2.167	0.460	0.141	-0.404	0.016	0.006	-0.669	0.044	0.024
sfarmer	0.017	0.491	0.071	1.006	0.144	0.041	-0.498	0.035	0.011	1.467	0.306	0.120	0.210	0.006	0.004
slawyer	0.024	0.407	0.030	0.569	0.152	0.019	0.031	0.000	0.000	-0.676	0.215	0.036	-0.289	0.039	0.009
sselfemp	0.051	0.714	0.076	-0.756	0.231	0.071	0.460	0.085	0.029	0.078	0.002	0.001	0.989	0.395	0.238
spetbour	0.021	0.931	0.102	-1.907	0.463	0.191	-1.878	0.449	0.205	-0.202	0.005	0.003	-0.338	0.015	0.012

are relatively low. The similarities within, as well as the contrast between, these two clusters are the major ones for describing the occupational structure of the members of 'Frankfurter Nationalversammlung'.

If the father was in the military, there was a relatively high inter-generation mobility of the three occupations of 'integration' also. In other words, if the father was a soldier, in May 1848 the son relatively often had a job in the system of 'social control'. The relatively low contribution of inertia of 'fmilit' (INR1 = 0.011) can be interpreted as a second-order effect of this variable category: the job of the father in the military is relatively highly associated with a job in 'integration' of the son when he entered the labour market. The primary effect should be the high occupational stability within the cluster 'integration'.

Referring to the first axis, for 'clergy' a low intra-generation mobility can be reported only: while 'sclergy' determines the geometric orientation of the first axis to a relatively high extent (QCOR1 = 0.176, INR1 = 0.119), the category 'fclergy' does not load on it (QCOR1 = 0.017, INR1 = 0.000). The association of 'sclergy' with the first axis can be interpreted as occupational constancy: persons who start with a job in the clergy changed their employer relatively seldom. Because the correlations are relatively low (less than 20% of explained variance by the first axis) the given conclusion has to be verified by the description of higher-order dimensions.

Comparing both the influences of 'father's occupation' and the influence of 'occupation when entering the labour market' on 'main occupation in May 1848', one should concentrate on interpreting the contribution of inertia. In the case where the amount of missing data in both variables is nearly the same (the proportion of available data is 48.1 to 52.0, 'occupation of the father' versus 'occupation when entering the labour market'), the influence of the masses can be neglected. Adding the contributions of inertia from the ten categories of the two row-variables separately, the proportion is 224 to 776. Thus the different amount of variation in the two tables of input data is reconfirmed: the influence of the variable 'occupation when entering the labour market' on the geometric orientation of the first axis is much higher than the influence of the variable 'occupation of the father'. Although both variables are highly associated with the variable 'main occupation in May 1848', with knowledge of the 'occupation when entering the labour market' a better prediction of the main occupation in May 1848 is available than with knowledge of the 'occupation of the father'. In other words: the profiles of the variable-categories 'main occupation in May 1848' are more identical with the profiles of 'occupation when entering the labour market' than with the profiles of 'occupation of the father'.

The second axis explains 22.4% of the total variation in the data. It mirrors the contrast between the categories of 'adaption' and 'pattern maintenance' whereby the three categories of '−selfemp' have to be neglected. They load on the right part only, but they do not have an influence on the geometric orientation of the second axis to be taken into account (see column INR2).

While on the first axis the categories of '−clergy' and '−busin' load on the same part, on the second axis both are strongly negatively associated. With the knowledge that there was relatively little mobility between both occupations, the reported positive association on the first axis has to be interpreted as common contrast of '−clergy' and '−busin' against the occupations of 'integration'.

As on the first axis, the influence of the variable 'occupation when entering the labour market' on the geometric orientation of the second axis is much stronger than the influence of the variable 'occupation of the father': the proportion of the contributions of inertia is 0.877 to 0.124. This means that the contrast of 'adaption' versus 'pattern maintenance' is mainly concentrated on their (missing) mobility between the 'occupation when entering the labour market' and the 'main occupation in May 1848'. In particular, the job stability within the 'clergy' as well as the job stability within the cluster 'adaption' is relatively high. In the case of 'clergy' this solution is a verification of the conclusion made by interpreting the first axis. Between both categories of 'adaption' there should be several movements from 'petit bourgeoisie' to 'business' (in addition to the squared correlations, the location, and the contributions of inertia, see the two values in the column MASS) which can be interpreted as ascent to a better job during the time from 'entering the labour market' up to 'May 1848'.

The third axis explains 18.0% of the total inertia. This axis mirrors a contrast between 'justice' and − to a lesser degree − 'lawyer' (both belong to 'integration') and 'goal attainment'. In particular, both categories of 'goal attainment' have a relatively high job stability; in the case of the farmer there is also a relatively high occupational heritage.

The fourth axis explains 12.6% of the total variation. This axis mirrors a contrast within Parsons' main function 'pattern maintenance': 'clergy' versus 'educ' and 'selfemp'. According to this axis, between the categories 'educ' and 'selfemp', there are relatively many movements but relatively few movements between these two occupations and 'clergy'. As reported before, a job in the church resulted in a change relatively seldom.

The fifth axis (not shown here) which explains 10.7% of the total variation displays a contrast within the cluster 'goal attainment'. Axes six to eight also reflect variation within one of the clusters.

Overall, it can be shown that the greatest distinctions are between Parsons' four main functions and that there are relatively many movements between the single jobs within each cluster. For example, if we focus on 'pattern maintenance', there is a relatively high mobility within the cluster, e.g. from 'selfemp' to 'educ'. Furthermore, changes from one of these jobs to a job of the cluster 'integration' are relatively few − the categories belonging to each of them are loading on opposite sides of the first axis. Concentrating on the variation within 'pattern maintenance', the fourth axis mirrors the contrast between the categories: there is a relatively low mobility between 'clergy' and 'educ' as well as 'selfemp'. Analyzing several dimensions simultaneously shows that the mobilities between the three categories are not so simply structured as one could assume by regarding the first two axes only.

When comparing our solutions with the solutions of Best (1990) no major differences can be reported for the first two axes. But in contrast to the application of MDS, CA provides the squared factor loadings as well as the contributions of inertia for each variable category. In particular, with the contributions of inertia we could describe how strong the differences are (due to inter-generational and intra-generation mobility) between and within each of the ten categories. Furthermore, the usefulness of the coefficients available using CA will be important especially if there are solutions in more than two dimensions.

2.5 DISCUSSION

As shown, CA is a powerful instrument for exploratory analyses of categorical data. The easily interpretable display leads to a first heuristic exploration of the structure of the data, but one has always to be aware that the interpretation of the maps can have pitfalls. When using the symmetric presentation as we have, and as is usual in the French School (Chapter 6), one is not allowed to interpret the distances between rows and columns (see Greenacre 1989, 1993, as well as chapters 1 and 3). To avoid misinterpretations and to obtain coefficients like those well-known in PCA we suggested using numerical diagnostics to a large extent. When using the asymmetric map one compares profiles and vertices, whereby the subspace is optimized for the profiles only (Chapters 1 and 3). When the variation is not high (say, for example, Cramer's $V < 0.50$) and the number of profiles is not low (say, for example, > 15), there are many overlappings of the profiles in the asymmetric display. In contrast to the symmetric presentation the readability of the map deteriorates very quickly with an increasing number of profiles.

It cannot be mentioned too often that the profit as well as the danger of the method lies in the graphical display. Even the frequently used symmetric map leads to interpreting the distances between rows and columns, but this kind of interpretation is allowed only when using the asymmetric scaling (where one meets the earlier-mentioned problems). On the one hand, even the graphical display can be used (as heuristic instrument) very well. For example, using the maps it was easy to show that CA produces highly stable solutions. In the case of 'cultural competence' there were only small changes in the two-dimensional maps when adding further variables, or when adding an artificial third dimension. But on the other hand, not all changes were as small as they looked on the displays, for example, comparing the first and the third examples we obtained a new (third) dimension also described by column categories which, in the first analysis, loads highly on the first or second axes.

In general, when interpreting correspondence analysis solutions one has to take into account the three main concepts: the chi-squared distances, the profiles, and the masses. With the help of the chi-square distances one can decide if there is enough variation for further analyses or if there is white noise only. Including the masses means that the rows and columns are weighted by their marginals with respect to the sample size of the (stacked) contingency table. Finally, by using CA, one always interprets profiles instead of absolute values. Interpreting variable-categories within CA solutions means interpreting the categories with respect to all other categories included in the analysis. This kind of interpretation must have its feedback in the language: when interpreting CA solutions one has to avoid expressions such as 'most of ...' and rather use terms such as 'above the average ...'. Using this information and considering the several coefficients of the numerical output as well as the maps simultaneously, the method provides detailed results which allow a very detailed interpretation of categorical data.

Computation of Correspondence Analysis

Jörg Blasius and Michael Greenacre

3.1 INTRODUCTION

In this chapter we give a step-by-step description of the computations involved in performing simple correspondence analysis (CA). The theory of each step will be given using matrix and scalar notation and the results of each step will be illustrated by the first data matrix given as an example in Chapter 1, the 7×4 contingency table in Figure 1.1. This table is a cross-tabulation of age (seven categories) and the question if civil rights people are pushing too fast (four categories). The source of the table is the American Social Attitudes Sourcebook (Converse *et al*. 1980).

There are a number of ways to compute the solution to CA, for example using the singular value decomposition, the eigenvalue decomposition or by an algorithm known as 'reciprocal averaging'. Of these methods, we have chosen to present the computations in terms of the singular value decomposition (SVD), which is a generalization of the well-known eigenvalue decomposition (EVD). Whereas the EVD is applicable to square matrices, which are usually symmetric in statistical applications, for example to decompose the correlation matrix in principal components analysis (PCA), the SVD is used to decompose any rectangular matrix. The SVD (Eckart and Young 1936) is one of the most useful tools in multivariate analysis and is available in many matrix programming languages such as SAS/IML, S-PLUS, and GAUSS. There are many similarities between CA and PCA, which was a reason for the French to label correspondence analysis as *analyse factorielle des correspondences*. We first review the definition of the EVD before introducing the SVD.

The eigenvalue decomposition of a symmetric matrix $\mathbf{R}$ can be defined as the following product of matrices: $\mathbf{R} = \mathbf{B} \mathbf{\Lambda} \mathbf{B}^{\mathrm{T}}$, where $\mathbf{B}$ contains the eigenvectors of $\mathbf{R}$ in its columns and Λ is a diagonal matrix containing the eigenvalues in descending order: $\lambda_1 > \lambda_2 > ... > \lambda_m$. The eigenvectors are normalized to have length 1 and are orthogonal, which means that they satisfy the following condition: $\mathbf{B}^T\mathbf{B} = \mathbf{I}$. In the case of PCA, $\mathbf{R}$ is either the covariance matrix or correlation matrix of the m variables.

Principal component analysis starts with the idea that each of a set of variables can be described as a different linear combination of a smaller number of common factors. In practice this means that the $n \times m$ data matrix can be written as:

$$\mathbf{Z} = \mathbf{SL} \tag{3.1}$$

where $\mathbf{Z}$ is the cases $\times$ variables matrix of mean-centered data (when the covariance matrix is analyzed) or both mean-centered and normalized (i.e. 'z-transformed') data (when the correlation matrix is analyzed). The matrix $\mathbf{L}$ contains the well-known factor loadings, and $\mathbf{S}$ the factor scores. Equation (3.1) is the complete solution where the number of factors (i.e. the number of columns of $\mathbf{L}$) is equal to the number of variables, in which case $\mathbf{Z}$ can be reproduced without error. For a reduced number of factors, dimensions which account for the least variance (the last rows of $\mathbf{L}$ and corresponding last columns of $\mathbf{S}$) are omitted, leading to an approximation to the data which can be shown to be a least-squares fit. This is different to factor analysis in which the same reduced-rank version of (3.1) is postulated but a specific error model is incorporated, i.e. $\mathbf{Z} = \mathbf{SL} + \mathbf{U}$, with distributional assumptions on the error term $\mathbf{U}$ and unknown scores $\mathbf{S}$.

Let us consider the PCA of the correlation matrix $\mathbf{R}$, which is related to the centered and normalized $\mathbf{Z}$ as follows: $\mathbf{R} = (1/n)\mathbf{Z}^T\mathbf{Z}$. The orthonormality of the factor scores is conventionally in the following form: $(1/n)\mathbf{S}^T\mathbf{S} = \mathbf{I}$. From (3.1) it follows that

$$\mathbf{R} = (1/n)\mathbf{L}^T\mathbf{S}^T\mathbf{SL} = \mathbf{L}^T\mathbf{L} \tag{3.2}$$

Again this is the complete solution when all dimensions are included and will be an approximation when less factors are used.

Now if we rewrite the eigenvalue-decomposition of $\mathbf{R}$: $\mathbf{R} = \mathbf{B}\Lambda\mathbf{B}^T$, as $\mathbf{R} = \mathbf{B}\Lambda^{1/2}(\mathbf{B}\Lambda^{1/2})^T$, then it can be shown that the matrix of factor loadings can be simply computed by $\mathbf{L} = (\mathbf{B}\Lambda^{1/2})^T$. Because the columns of $\mathbf{B}$ are normalized, the sum of squares of the elements in each row of $\mathbf{L}$ is the corresponding eigenvalue, the sum of the complete set of eigenvalues is equal to the trace of $\mathbf{R}$, which is the number of variables. Expressing each eigenvalue as a percentage of their sum provides a measure of explained variance of each factor. These explained variances are the basis for interpreting a PCA solution in any social science application. The sum of squares in each column of $\mathbf{L}$ is equal to the explained variance of the corresponding variable, which is exactly one in the complete solution.

In the reduced solution, for example a two-dimensional solution, only the first two rows of **L** are retained. If we again use the squared factor loadings and sum them up over the two rows in each column, the explained variance we obtain is known as the 'communality' of the respective variable. The eigenvalues (or, more commonly, the explained variances of the eigenvalues), the factor loadings and the communalities are the coefficients used for interpretation of a PCA solution. As shown in this chapter, all three diagnostic instruments have their equivalents in correspondence analysis.

Whereas the eigenvalue-eigenvector decomposition applies to square matrices, the singular-value decomposition (SVD) applies to rectangular matrices. The SVD is defined as the decomposition of an $I \times J$ matrix **A** as the product of three matrices:

$$\mathbf{A} = \mathbf{U}\boldsymbol{\Gamma}\mathbf{V}^{\mathrm{T}} \qquad (3.3)$$

where the matrix $\boldsymbol{\Gamma}$ is a diagonal matrix of positive numbers in descending order: $\gamma_1 \geqslant \gamma_2 \geqslant \ldots \geqslant \gamma_K > 0$, K is the rank of **A**, and the columns of the matrices **U** and **V** are orthonormal, that is $\mathbf{U}^{\mathrm{T}}\mathbf{U} = \mathbf{V}^{\mathrm{T}}\mathbf{V} = \mathbf{I}$. The **K** numbers $\gamma_1, \gamma_2, \ldots$ are called 'singular values', the K columns of **U** are called 'left singular vectors' and the K columns of **V** are called 'right singular vectors'. The close connection between the SVD and the eigenvalue-eigenvector decomposition is embodied in the following two results which are obtainable directly from the definition (3.1) of the SVD:

$$\mathbf{A}^{\mathrm{T}}\mathbf{A} = \mathbf{V}\boldsymbol{\Gamma}^2\mathbf{V}^{\mathrm{T}} \qquad (3.4)$$

$$\mathbf{A}\mathbf{A}^{\mathrm{T}} = \mathbf{U}\boldsymbol{\Gamma}^2\mathbf{U}^{\mathrm{T}} \qquad (3.5)$$

Equation (3.4) shows that the right singular vectors of **A** are identical to the eigenvectors of $\mathbf{A}^{\mathrm{T}}\mathbf{A}$ and the singular values of **A**, $\gamma_1^2, \gamma_2^2, \ldots \gamma_k^2$, are the square roots of the eigenvalues of $\mathbf{A}^{\mathrm{T}}\mathbf{A}$. Equation (3.5) shows the similar connection to the eigenvalues and eigenvectors of $\mathbf{A}\mathbf{A}^{\mathrm{T}}$.

The usefulness of the SVD in our present context is as follows. We consider the general case of a set of I points in J-dimensional space, where coordinates are in the rows of the matrix **Y**, with masses $w_1, w_2, \ldots, w_I$ assigned to the respective points and where the space is structured by the weighted Euclidean with dimension weights $q_1, q_2, \ldots, q_J$ associated with the respective dimensions (i.e. the distance between any two points, say **x** and **y**, is equal to $\sqrt{(\mathbf{x} - \mathbf{y})^{\mathrm{T}} \mathbf{D}_q(\mathbf{x} - \mathbf{y})}$, see Chapter 1). Let $\mathbf{D}_w$ and $\mathbf{D}_q$ be the diagonal matrices of point masses and dimension weights respectively, and let **w** be the vector of point masses, where we assume that the point masses add up to 1: $\Sigma_i w_i = 1$, or in matrix notation: $\mathbf{1}^{\mathrm{T}}\mathbf{w} = 1$, where **1** is the vector of ones. Then the general result is that any low-dimensional map of the points can be derived directly from the SVD of the following matrix:

$$A = D_w^{1/2}(Y - 1\bar{y}^T)D_q^{1/2} \tag{3.6}$$

where $\bar{y}^T$ is the centroid of the rows of Y. If we write the SVD as before:

$$A = U\Gamma V^T \tag{3.7}$$

then the principal coordinates of the row points, i.e. the projections of the row profiles onto principal axes, are contained in the following matrix:

$$F = D_w^{-1/2}U\Gamma \tag{3.8}$$

for example, the coordinates of the points in an optimal two-dimensional display are contained in the first two columns of F. The principal axes of the display are contained in the matrix:

$$A = D_q^{-1/2}V \tag{3.9}$$

for example, the first two principal axes are the first two column vectors of A. This general result is proved and illustrated in detail by Greenacre (1984, Chapter 2 and Appendix A).

The simplest illustration of this result is the geometric definition of PCA itself, where Y is the raw data matrix, $D_w = (1/n)I$ (equal masses for all cases) and D_q is either the identity matrix (no differential weighting of the variables) or the diagonal matrix of the inverses of the variances (when variables are normalized).

Coming now to CA, we have two special cases of the above general result, namely the row and column problems where we aim to reduce the dimensionality of the row profiles and column profiles, respectively, where each set of points has its own set of associated masses and weighted Euclidean metric. As we demonstrate in this chapter, both these problems reduce to the SVD of the same matrix of so-called 'standardized residuals' of the contingency table.

At this point it is convenient to start the description of the computations. In the course of the description all the relevant notation will be defined. All the results which accompany the theoretical description have been obtained using SAS/IML (SAS Institute Inc. 1985).

3.2 COMPUTATIONS

3.2.1   We start with the contingency table N ($I \times J$). In this case N is the table given in Figure 1.1 and $I = 7$, $J = 4$ (see Table 3.1).

TABLE 3.1
Input data.

Age group	Too fast	About right	Too slow	Don't know
18–24	69	37	7	5
25–34	148	45	14	22
35–44	170	65	12	29
45–54	159	57	12	28
55–64	122	26	6	18
65–74	106	21	5	23
75+	40	7	1	14

3.2.2 First the row sums $n_{1+}, n_{2+}, \ldots, n_{I+}$ of **N** are calculated; e.g. $n_{2+} = \Sigma_j n_{2j} = 229$ (see Table 3.2).

TABLE 3.2
Row sums.

18–24	118
25–34	229
35–44	276
45–54	256
55–64	172
65–74	155
75+	62

3.2.3 The column sums $n_{+1}, n_{+2}, \ldots, n_{+J}$; e.g. $n_{+4} = \Sigma_i n_{i4} = 139$ (see Table 3.3).

TABLE 3.3
Column sums.

Too fast	About right	Too slow	Don't know
814	258	57	139

3.2.4 The grand total of the table is $n = \Sigma_i \Sigma_j n_{ij} = 1268$.

3.2.5 The row masses $r_1, r_2, \ldots, r_I$ are the row sums divided by the grand total, $r_i = n_{i+}/n$; e.g. $r_2 = 229/1268 = 0.1806$. We denote the vector of row masses by **r** (see Table 3.4).

TABLE 3.4
Row masses.

18–24	0.0931
25–34	0.1806
35–44	0.2177
45–54	0.2019
55–64	0.1356
65–74	0.1222
75 +	0.0489

3.2.6 The column masses $c_1, c_2, ..., c_J$ are the column sums divided by the grand total, $c_j = n_{+j}/n$; e.g. $c_4 = 139/1268 = 0.1096$. We denote the vector of row masses by **c** (see Table 3.5).

TABLE 3.5
Column masses.

Too fast	About right	Too slow	Don't know
0.6420	0.2035	0.0450	0.1096

3.2.7 The correspondence matrix **P** is defined as the original table **N** divided by the grand total n, $\mathbf{P} = (1/n)\mathbf{N}$; e.g. $p_{24} = 22/1268 = 0.0174$. Notice that the row and column sums of **P** are the row and column masses respectively; e.g. $r_2 = 0.1167 + 0.0355 + 0.0110 + 0.0174 = 0.1806$ (see Table 3.6).

TABLE 3.6
Correspondence matrix.

Age group	Too fast	About right	Too slow	Don't know
18–24	0.0544	0.0292	0.0055	0.0039
25–34	0.1167	0.0355	0.0110	0.0174
35–44	0.1341	0.0513	0.0095	0.0229
45–54	0.1254	0.0450	0.0095	0.0221
55–64	0.0962	0.0205	0.0047	0.0142
65–74	0.0836	0.0166	0.0039	0.0181
75 +	0.0315	0.0055	0.0008	0.0110

3.2.8 The row profiles are the rows of the original table **N** divided by their respective row totals; e.g. the first row profile consists of the first row of **N**

[69 37 7 5], divided by its total, 118. Equivalently, the matrix of row profiles can be defined as the rows of the correspondence matrix **P** divided by their respective row sums (i.e. row masses), which can be written as $D_r^{-1}P$, where D_r is the diagonal matrix of row masses; e.g. $p_{11}/r_1 = 0.0544/0.0931 = 0.5847$ (see Table 3.7).

TABLE 3.7
Row profiles.

Age group	Too fast	About right	Too slow	Don't know
18–24	0.5847	0.3136	0.0593	0.0424
25–34	0.6463	0.1965	0.0611	0.0961
35–44	0.6159	0.2355	0.0435	0.1051
45–54	0.6211	0.2227	0.0469	0.1094
55–64	0.7093	0.1512	0.0349	0.1047
65–74	0.6839	0.1355	0.0323	0.1484
75 +	0.6452	0.1129	0.0161	0.2258

3.2.9 The column profiles are the columns of the original table **N** divided by their respective totals; e.g. the first column profile consists of the first column of **N** [69 148 ... 40], divided by its total, 814. Equivalently, the matrix of column profiles consists of the columns of the correspondence matrix **P** divided by their respective column sums: PD_c^{-1}, where D_c is the diagonal matrix of column masses; e.g. $p_{11}/c_1 = 0.0544/0.6420 = 0.0848$ (see Table 3.8).

TABLE 3.8
Column profiles.

Age group	Too fast	About right	Too slow	Don't know
18–24	0.0848	0.1434	0.1228	0.0360
25–34	0.1818	0.1744	0.2456	0.1583
35–44	0.2088	0.2519	0.2105	0.2086
45–54	0.1953	0.2209	0.2105	0.2014
55–64	0.1499	0.1008	0.1053	0.1295
65–74	0.1302	0.0814	0.0877	0.1655
75 +	0.0491	0.0271	0.0175	0.1007

3.2.10 Equivalence of row and column problems. At this point we have assembled all the entities we require to define both the row and column problems. Let us consider the row problem, for example, which consists of a set of I profiles (in the rows of $D_r^{-1}P$, with masses **r** in the diagonal matrix D_r, in a space with distance defined by the diagonal matrix D_c^{-1}). We need to derive the centroid of the row profiles, which is: $r^T D_r^{-1} P = 1^T P = c^T$, the row vector of column masses. The matrix A in (3.6) is thus:

$$A = D_r^{1/2}(D_r^{-1}P - 1c^T)D_c^{-1/2} \tag{3.10}$$

which can be rewritten as:

$$A = D_r^{-1/2}(P - rc^T)D_c^{-1/2}$$

On the other hand, the column problem consists of a set of J profiles in the columns of PD_c^{-1}, with masses c in the diagonal of D_c, in a space with distance defined by the diagonal matrix D_r^{-1}. The difference here is that the profiles are in the columns, not the rows as assumed by the general theory in section 3.1, so we simply transpose the matrix PD_c^{-1} of column profiles to obtain $D_c^{-1}P^T$. The centroid of these profiles is $c^T D_c^{-1} P^T = 1^T P^T = r^T$, the row vector of row masses. The matrix in (3.6) is thus:

$$A = D_c^{1/2}(D_c^{-1}P^T - 1r^T)D_r^{-1/2}$$
$$= D_c^{-1/2}(P^T - cr^T)D_r^{-1/2} \tag{3.11}$$

which is just the transpose of the matrix derived for the row problem above. It follows that both the row and column problems are solved by computing the SVD of the same matrix, called the matrix of standardized residuals (Table 3.9):

$$A = D_r^{-1/2}(P - rc^T)D_c^{-1/2} \tag{3.12}$$

i.e. the $I \times J$ matrix with elements

$$a_{ij} = (p_{ij} - r_i c_j)/\sqrt{r_i c_j} \tag{3.13}$$

e.g.

$$a_{11} = (0.0544 - 0.0931 \times 0.6420)/\sqrt{0.0931 \times 0.6420} = -0.0218$$

TABLE 3.9
Standardized residuals.

Age group	Too fast	About right	Too slow	Don't know
18–24	− 0.0218	0.0745	0.0207	− 0.0620
25–34	0.0023	− 0.0066	0.0324	− 0.0174
35–44	− 0.0151	0.0331	− 0.0032	− 0.0064
45–54	− 0.0117	0.0191	0.0041	− 0.0003
55–64	0.0310	− 0.0427	− 0.0175	− 0.0055
65–74	0.0183	− 0.0527	− 0.0209	0.0409
75+	0.0009	− 0.0444	− 0.0301	0.0776

Let us look at formula (3.13) more closely. In the brackets we find the difference between the observed proportion p_{ij} in the (ij)th cell and the expected proportion $r_i c_j$ calculated from the product of the corresponding marginal proportions. If we multiplied this expression by the sample size n, we would obtain the observed frequency n_{ij} and expected frequency $(n_{i+}n_{+j})/n$, respectively.

3.2.11   In Tables 3.10, 3.11 and 3.12 we show the singular values as well as the left and right singular vectors of the SVD (3.3) of $\mathbf{A}$:

$$\mathbf{A} = \mathbf{U}\boldsymbol{\Gamma}\mathbf{V}^{\mathrm{T}}$$

TABLE 3.10
Singular values.

0.1611
0.0617
0.0324
0.0000

TABLE 3.11
Left singular vectors.

− 0.6267	0.0888	− 0.2293	− 0.3051
− 0.0937	− 0.3761	0.7776	− 0.1857
− 0.1815	0.3513	− 0.2252	− 0.3450
− 0.1059	0.2401	0.0882	0.0107
0.2331	− 0.6108	− 0.5227	0.1302
0.4470	− 0.0937	− 0.0611	− 0.8462
0.5478	0.5364	0.0853	0.1420

TABLE 3.12
Right singular vectors.

0.2067	− 0.5036	− 0.2485	0.8012
− 0.6946	0.5269	− 0.1910	0.4511
− 0.2839	− 0.2269	0.9072	0.2120
0.6279	0.6460	0.2807	0.3311

3.2.12   If we square the standardized residuals (3.13), then sum them up over the $I \times J$ cells of the table and multiply the result by the total n, we obtain the chi-square statistic for the contingency table:

$$\chi^2 = n \sum_i \sum_j \frac{(p_{ij} - r_i c_j)^2}{r_i c_j} = 39.09$$

Thus the decomposition of the matrix $\mathbf{A}$ of standardized residuals implicitly involves a decomposition of the chi-square statistic, which is proportional to the sum of squares of the elements of $\mathbf{A}$, i.e. trace $(\mathbf{A})$.

3.2.13 From 3.2.12 it follows that the chi-square statistic can be decomposed into $I \times J$ components of the form:

$$n \times \frac{(p_{ij} - r_i c_j)^2}{r_i c_j}$$

e.g. for $i = j = 1$, $= 1268 \times (0.0544 - 0.0931 \times 0.6420)^2/(0.0931 \times 0.6420)$
$= 1268 \times (-0.0218)^2 = 0.600$ (cf. a_{11} in Table 3.9).

These are identical to the usual terms in the chi-square calculation of the form:

$$\frac{(n_{ij} - n_{i+}n_{+j}/n)^2}{(n_{i+}n_{+j}/n)}$$

where $n_{i+}n_{+j}/n$ is the well-known 'expected frequency', e.g. for $i = j = 1$, $(69 - 118 \times 814/1268)^2/(118 \times 814/1268) = 0.600$ (see Table 3.13).

TABLE 3.13
Chi-square components.

Age group	Too fast	About right	Too slow	Don't know
18–24	0.60	7.03	0.54	4.87
25–34	0.01	0.05	1.33	0.38
35–44	0.29	1.39	0.01	0.05
45–54	0.17	0.46	0.02	0.00
55–64	1.22	2.31	0.39	0.04
65–74	0.42	3.52	0.56	2.12
75+	0.00	2.50	1.15	7.63

3.2.14 The sum of the squares of the elements of $\mathbf{A}$ is the 'total inertia' of the contingency table:

$$\text{total inertia} = \sum_i \sum_j \frac{(p_{ij} - r_i c_j)^2}{r_i c_j}$$

which from (3.2.12) is the chi-square statistic divided by n:

$$\text{total inertia} = \chi^2/n = 39.09/1268 = 0.0308$$

The total inertia of 0.0308 in this case is quite small, in other words the degree of row–column association is low (see Figure 1.2 in Chapter 1).

3.2.15 There are $K = \min\{I - 1, J - 1\}$ dimensions in the solution, in the present example $K = \min\{6, 3\} = 3$. The squares of the singular values of $\mathbf{A}$ in 3.2.10, i.e. the eigenvalues of $\mathbf{A}^T\mathbf{A}$ or $\mathbf{A}\mathbf{A}^T$, also decompose the total inertia. These are denoted by $\lambda_1, \lambda_2, ..., \lambda_K$ and are called the principal inertias. Like in PCA, the principal inertias are often expressed as percentages of the total inertia, e.g. the first principal inertia of 0.02597 is 84.2% of the total inertia of 0.03083 (see Table 3.14).

TABLE 3.14
Principal inertias and percentages

	Principal inertia	Explained inertia (%)
Axis 1	0.0260	84.2
Axis 2	0.0038	12.4
Axis 3	0.0011	3.4
Axis 4	0.0000	0.0

3.2.16 The principal coordinates of the rows are obtained using (3.8), for the row problem, i.e.:

$$\mathbf{F} = \mathbf{D}_r^{-1/2}\mathbf{U}\boldsymbol{\Gamma}$$

or in scalar notation:

$$f_{ik} = u_{ik}\gamma_k/\sqrt{r_i}$$

e.g. $f_{11} = -0.6267 \times 0.1611/\sqrt{0.0931} = -0.3310$ (see Table 3.15).

TABLE 3.15
Principal coordinates of rows.

Age group	Axis 1	Axis 2	Axis 3
18–24	−0.3310	0.0180	−0.0244
25–34	−0.0355	−0.0546	0.0594
35–44	−0.0627	0.0465	−0.0157
45–54	−0.0380	0.0330	0.0064
55–64	0.1020	−0.1024	−0.0460
65–74	0.2060	−0.0165	−0.0057
75+	0.3992	0.1497	0.0125

3.2.17 The principal coordinates of the columns are obtained using (3.8), for the column problem, i.e.:

$$\mathbf{G} = \mathbf{D}_c^{-1/2}\mathbf{V}\boldsymbol{\Gamma}$$

i.e.

$$g_{ij} = v_{jk}\gamma_k/\sqrt{c_j}$$

e.g. $g_{42} = 0.6460 \times 0.0617/\sqrt{0.1096} = 0.1204$ (see Table 3.16).

TABLE 3.16
Principal coordinates of columns.

Response	Axis 1	Axis 2	Axis 3
Too fast	0.0416	− 0.0388	− 0.0101
About right	− 0.2481	0.0721	− 0.0137
Too slow	− 0.2157	− 0.0660	0.1388
Don't know	0.3056	0.1204	0.0275

3.2.18 The standard coordinates of the rows are the principal coordinates divided by their respective singular values, in other words the standard coordinates are

$$\mathbf{X} = \mathbf{F}\mathbf{\Gamma}^{-1} = \mathbf{D}_r^{-1/2}\mathbf{U}$$

i.e.

$$x_{ik} = f_{ik}/\gamma_k$$

e.g. $x_{11} = -0.3310/0.1611 = -2.0545$ (see Table 3.17).

TABLE 3.17
Standard coordinates of rows.

Age group	Axis 1	Axis 2	Axis 3
18–24	− 2.0545	0.2912	− 0.7518
25–34	− 0.2205	− 0.8849	1.8298
35–44	− 0.3891	0.7530	− 0.4827
45–54	− 0.2356	0.5345	0.1962
55–64	0.6328	−1.6584	−1.4193
65–74	1.2785	− 0.2680	− 0.1749
75 +	2.4776	2.4259	0.3858

3.2.19 The standard coordinates of the columns are the principal coordinates divided by their respective singular values, in other words the standard coordinates are

$$\mathbf{Y} = \mathbf{G}\mathbf{\Gamma}^{-1} = \mathbf{D}_c^{-1/2}\mathbf{V}$$

i.e.

$$y_{ik} = g_{ik}/\gamma_k$$

e.g. $y_{22} = 0.0721/0.0617 = 1.1682$ (see Table 3.18).

TABLE 3.18
Standard coordinates of columns.

Response	Axis 1	Axis 2	Axis 3
Too fast	0.2579	-0.6285	-0.3102
About right	-1.5398	1.1682	-0.4234
Too slow	-1.3389	-1.0701	4.2788
Don't know	1.8966	1.9511	0.8477

Note that the first map in Figure 1.1 shows the joint plot of the rows and columns, both in terms of their first two principal coordinates, i.e. it is a map constructed using the first two columns of $\mathbf{F}$ and the first two columns of $\mathbf{G}$, where in each case the first column is used for the horizontal (x) axis and the second column for the vertical (y) axis.

3.2.20   Each principal inertia λ_k is decomposed into components $r_i f_{ik}^2$ for each row i:

$$\lambda_k = \sum_i r_i f_{ik}^2$$

or in matrix notation:

$$\mathbf{D}_\lambda = \mathbf{F}^T \mathbf{D}_r \mathbf{F}$$

The values of the inertia components $r_i f_{ik}^2$ are given in Table 3.19, e.g. for the first row and first column: $0.0931 \times (-0.3310)^2$. Summing up each column of this matrix gives the eigenvalues, e.g. $\lambda_1 = 0.01020 + \ldots + 0.00779 = 0.0260$.

TABLE 3.19
Inertia of each axis by rows.

Age group	Axis 1	Axis 2	Axis 3
18–24	0.01020	0.00003	0.00006
25–34	0.00023	0.00054	0.00064
35–44	0.00086	0.00047	0.00005
45–54	0.00029	0.00022	0.00001
55–64	0.00141	0.00142	0.00029
65–74	0.00519	0.00003	0.00000
75+	0.00779	0.00110	0.00001

3.2.21 The contributions of the rows to the principal inertia are conventionally defined as the inertia components relative to their total (the principal inertia λ_k):

$$r_i f_{ik}^2 / \lambda_k$$

e.g. for the first row and first column: $0.01020/0.02597 = 0.3928$ (see Table 3.20).

TABLE 3.20
Explained inertia of axis by rows.

Age group	Axis 1	Axis 2	Axis 3
18–24	0.3928	0.0079	0.0526
25–34	0.0088	0.1414	0.6047
35–44	0.0329	0.1234	0.0507
45–54	0.0112	0.0577	0.0078
55–64	0.0543	0.3731	0.2732
65–74	0.1998	0.0088	0.0037
75+	0.3001	0.2877	0.0073

The interpretation of the row contributions is opposite to that of the squared factor loadings. Whereas the squared factor loadings tell to what extent each row category and each column category is described by the axes, the contributions of inertia show to what extent the geometric orientation of an axis is determined by the single variable categories.

3.2.22 For the ith row the inertia components for all K axes sum up to the 'row inertia' of the ith row, which is defined as the mass × squared distance of the row profile to the centroid:

$$\sum_j (p_{ij}/r_i - c_j)^2 / c_j = \sum_k r_i f_{ik}^2$$

e.g. for the last row: $0.00779 + 0.00110 + 0.00001 = 0.00890$. The row inertia on the left-hand side is identical to the sum of squared elements in the ith row of **A** (cf. (3.13)):

$$\sum_j s_{ij}^2 = \sum_j (p_{ij} - r_i c_j)^2 / (r_i c_j)$$

e.g. for the last row: $0.0009^2 + 0.0444^2 + 0.0301^2 + 0.0776^2 = 0.00890$ (see Table 3.21).

TABLE 3.21
Inertias of rows.

18–24	0.01028
25–34	0.00140
35–44	0.00138
45–54	0.00052
55–64	0.00312
65–74	0.00523
75+	0.00890

3.2.23 The squared correlations of the rows with the principal axes are the inertia components $r_i f^2_{ik}$ expressed relative to the row inertia:

$$r_i f^2_{ik} / \sum_j s^2_{ij}$$

e.g. first row and first column: $0.01020/0.01028 = 0.9917$. These can be interpreted geometrically as the squared cosines of the angles between each row profile and each principal axis (see Table 3.22).

TABLE 3.22
Contributions of principal axes to rows.

Age group	Axis 1	Axis 2	Axis 3
18–24	0.9917	0.0029	0.0054
25–34	0.1625	0.3840	0.4535
35–44	0.6204	0.3409	0.0387
45–54	0.5608	0.4234	0.0158
55–64	0.4522	0.4556	0.0922
65–74	0.9928	0.0064	0.0008
75+	0.8759	0.1232	0.0009

3.2.24 In the reduced K^*-dimensional space, the explained inertia can be summed over the K^* axes to obtain a measure of quality of display for each row:

$$\text{quality of } i\text{th row} = \sum_{k=1}^{K^*} r_i f^2_{ik} / \sum_j s^2_{ij}$$

e.g. the quality of the second row in the two-dimensional map is: $0.1625 + 0.3840 = 0.5465$ (in other words, about 55% of the inertia is contained in the two-dimensional map, and 45% in the remaining third dimension, so that this is not a well-represented point, whereas the qualities of the other points are very high (>90%) in the map, Table 3.23). Geometrically the qualities can also be interpreted as the squared cosines of the angle between each row profile and the subspace defined by the first K^* axes. The qualities are equivalent to the communalities in PCA.

TABLE 3.23
Quality of rows in 2-D
map.

18–24	0.9946
25–34	0.5465
35–44	0.9613
45–54	0.9842
55–64	0.9078
65–74	0.9992
75 +	0.9991

3.2.25 The square root of the quantities in Table 3.22 are correlations between each row profile and each principal axis. By attaching the sign of the corresponding coordinate to each correlation, we obtain the equivalent of the factor loadings in PCA. For example, the first row (18–24) has a negative first coordinate (-0.3310, see Table 3.15), so the correlation between this row and the first principal axis is: $-\sqrt{0.9917} = -0.9958$ (see Table 3.24).

TABLE 3.24
Correlation of rows with axes.

Age group	Axis 1	Axis 2	Axis 3
18–24	−0.9958	0.0541	−0.0734
25–34	−0.4031	−0.6196	0.6735
35–44	−0.7876	0.5839	−0.1967
45–54	−0.7489	0.6507	0.1255
55–64	0.6725	−0.6750	−0.3036
65–74	0.9964	−0.0800	−0.0274
75 +	0.9359	0.3510	0.0293

Notice that the following sections, 3.2.26–3.2.31, are similar to 3.2.20–3.2.26 applied to the columns.

3.2.26 Each principal inertia λ_k is decomposed into components $c_j g_{jk}^2$ for each row j:

$$\lambda_k = \sum_j c_j g_{jk}^2$$

i.e.

$$\mathbf{D}_\lambda = \mathbf{G}^\mathrm{T}\mathbf{D}_c\mathbf{G}$$

The values of the inertia components $c_j g_{jk}^2$ are given in Table 3.25, e.g. second row, first column:

$$0.2035 \times (-0.2481)^2 = 0.01253$$

As in section 3.2.20, the column sums of this matrix are the eigenvalues, e.g.

$$\lambda_1 = 0.00111 + 0.01253 + 0.00209 + 0.01024 = 0.0260$$

TABLE 3.25
Inertia of each axis by columns.

Response	Axis 1	Axis 2	Axis 3
Too fast	0.00111	0.00097	0.00007
About right	0.01253	0.00106	0.00004
Too slow	0.00209	0.00020	0.00087
Don't know	0.01024	0.00159	0.00008

3.2.27 The contributions of the columns to the principal inertia are conventionally defined as the inertia components relative to their total (the principal inertia λ_k):

$$c_j g_{jk}^2 / \lambda_k$$

e.g. second row, first column: $0.01253/0.0260 = 0.4824$ (see Table 3.26).

TABLE 3.26
Explained inertia of each axis by columns.

Response	Axis 1	Axis 2	Axis 3
Too fast	0.0427	0.2536	0.0618
About right	0.4824	0.2777	0.0365
Too slow	0.0806	0.0515	0.8230
Don't know	0.3943	0.4173	0.0788

3.2.28   For the jth column the inertia components for all K axes sum up to the column inertia of the jth column (again the mass × squared distance of the column profile to the centroid):

$$\sum_i (p_{ij}/r_i - c_j)^2/c_j = \sum_k c_j g_{jk}^2$$

e.g. for the second column (About right):

$$0.01253 + 0.00106 + 0.00004 = 0.01362$$

(See Table 3.27.)

TABLE 3.27
Inertias of columns.

Too fast	0.00214
About right	0.01362
Too slow	0.00315
Don't know	0.01191

The column inertia is identical to the sum of squared elements in the jth column of **A** (cf. (3.13)):

$$\sum_i s_{ij}^2 = \sum_i (p_{ij} - r_i c_j)^2/(r_i c_j)$$

e.g. for the second column:

$$(0.0745)^2 + (-0.0066)^2 + \ldots + (-0.0444)^2 = 0.01362$$

3.2.29   The squared correlations of the columns with the principal axes are the inertia components $c_j g_{jk}^2$ expressed relative to the column inertia:

$$c_j g_{jk}^2 / \sum_i s_{ij}^2$$

e.g. the squared correlation between the second column (About right) and the first axis: $0.01253/0.01362 = 0.9195$ (see Table 3.28).

TABLE 3.28
Contributions of principal axes to columns.

Response	Axis 1	Axis 2	Axis 3
Too fast	0.5182	0.4514	0.0304
About right	0.9195	0.0776	0.0028
Too slow	0.6633	0.0622	0.2745
Don't know	0.8596	0.1335	0.0070

3.2.30 In the reduced K^*-dimensional space, the explained inertia can be summed over the K^* axes to obtain a measure of quality of display for each column:

$$\text{quality of } j\text{th column} = \sum_{k=1}^{K^*} c_j g_{jk}^2 \Big/ \sum_i s_{ij}^2$$

e.g. the quality of the first column (Too fast) in the two-dimensional map is: $0.5182 + 0.4514 = 0.9696$, i.e. 97% of the inertia of this profile is displayed in the map (Table 3.29). Again this value can be interpreted as a communality and, geometrically, as the squared cosine of the angle between the profile and the two-dimensional map, hence the angle is quite small and the profile vector almost coincides with the map.

TABLE 3.29
Quality of columns in 2-D
map.

Too fast	0.9696
About right	0.9971
Too slow	0.7255
Don't know	0.9931

3.2.31 The square root of the quantities in Table 3.28 are correlations between each column profile and each principal axis. By attaching the sign of the corresponding coordinate to each correlation, we obtain the equivalent of the factor loadings in PCA. For example, the first column has a positive first coordinate (0.0416, see Table 3.16), so the correlation between this column and the first principal axis is: $+\sqrt{0.5182} = 0.7199$ (see Table 3.30).

TABLE 3.30
Correlations of columns with axes.

Response	Axis 1	Axis 2	Axis 3
Too fast	0.7199	−0.6719	−0.1743
About right	−0.9589	0.2786	−0.0531
Too slow	−0.8144	−0.2493	0.5240
Don't know	0.9271	0.3653	0.0834

3.2.32 The row profiles are situated in a K-dimensional space (e.g. $K = 3$ in this example). The chi-square distances between the row points can be calculated exactly in this full space. For the ith and i'th rows, the distance between them is:

$$\sqrt{\sum_j \left(\frac{p_{ij}}{r_i} - \frac{p_{i'j}}{r_{i'}}\right)^2 / c_j}$$

e.g. the distance between the first and second rows is the square root of:

$(0.0544/0.0931 - 0.1167/0.1806)^2/0.6402 + \dots$

$+ (0.0039/0.0931 - 0.0174/0.1806)^2/0.1096$

which is equal to 0.3156 (see Table 3.31). These distances are also the usual Euclidean distances between the principal coordinates in all K dimensions:

$$\sqrt{\sum_{k=1}^{K} (f_{ik} - f_{i'k})^2}$$

e.g. between the first and second rows:

$$\sqrt{(-0.3310 + 0.0355)^2 + (0.0180 + 0.0546)^2 + (-0.0244 - 0.0594)^2} = 0.3156$$

TABLE 3.31
Distances between row profiles.

Age group	18–24	25–34	35–44	45–54	55–64	65–74	75+
18–24	0.0000	0.3156	0.2700	0.2951	0.4499	0.5385	0.7430
25–34	0.3156	0.0000	0.1288	0.1024	0.1797	0.2530	0.4827
35–44	0.2700	0.1288	0.0000	0.0358	0.2240	0.2762	0.4742
45–54	0.2951	0.1024	0.0358	0.0000	0.2016	0.2492	0.4525
55–64	0.4499	0.1797	0.2240	0.2016	0.0000	0.1408	0.3941
65–74	0.5385	0.2530	0.2762	0.2492	0.1408	0.0000	0.2555
75+	0.7430	0.4827	0.4742	0.4525	0.3941	0.2555	0.0000

3.2.33 In the low-dimensional map, or reduced space, of dimensionality K^* (e.g. $K^* = 2$ for a planar map, which is the most frequent case), the distances between row points can be calculated as the usual Euclidean distance between their principal coordinates:

$$\sqrt{\sum_{k=1}^{K^*} (f_{ik} - f_{i'k})^2}$$

e.g. between the first and second rows in $K^* = 2$ dimensions:

$$\sqrt{(-0.3310 + 0.0355)^2 + (0.0180 + 0.0546)^2} = 0.3043$$

(See Table 3.32.)

Being partial sums of the distances in Table 3.31, these distances in the reduced space are all less than or equal to the distances in the full space.

TABLE 3.32
Distances between row profiles in two dimensions.

Age group	18–24	25–34	35–44	45–54	55–64	65–74	75+
18–24	0.0000	0.3043	0.2699	0.2935	0.4494	0.5382	0.7421
25–34	0.3043	0.0000	0.1047	0.0876	0.1455	0.2445	0.4804
35–44	0.2699	0.1047	0.0000	0.0282	0.2220	0.2760	0.4733
45–54	0.2935	0.0876	0.0282	0.0000	0.1947	0.2490	0.4525
55–64	0.4494	0.1455	0.2220	0.1947	0.0000	0.1349	0.3897
65–74	0.5382	0.2445	0.2760	0.2490	0.1349	0.0000	0.2549
75+	0.7421	0.4804	0.4733	0.4525	0.3897	0.2549	0.0000

3.2.34 The column profiles are also situated in a K-dimensional space (e.g. $K = 3$ in this example). The chi-square distances between the column points can be calculated exactly in this full space, as for the rows. For the jth and j'th rows, the distance between them is:

$$\sqrt{\sum_i \left(\frac{p_{ij}}{c_j} - \frac{p_{ij'}}{c_{j'}}\right)^2 / r_i}$$

e.g. the distance between the first and third columns (Too fast and Too slow) is the square root of:

$$(0.0544/0.6420 - 0.0055/0.2035)^2/0.0931 + \ldots$$
$$+ (0.0315/0.6420 - 0.0008/0.2035)^2/0.0489$$

which is equal to 0.2985 (see Table 3.33). These distances are also the usual Euclidean distances between the principal coordinates of the columns in all K dimensions:

$$\sqrt{\sum_{k=1}^{K} (g_{jk} - g_{j'k})^2}$$

e.g. between the first and third columns:

$$\sqrt{(0.0416 + 0.2157)^2 + (-0.0388 + 0.066)^2 + (-0.0101 - 0.1388)^2} = 0.2985$$

TABLE 3.33
Distances between column profiles.

Response	Too fast	About right	Too slow	Don't know
Too fast	0.0000	0.3102	0.2985	0.3106
About right	0.3102	0.0000	0.2083	0.5573
Too slow	0.2985	0.2083	0.0000	0.5648
Don't know	0.3106	0.5573	0.5648	0.0000

3.2.35 In the low-dimensional map, or reduced space, of dimensionality K^* (e.g. $K^* = 2$), the distances between projected row points can be calculated as the usual Euclidean distance between their principal coordinates:

$$\sqrt{\sum_{k=1}^{K^*} (g_{jk} - g_{j'k})^2}$$

e.g. between the first and third columns:

$$\sqrt{(0.0416 + 0.2157)^2 + (-0.0388 + 0.066)^2} = 0.2587$$

(See Table 3.34.)

TABLE 3.34
Distances between column profiles in two dimensions.

Response	Too fast	About right	Too slow	Don't know
Too fast	0.0000	0.3102	0.2587	0.3083
About right	0.3102	0.0000	0.1419	0.5558
Too slow	0.2587	0.1419	0.0000	0.5537
Don't know	0.3083	0.5558	0.5537	0.0000

3.2.36 In the full space, the distances between the set of profiles (of the rows, say) and the other set of vertices (of the columns) can be related to the profile elements. To calculate the profile-to-vertex distances, we compute the Euclidean distance between the row principal coordinates (Table 3.15) and column standard coordinates (Table 3.18):

$$\sqrt{\sum_{k=1}^{K} (f_{ik} - y_{jk})^2}$$

e.g. between the first row and the first column:

$$\sqrt{(0.3310 + 0.2579)^2 + (0.0180 - 0.2912)^2 + (-0.0244 + 0.7518)^2} = 0.9200$$

Notice that, as a general rule, the smaller the distance between a profile point and a vertex point, the higher is the corresponding profile element. For example, the first row profile (Table 3.7) has elements 0.5847, 0.3136, 0.0593 and 0.0424, whereas the profile-to-vertex distances for the first row are 0.9200, 1.7156, 4.5516 and 3.0757, showing that the row profile is closer to the column vertices for which the profile elements are greater (Table 3.35).

TABLE 3.35
Distances between row profiles and column vertices.

Age group	Too fast	About right	Too slow	Don't know
18–24	0.9200	1.7156	4.5516	3.0757
25–34	0.7430	1.9977	4.5314	2.8944
35–44	0.8032	1.8990	4.6171	2.8656
45–54	0.7908	1.9310	4.6003	2.8512
55–64	0.6090	2.1099	4.6601	2.8698
65–74	0.6855	2.1508	4.6747	2.7309
75+	0.8542	2.2331	4.7655	2.4869

3.2.37 The profile-to-vertex distances in the reduced space are calculated as in the full space, but are restricted to the first K^* dimensions:

$$\sqrt{\sum_{k=1}^{K^*} (f_{ik} - y_{jk})^2}$$

e.g. between the first row and the first column in two dimensions (Table 3.36):

$$\sqrt{(0.3310 + 0.2579)^2 + (0.0180 - 0.2912)^2} = 0.8745$$

A similar interpretation of these distances in terms of profile elements is possible in the reduced space, though it should be noted that the vertex points are not as well represented as the profile points. Vertex points which project towards the centre of the map are poorly represented, so the profile-to-vertex distances for such points are not worth considering.

TABLE 3.36
Distances between row profiles and column vertices in two dimensions.

Age group	Too fast	About right	Too slow	Don't know
18–24	0.8745	1.6685	1.4832	2.9494
25–34	0.6446	1.9385	1.6523	2.7849
35–44	0.7472	1.8547	1.6957	2.7324
45–54	0.7246	1.8826	1.7057	2.7242
55–64	0.5488	2.0759	1.7357	2.7271
65–74	0.6141	2.1098	1.8700	2.5941
75+	0.7909	2.1902	2.1235	2.3424

3.2.38 The original correspondence matrix (Table 3.6) can be reconstructed from the row and column masses (Tables 3.4 and 3.5), the principal inertias (denoted by λ_k in section 3.2.15, the squares of the singular values γ_k in section 3.2.11) and the row and column coordinates in either the principal or standard form (sections 3.2.16–3.2.19). This 'reconstitution formula' is derived directly from the SVD of the matrix of standardized residuals given by formula (3.12) and the definitions of the principal and standard coordinates in terms of the singular vectors and singular values. Depending on whether the principal or standard coordinates are used, this formula can be expressed in four equivalent ways:

$$\mathbf{P} = \mathbf{rc}^T + \mathbf{D}_r\mathbf{F}\mathbf{\Gamma}^{-1}\mathbf{G}^T\mathbf{D}_c$$
$$\mathbf{P} = \mathbf{rc}^T + \mathbf{D}_r\mathbf{F}\mathbf{Y}^T\mathbf{D}_c$$
$$\mathbf{P} = \mathbf{rc}^T + \mathbf{D}_r\mathbf{X}\mathbf{G}^T\mathbf{D}_c$$
$$\mathbf{P} = \mathbf{rc}^T + \mathbf{D}_r\mathbf{X}\mathbf{\Gamma}\mathbf{Y}^T\mathbf{D}_c$$

The scalar equivalents of these formulae are as follows (where we use the fact that $\gamma_k = \sqrt{\lambda_k}$:

$$p_{ij} = r_i c_j \left(1 + \sum_{k=1}^{k} f_{ik}g_{jk}/\sqrt{\lambda_k}\right)$$

$$p_{ij} = r_i c_j \left(1 + \sum_{k=1}^{k} f_{ik}y_{jk}\right)$$

$$p_{ij} = r_i c_j \left(1 + \sum_{k=1}^{k} x_{ik}g_{jk}\right)$$

$$p_{ij} = r_i c_j \left(1 + \sum_{k=1}^{k} \sqrt{\lambda_k}x_{ik}y_{jk}\right)$$

e.g. using the first version of the formula:

$$p_{11} = 0.0931 \times 0.6420 \times (1 + (-0.3310) \times (-0.0416)/0.1611 + \ldots \text{etc}) = 0.0544$$

Multiplying the elements of the reconstructed correspondence matrix by the total n (= 1268, in this example) gives the reconstructed data matrix of frequencies in Table 3.1, e.g.

$$n_{11} = 1268 \times 0.0544 = 69$$

(See Table 3.37.)

TABLE 3.37
Recomputed input data.

Age group	Too fast	About right	Too slow	Don't know
18–24	69	37	7	5
25–34	148	45	14	22
35–44	170	65	12	29
45–54	159	57	12	28
55–64	122	26	6	18
65–74	106	21	5	23
75+	40	7	1	14

3.2.39 When restricted to the K^* dimensions of the reduced space solution, the formulae in section 3.2.38 give approximate values of the original correspondence matrix or (when multiplied by n) of the contingency table. For example, the values in Table 3.38 were calculated from the formula:

$$n_{ij} = nr_ic_j\left(1 + \sum_{k=1}^{K^*} f_{ik}g_{jk}/\sqrt{\lambda_k}\right)$$

and then rounded to the nearest whole number.

TABLE 3.38
Recomputed input data in two dimensions.

Age group	Too fast	About right	Too slow	Don't know
18–24	68	37	8	5
25–34	151	46	11	21
35–44	169	65	13	29
45–54	159	57	12	28
55–64	120	25	8	19
65–74	106	21	5	23
75 +	40	7	1	14

and then rounded to the nearest whole number.

3.2.40 Finally, the essential numerical results of CA are the table of eigenvalues (principal inertias) and percentages of inertia, given in Table 3.14, and the complete numerical results for the rows and the columns are shown in Table 3.39.

TABLE 3.39
General statistics.

	MASS	SQCOR	INR	LOC1	QCOR1	INR1	LOC2	QCOR2	INR2
18 – 24	0.093	0.995	0.334	− 0.331	0.992	0.393	0.018	0.003	0.008
25 – 34	0.181	0.546	0.046	− 0.036	0.162	0.009	− 0.055	0.384	0.141
35 – 44	0.218	0.961	0.045	− 0.063	0.620	0.033	0.046	0.341	0.123
45 – 54	0.202	0.984	0.017	− 0.038	0.561	0.011	0.033	0.423	0.058
55 – 64	0.136	0.908	0.101	0.102	0.452	0.054	− 0.102	0.456	0.373
65 – 74	0.122	0.999	0.170	0.206	0.993	0.200	− 0.017	0.006	0.009
75 +	0.049	0.999	0.289	0.399	0.876	0.300	0.150	0.123	0.288
Too fast	0.642	0.970	0.069	0.042	0.518	0.043	− 0.039	0.451	0.254
About right	0.203	0.997	0.442	− 0.248	0.920	0.482	0.072	0.078	0.278
Too slow	0.045	0.725	0.102	− 0.216	0.663	0.081	− 0.066	0.062	0.051
Don't know	0.110	0.993	0.386	0.306	0.860	0.394	0.120	0.133	0.417

The values in each column are as follows:

- MASS – mass [see sections 3.2.5 and 3.2.6].
- SQCOR – squared correlation (of profiles with subspace, in this case the solution subspace is two-dimensional), or qualities [see sections 3.2.24 and 3.2.30].
- INR – inertia of each profile point [see sections 3.2.22 and 3.2.28].
- LOC1 – principal coordinate on axis 1 [see sections 3.2.16 and 3.2.17].
- QCOR1 – squared correlation (of profile with axis 1) [see sections 3.2.23 and 3.2.29].
- INR1 – proportion of inertia on axis 1 explained by each profile [see sections 3.2.21 and 3.2.27].
- LOC2, QCOR2 and INR2 – corresponding quantities for principal axis 2.

The above output is in the same format as the SimCA and BMDP output, except in SimCA all quantities are multiplied by 1000 and written as integers to save space in the tables. Notice that the column SQCOR (quality) is the sum of the columns QCOR1 and QCOR2. If the solution was three-dimensional, then SQCOR would be the sum of QCOR1, QCOR2 and QCOR3, and so on. In the full solution, with K axes, SQCOR $= 1$.

Correspondence Analysis and Contingency Table Models

Peter G. M. van der Heijden, Ab Mooijaart and Yoshio Takane

4.1 INTRODUCTION

Sometimes correspondence analysis (CA) is presented as a model-free technique (see Benzecri *et al.* 1973). The idea is that CA is helpful when we want 'to let the data speak for itself'. The idea is that no assumptions are made about the distribution which yielded the values in the matrix to be studied. No prejudices of the researcher lead the analysis of the data. CA is simply a tool to make a graphical representation of the data. The generalized singular value decomposition performed in the computation of CA is helpful in projecting most information to the first few dimensions of the full-dimensional solution.

We have some difficulty with calling correspondence analysis 'model-free'. We rather adopt another definition of a model, namely that a model is a nonlinear projection of the data on a (usually low-dimensional) parameter space (see discussion of van der Heijden *et al.* 1989; see also de Leeuw 1988). This nonlinear projection can be optimal in terms of some criterion, and often used criteria are least squares, generalized least squares, or maximum likelihood (ML). In earlier chapters CA is estimated using a singular value decomposition that has optimality properties in a least squares sense.

We prefer this definition because we find that by choosing CA to represent the data graphically an explicit choice is made to emphasize certain aspects of the data. One aspect is to emphasize the association between the row and the column variable instead of the margins. A second choice is to study the association by using certain metrics for the row space and the column space, and certain weights for the row points and the column points. A term like 'model-free' suggests that no explicit choices are made in studying the data, and we do not agree with this. For a related discussion we refer to the discussion of the paper of van der Heijden *et al.* (1989).

This leads us to a presentation of CA as a model that can be fitted using different criteria. In our view CA as presented in earlier chapters is optimized in terms of a least-squares criterion for a transformation of the observed proportions. A different approach is to optimize CA by ML. This will be discussed in section 4.2. In section 4.3 we will show the close relation that exists between CA and the latent class model, a model that assumes that the manifest variables are independent given the level of a latent variable. It turns out that latent class analysis and CA are models that are often equivalent. In section 4.4 we describe some relations between CA and log–linear models, models having log–bilinear terms, and ideal point discriminant analysis (IPDA). In this section it is shown that CA can be interpreted as a tool for residual analysis. It is also shown that if certain conditions are fulfilled, the estimates obtained with CA will be very similar to the estimates obtained with models having log–bilinear terms. We end with a discussion about the relative merits of CA and other statistical models.

Throughout this paper the same data matrix will be analyzed with different models. This matrix is displayed in Table 4.1. The matrix is concerned with results of an election held in The Netherlands in 1986 for the so-called Second Chamber (comparable with the House of Commons). The rows of the matrix are six types of city, namely rural, industrialized rural, commuter, small, middle large and large. The political parties are subdivided into six categories, namely PvdA (labour party), CDA (christian democrats), VVD (right-winged liberals), D'66 (left-winged liberals), small left-winged parties and small right-winged parties (mostly religious). We will consider this table to be a one in a

TABLE 4.1

Voters for the 1986 elections in The Netherlands. In the rows city types are found, and in the column political parties are given. For more details, see the text.

City type	PvdA	CDA	Political party VVD	D'66	Left	Right
Rural	285	482	186	49	21	60
Rural, industrialized	620	914	308	102	42	97
Commuter	355	460	347	104	36	47
Small city	336	337	168	62	27	46
Middle large city	548	455	233	91	47	43
Large city	903	516	343	153	110	37

thousand random sample of voters (see Central Bureau for Statistics 1987, for more details).

Since this book focuses on correspondence analysis approximated by least squares, we present here the results of the correspondence analysis of our table (see also van de Geer 1989). In Figure 4.1(a) and 4.1(b) the graphical representation is given of the first two dimensions of the five-dimensional space.

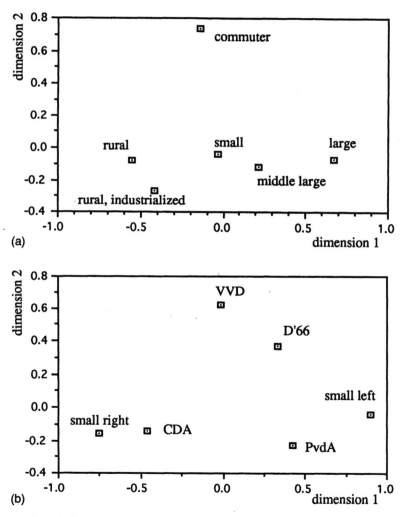

FIGURE 4.1 (a) Correspondence analysis solution in two dimensions, row points. See text for explanation. (b) Correspondence analysis solution in two dimensions, column points. See text for explanation.

The singular values are 0.19 and 0.10, displaying 0.77 and 0.22 of the total inertia. In both figures the points are represented using principal coordinates (see Greenacre, Chapter 1). Roughly, we find rural towns in the bottom left corner, larger cities in the bottom right corner, and commuters at the top. It turns out that in the rural towns people vote more than average for religious parties (small right and CDA), in large cities they vote more than average for left-winged parties, and in commuter towns they vote more than average for the VVD. Both in large cities and in commuter cities people vote more than average for D'66.

4.2 CORRESPONDENCE ANALYSIS

CA approximated by least squares

Our starting point in a comparison of CA with other contingency table models is to show how the observed proportions p_{ij} are decomposed in CA. The reason for this starting point is that statistical models are usually defined in terms of theoretical probabilities π_{ij} that are assumed to hold in some population. Thus by starting with the observed proportions, a later comparison with other statistical models becomes easier.

The reconstitution formula can be written as

$$p_{ij} = p_{i+}p_{+j} \left(1 + \sum_{k=1}^{K} \mu_k r_{ik} c_{jk} \right)$$

where $\mu_k = \sqrt{\lambda_k}$ is the square root of the kth principal inertia and the standard coordinates r_{ik} and c_{jk} are normalized in the usual way. The graphical representations such as in Figure 4.1 are usually made with principal coordinates, which are given by $f_{ik} = r_{ik}\mu_k$ and $g_{jk} = c_{jk}\mu_k$, in which case the reconstitution formula can be written equivalently as:

$$p_{ij} = p_{i+}p_{+j} \left(1 + \sum_{k=1}^{K} \mu_k^{-1} f_{ik} g_{jk} \right)$$

In earlier chapters CA is estimated by means of the singular value decomposition of the matrix of standardized residuals with elements

$$s_{ij} = \frac{(p_{ij} - p_{i+}p_{+j})}{\sqrt{p_{i+}p_{+j}}}$$

This decomposition is optimal in terms of least squares: by using only K^* of the K dimensions, i.e. by using only the first K^* terms in the reconstitution formula we obtain a weighted least-squares approximation of the proportions p_{ij}.

It may be noted, in passing, that it is relatively straightforward to incorporate linear constraints in correspondence analysis (see Chapter 5). They come in different guises, redundancy of qualitative data (Israëls, 1984), canonical correspondence analysis (ter Braak 1986), canonical analysis with linear constraints (Böckenholt and Böckenholt 1990). These techniques use external information to limit the space in which rows and columns of contingency tables are represented. Analogous developments in modelling approaches to contingency tables will be presented later in this chapter.

As we said before, the correspondence analysis solution discussed in section 4.1 for Table 4.1 is presented with principal coordinates f_{ik} and g_{jk}. So in Figure 4.1(a) for row point i on dimension k, f_{ik} is used as coordinate, and in Figure 4.1(b) for column point j on dimension k, g_{jk} is used as coordinate. In the tradition to see correspondence analysis as a tool to represent the matrix graphically (see section 4.1) this is quite useful, because thus the Euclidean distances in Figures 4.1(a) and 4.1(b) are approximations of the so-called chi-squared distances between the rows (columns) of the matrix (see the earlier chapters). In the modelling tradition it is more standard to show the estimates for the parameters r_{ik}, c_{jk} and μ_k. These estimates are given in panel A of Table 4.2. We will now compare these estimates with estimates of models approximated by ML.

CA approximated by ML

Correspondence analysis approximated using least squares has quite a long history. A much shorter tradition exists since 1985, when Goodman (1985, 1986) started studying CA as a model to be estimated by ML (see also Gilula and Haberman 1986, 1988). In this tradition there are population probabilities π_{ij} that follow a statistical model, in this case CA:

$$\pi_{ij} = \alpha_i \beta_j \left(1 + \sum_{k=1}^{K} \mu_k r_{ik} c_{jk} \right)$$

Under the assumption that the frequencies follow a (product) multinomial or a Poisson distribution, a likelihood function is set up that is maximized over the unknown parameters α_i, β_j, μ_k, r_{ik} and c_{jk}. If the parameter estimates r_{ik} and c_{jk} are unconstrained, then it turns out that $\alpha_i = p_{i+}$ and $\beta_j = p_{+j}$ (see Goodman 1985, Siciliano *et al.* 1990).

The CA model defined above is not restrictive because $K = \min(I - 1, J - 1)$. Therefore the CA estimated by least squares and the CA estimated by ML are identical for $K = \min(I - 1, J - 1)$. The model becomes restrictive if either restrictions are imposed on the parameters r_{ik} and/or c_{jk}, or on the dimensionality of the solution. For example, in the model

$$\pi_{ij} = \alpha_i \beta_j \left(1 + \sum_{k=1}^{K^*} \mu_k r_{ik} c_{jk} \right)$$

where $0 \leqslant K^* < K$, for K^* dimensions $\mu_k \geqslant 0$. If this model fits adequately, then it is not necessary to study more than K^* dimensions. It is also possible

TABLE 4.2

Parameter estimates for the data in Table 4.1. In panel A we find the estimates for correspondence analysis approximated by least squares; in panel B estimates for correspondence analysis approximated by maximum likelihood; in panel C estimates for the RC(2) association model; in panel D estimates for the latent budget model with three latent budgets; in panel E estimates for ideal point discriminant analysis; in panel F rescaled estimates for ideal point discriminant analysis (see text).

	Panel A		Panel B		Panel C		Panel D			Panel E		Panel F	
	0.191	0.102	0.190	0.102	0.194	0.093	0.458	0.294	0.248			0.097	0.046
Rows													
1	−1.274	−0.241	−1.262	−0.248	−1.272	−0.094	0.659	0.103	0.238	−0.917	−0.055	−1.287	−0.082
2	−0.955	−0.825	−0.964	−0.824	−0.937	−0.883	0.627	0.163	0.210	−0.665	−0.593	−0.933	−0.881
3	−0.317	2.304	−0.321	2.305	−0.339	2.272	0.450	0.188	0.362	−0.242	1.529	−0.340	2.271
4	−0.067	−0.121	−0.066	−0.125	−0.076	−0.149	0.471	0.286	0.242	−0.049	−0.106	−0.069	−0.157
5	0.501	−0.375	0.514	−0.366	0.483	−0.458	0.388	0.383	0.229	0.350	−0.319	0.491	−0.474
6	1.528	−0.232	1.525	−0.234	1.541	−0.160	0.228	0.537	0.234	1.094	−0.103	1.535	−0.153
Columns													
1	0.973	−0.704	0.984	−0.703	0.957	−0.790	0.295	0.697	0.000	0.133	−0.047	0.980	−0.689
2	−1.059	−0.441	−1.058	−0.437	−1.059	−0.398	0.623	0.164	0.078	−0.143	−0.031	−1.053	−0.455
3	−0.037	1.945	−0.037	1.949	0.015	1.875	0.000	0.000	0.711	−0.007	0.133	−0.052	1.951
4	0.772	1.167	0.784	1.150	0.810	1.297	0.000	0.062	0.179	0.104	0.079	0.766	1.159
5	2.070	−0.111	1.959	−0.114	2.004	0.248	0.002	0.077	0.032	0.281	−0.001	2.070	−0.015
6	−1.732	−0.493	−1.778	−0.536	−1.848	−0.320	0.080	0.000	0.000	−0.235	−0.034	−1.731	−0.499

to test whether some parameters are fixed, or equal to other parameters. For example, if the parameters $r_{ik} = 0$ for $k = 1, ..., K^*$, then the point i falls into the origin, and the result is that $\pi_{ij}/\pi_{i+} = \pi_{+j}$ for point i. As another example, if $r_{ik} = r_{i'k}$ for $k = 1, ..., K^*$, then the points i and i' are located in the same place. A third possible restriction that can be tested is whether the scores are equidistant, for example, whether the scores r_{ik} are $(-3, -1, 1, 3)$ if $I = 4$, or, whether $r_{ik} = c_{ik}$ if we deal with the analysis of a square table. See Gilula and Haberman (1988) for other kinds of restrictions that may be imposed.

The correspondence analysis model with constraints on the dimensionality of the solution only, has $(I - K^* - 1)(J - K^* - 1)$ degrees of freedom. Thus we can test whether one or two dimensions are sufficient to describe the association in Table 4.1. For Table 4.1 we fit the model with $K^* = 0$, $K^* = 1$ and $K^* = 2$ dimensions. For $K^* = 0$ the correspondence analysis model comes down to the independence model: $\pi_{ij} = \pi_{i+}\pi_{+j}$. This model has a likelihood ratio chi-square statistic $G^2 = 420.6$ for $(I - 1)(J - 1) = 25$ degrees of freedom. The model with one dimension only $(K^* = 1)$ has a $G^2 = 95.1$ for 16 degrees of freedom. The model with two dimensions $(K^* = 2)$ has a $G^2 = 6.7$ for 9 degrees of freedom.

It turns out that the model with two dimensions is not significantly different from the saturated model where $K^* = K = 5$. The estimates for r_{ik}, c_{jk} and μ_k are given in panel B of Table 4.2. It can be seen that the estimates found in the ML solution are very similar to the estimates found in the least squares solution found in panel A. In fact, for this example there is hardly any difference.

Comparison

It is much easier to estimate CA by least squares than by ML, so one should have good reasons to do the latter instead of the former. Compared with CA approximated by least squares, it is a great advantage of the version of CA estimated by ML that restrictions on the model parameters can be tested. Very natural questions can be answered about a CA solution, such as 'how many dimensions should I interpret?'; 'is the distance between two row points or two column points in my graphical display significantly different from zero?'; 'is the distance of a point to the origin significantly different from zero?'; and 'are the scores for the rows and/or the columns equidistant?'.

It should be emphasized, however, that it is only useful to perform such tests when the assumption is fulfilled that the data stem from a (product)-multinomial or a Poisson distribution. This assumption can be violated if the observations leading to the frequencies are dependent, for example, if the answers of respondents in households are related. Another case in which the usefulness of this approach is doubtful is if the data are not a random sample from the population one had in mind by setting up the study. This often happens in surveys. If the non-response is selective in the sense that it is larger in some cells than in other cells, then the test results only tell us how to generalize from a sample to some unclear population.

Often the results from CA approximated by least squares and those from CA approximated by ML will be very similar (see, for example, Goodman 1985, 1986). We also found this for the example discussed above (see panel A and B in Table 4.2). We already indicated that if $K^* = K$, then these approximations are identical because they do not impose restrictions on the data. However, there are also situations in which they may be rather different. One such situation is when in the CA approximated by least squares

$$\left(1 + \sum_{k=1}^{K^*} \mu_k r_{ik} c_{jk}\right)$$

is negative. This might occur when, for example, $K^* = 1$ and μ_1 is large. Then for a one-dimensional solution it is most likely that there will be reconstituted proportions that are smaller than zero. In the one-dimensional CA solution approximated by ML the probabilities π_{ij} are always positive, and therefore either the estimates for μ_1 or for some parameters r_{i1} or c_{j1} will be quite different from those found in the CA approximated by least squares.

A question that remains is how the ML estimates of the scores should be interpreted. One answer is that there is no difference in interpretation between the ML estimates and the least squares estimates, but that these estimates have other properties (see the statistical literature). One could also argue as follows: when CA is performed by least squares (as explained and illustrated in Chapters 1 to 3), we have a geometric interpretation of the results in the form of profiles, masses, projections of profiles onto optimal subspaces, etc. When CA is estimated by ML a similar but different interpretation is found. The key to the solution of the interpretation problem is that now the matrix with elements $\hat{\pi}_{ij}$ is the starting point for the interpretation, and not the matrix with observed proportions p_{ij}. For the ML solution the points in a representation like Figure 4.1(a) do not represent *observed* profiles with elements p_{ij}/p_{i+}, but estimates of *expected* profiles with elements $\hat{\pi}_{ij}/\hat{\pi}_{i+}$. Masses of the points are the same, and the metric of the row and column spaces are also the same. However, whereas for the least squares solution in Table 4.2 the scores represent an optimal least squares projection to a two-dimensional subspace, for the ML solution the matrix with elements $\hat{\pi}_{ij}$ is perfectly represented in a two-dimensional space. The fit of the model, $G^2 = 6.7$ for 9 degrees of freedom in this case, helps us to evaluate whether the matrix with elements p_{ij} may be represented by the matrix with elements $\hat{\pi}_{ij}$. The principle of ML tells us that, if there is a true two-dimensional graphical display for the population, the parameter estimates in panel B of Table 4.2 make the observed data most likely. So the scores in panel B of Table 4.2 are not only useful for plots by analogy with CA, they can be understood in terms of profile points and the like in their own right.

4.3 SOME RELATIONS BETWEEN CA, LATENT BUDGET ANALYSIS AND LATENT CLASS ANALYSIS

4.3.1 Latent budget analysis

Consider the matrix in Table 4.3. The elements of this matrix are derived from the matrix in Table 4.1 as $p_{j|i} = n_{ij}/n_{i+}$. The proportion $p_{j|i}$ is a conditional proportion: it is the proportion of voters that have voted for political party j given that they live in city type i. It is easy to study a matrix as in Table 4.3 by comparing the conditional proportions $p_{j|i}$ with the marginal proportions p_{+j}, that are given in the last line of Table 4.3. It shows that PvdA is voted more than average (i.e. 0.340) in large cities (i.e. 0.387), CDA is voted more than average in rural cities (0.445, 0.439 versus 0.353), VVD is voted more than average in commuter towns (0.257 versus 0.177), and so on. All these results were already clear from Figures 4.1(a) and 4.1(b), because CA is based on an analysis of a matrix with elements $p_{j|i}$ and a matrix with elements $p_{i|j}$. The rows of conditional proportions $p_{j|i}$ in Table 4.3 are the row profiles. The average profile is given in the last line of the table. One might ask whether there exists a small number of *typical* profiles, say three, that have *generated* the observed profiles. Thus the observed profiles would be approximated by a weighted average of three unknown profiles. The question then is: what are these unknown profiles, and how do they approximate the observed profiles.

TABLE 4.3
Voters for the 1986 elections in The Netherlands. In the rows city types are found, and in the column political parties are given. Conditional proportions: given the type of city, what is the distribution of voters for the political parties?

	Political party					
City type	PvdA	CDA	VVD	D'66	Left	Right
Rural	0.263	0.445	0.172	0.045	0.019	0.055
Rural, industrialized	0.298	0.439	0.148	0.049	0.020	0.047
Commuter	0.263	0.341	0.257	0.077	0.027	0.035
Small city	0.344	0.345	0.172	0.064	0.028	0.047
Middle large city	0.387	0.321	0.164	0.064	0.033	0.030
Large city	0.438	0.250	0.166	0.074	0.053	0.018
Total sample	0.340	0.353	0.177	0.063	0.032	0.037

This solution is displayed in the lower part of panel D of Table 4.2. Here we find as a first typical profile 0.295, 0.623, 0.000, 0.000, 0.002, 0.080, where the CDA and small right-winged parties have much higher proportions than their average (0.623 versus 0.340; 0.080 versus 0.037). We call this the religious profile. The second typical profile is 0.697, 0.164, 0.000, 0.062, 0.077 and 0.000, showing that here the PvdA and small left-winged parties are much higher than their average. We call this the left-winged profile. The third typical profile is 0.000, 0.078, 0.711, 0.179, 0.032, 0.000, showing that here the liberal parties VVD and D'66 are higher than their average. We call this the liberal profile. Notice that this is very similar to the three clusters of parties that we found in the correspondence analysis plots in Figure 4.1(b).

In the top part of panel D of Table 4.2 we find how the observed profiles are approximated by the typical profiles. On average, the proportions of voters making use of the three profiles are 0.458, 0.294 and 0.248, so 0.458 of the voters are in the religious profile, 0.294 in the left-winged profile and 0.248 in the liberal profile. The top part of panel D shows that, for example, the rural city profile is built up for 0.659 of the religious profile, for 0.103 of the left-winged profile, and for 0.238 of the liberal profile. We see that, on the whole, the profiles of rural and industrialized rural are built up more than average by the typical religious profile (0.659, 0.627 versus the average 0.458); the profiles of middle large and large more than average by the typical left-winged profile; and the profile of commuter cities by the typical liberal profile.

After this informal introduction of the model, we now introduce some notation. Let us denote theoretical profile elements as $\pi_{j|i} \equiv \pi_{ij}/\pi_{i+}$. Let us index the typical profiles to be estimated by $x(x = 1, ..., X)$, so that we can denote the element j of typical profile x as $\pi_{j|x}$, since $\Sigma_j \pi_{j|x} = 1$ (see bottom part of panel D). Let the probabilities that relate the rows to the typical profile be denoted by $\pi_{x|i}$, since $\Sigma_x \pi_{x|i} = 1$ (see top part of panel D). Then the model is

$$\pi_{j|i} = \sum_{x=1}^{X} \pi_{x|i} \pi_{j|x}$$

This model is called the latent budget model, and for many more details we refer to de Leeuw *et al.* (1990) and van der Heijden *et al.* (1992). Here the term 'budget' is synonymous with 'profile', and the term 'latent' corresponds to the fact that unknown, typical profiles ('budgets') are sought.

The model is usually approximated by ML. In its unconstrained form it has $(I - X)(J - X)$ degrees of freedom (for constraints we refer to van der Heijden *et al.* 1992). For one typical budget the model comes down to the independence model, having only the 'typical' budget with elements p_{+j}. We saw earlier that for this model our example gives a test statistic $G^2 = 420.6$ (df = 25). For two latent budgets (typical profiles) the model has a chi-square of $G^2 = 95.1$ (df = 16) and for three latent budgets the model has a chi-square of $G^2 = 6.7$

(df = 9). Thus the solution that we discussed above corresponds with a model that fits the data.

The latent budget model is not identified, but it can be identified by imposing fixed value constraints upon the parameters. The identification problem is similar to that in factor analysis. For the example discussed above we have imposed six constraints to the elements of the latent budgets, namely that elements 3 and 4 of latent budget 1, element 3 and 6 of latent budget 2 and elements 1 and 6 of budget 3 are all constrained to zero. For more details we refer to de Leeuw *et al.* (1990). Van der Heijden *et al.* (1992) indicate how general classes of constraints can be imposed on the parameters $\pi_{x|i}$ and $\pi_{j|x}$. These constraints are particularly useful in the analysis of higher-way tables.

4.3.2 Equivalence of the latent class model and the latent budget model

The latent class model is a closely related model for the analysis of contingency tables. For example, if there are three variables A, B and C, then the basic latent class model assumes the existence of a categorical latent variable X that explains the relation between the observed variables. Given the level of X the observed variables A, B and C are independent (for more details, see Goodman 1974).

Less attention is given to the latent class model for a two-way table

$$\pi_{ij} = \sum_{x=1}^{X} \pi_x \pi_{i|x} \pi_{j|x}$$

(but see Clogg 1981). One of the reasons might be that this model is not identified, whereas the latent class model for three or more variables is identified (see Mooijaart 1982, Goodman 1987, de Leeuw and van der Heijden 1991).

This model is equivalent to the latent budget model $\pi_{j|i} = \Sigma_x \pi_{x|i} \pi_{j|x}$: we can derive $\pi_{x|i}$ from the latent class parameters π_x and $\pi_{i|x}$ as

$$\pi_{x|i} = \frac{\pi_x \pi_{i|x}}{\sum\limits_{z}^{X} \pi_z \pi_{i|z}}$$

For the practice of data analysis this implies that the latent class model with X latent classes provides the same estimates of expected frequencies as the latent budget model with X latent budgets. The parameter estimates from the latent class model can be used to obtain the estimates for the latent budget model, and vice versa. For a discussion of relative advantages of one model over the other we refer to van der Heijden *et al.* (1992).

It turns out that the latent budget model (and the latent class model) is also closely related to CA, because all models are reduced rank models for contingency tables. We follow here the presentation of de Leeuw and van der Heijden (1991), who describe this relation in terms of theoretical probabilities.

4.3.3 CA and LCA as reduced rank models

Let π_{ij} be the elements of a matrix Π with theoretical probabilities, where $\Sigma_i \Sigma_j \pi_{ij} = 1$. Then Π has rank R ($R \leqslant \min(I, J)$) if it is possible to decompose Π by $\Pi = \mathbf{XY}'$, where $\mathbf{X}$ is some matrix of order $I \times R$ and $\mathbf{Y}$ is some matrix of order $J \times R$. If $R = \min(I, J)$, then Π has full rank, and if $R < \min(I, J)$, then Π has a reduced rank R. Notice that in practice it is unlikely to encounter an observed matrix $\mathbf{P}$ having a reduced rank, that is a rank smaller than $\min(I, J)$.

We will now show that CA defining a rank R model is equivalent to the more general 'model' that some matrix has rank R. This is done by showing that both models imply each other.

If the matrix Π has rank R, then the matrix can always be decomposed by a CA with $R - 1$ dimensions (see Lancaster 1958; see also de Leeuw and van der Heijden 1991). The reverse is evident: if the matrix Π is decomposed by a CA with $R - 1$ dimensions, then the matrix Π has rank R. This can be easily seen by rewriting the reconstitution formula as

$$p_{ij} = p_{i+}p_{+j} \left(1 + \sum_{k=1}^{R-1} \mu_k r_{ik} c_{jk}\right) = p_{i+}p_{+j} + \sum_{k=1}^{R-1} (\mu_k^{1/2} p_{i+} r_{ik})(\mu_k^{1/2} p_{+j} c_{jk})$$

$$= p_{i+}p_{+j} + \sum_{k=1}^{R-1} r_{ik}^* c_{jk}^*$$

where $r_{ik}^* \equiv \mu_k^{1/2} p_{i+} r_{ik}$ and $c_{jk}^* \equiv \mu_k^{1/2} p_{+j} c_{jk}$. This shows that $p_{i+}p_{+j}$ is a rank 1 matrix, and the $R - 1$ terms defined by $r_{ik}^* c_{jk}^*$ constitute a rank $R - 1$ matrix. If we denote the matrix Π of rank R as Π_R, and CA decomposing a rank R matrix as CA_R, then we can say that Π_R and CA_R are equivalent.

From the equation for latent class analysis of two-way tables we immediately see that the latent class model is a reduced rank model. If the latent variable has R levels, then the latent class model defines a matrix of rank R. We denote this model as LCA_R.

It is evident that if LCA_R is true for some matrix Π, that for this matrix Π_R is true, and therefore CA_R is true. The reverse does not hold. The reason is that the parameters in LCA are all non-negative, whereas in CA the parameters may be both positive or negative. Therefore LCA is more restrictive than CA, and therefore if CA_R is true for some matrix Π, then it might be that LCA_R is not true. In other words, there are matrices Π of rank R for which LCA_R will not be true. However, it is obvious that for rank 1 (independence), both models are equivalent. Second, for full rank, i.e. $R = \min(I, J)$, both models are also equivalent. Third, in contrast with statements of Gilula (see, for example, Gilula and Haberman 1986), de Leeuw and van der Heijden

(1991) prove that for a matrix Π of rank 2 the latent class model is equivalent to CA.

4.3.4 Conclusions

Several conclusions follow from these results. First, although the parametrization from CA and LCA are very different, they extract similar information from the data.

Second, if both models are fit by ML (which is the usual criterion for latent class analysis but not for CA, but see section 4.3), then they will have the same fit for $R = 2$. For $R > 2$, the models CA_R and LCA_R may have the same fit (this is almost always the case in situations where we tried this out), but it may also be that CA_R has a better fit than LCA_R, because the latter is more restrictive than the former. For the example discussed in this section and in section 4.2 the both models yield equivalent results. This can be seen from the fact that, if we compare the latent budget model with the correspondence analysis approximated by ML (so that both models are optimized using the same criterion), both for rank 2 and for rank 3 both models have the same likelihood ratio chi-square. For rank 2 this is true for every possible matrix (since CA with one dimension and the latent budget model with two latent budgets are equivalent) but this need not be true for rank 3.

Third, since for rank 2 CA_R and LCA_R are equivalent, they yield the same estimates of expected frequencies. This result may be used in the ML estimation of CA: it is possible to estimate the latent class model as a first step, and then to decompose the estimates of expected frequencies with ordinary CA in order to obtain the ML estimates of the CA model with one dimension.

In situations where both models are fit by ML, and they yield the same fit, it is not possible to prefer one model over the other based on statistical considerations. However, the parameters of both models have different interpretations: the parameters of CA can be interpreted as optimal scores, whereas the parameters of LCA are conditional probabilities, and therefore a choice for either CA or the latent class model will depend on the substantive research question about the data. LCA has the problem that its solution is not identified. However, this identification problem seems to be well understood (see de Leeuw *et al.* 1990).

Goodman (1987) compares both models, and shows an empirical example. In van der Heijden *et al.* (1990) it is shown that CA and the latent class model can provide graphical representations that are closely related. We refer to van der Heijden (1992) for a comparison of CA and the latent class model in the context of square tables where interest goes out to the off-diagonal frequencies. It is shown in van der Heijden *et al.* (1990) that multiple correspondence analysis (discussed in part two of this book) and the latent class model are also closely related.

4.4　CORRESPONDENCE ANALYSIS AND LOG-LINEAR MODELS

In earlier sections we have discussed a ML version of CA, and the relation of (the ML version of) CA with the latent class model. In this paragraph we will discuss a specific way to view the relation of CA with log-linear models. We refer to van der Heijden *et al.* (1989) and the papers cited there for many more details concerning this approach. An approach that is slightly different, but basically the same, is discussed by Novak and Hoffman (1990).

4.4.1　CA as a tool for residual analysis

The starting point of the relationship of CA with log-linear models is the relation of CA with the independence model. This relation can easily be derived from both the reconstitution formula as well as from the relation between the Pearson chi-square with the sum of the squared singular values. The reconstitution formula given in section 2 can be rewritten as

$$p_{ij} = p_{i+}p_{+j}\left(1 + \sum_{k=1}^{K} \mu_k r_{ik} c_{jk}\right) = p_{i+}p_{+j} + p_{i+}p_{+j}\sum_{k=1}^{K} \mu_k r_{ik} c_{jk}$$

which shows that in CA the (scaled) residuals $(p_{ij} - p_{i+}p_{+j})/p_{i+}p_{+j}$ are decomposed into K dimensions, where $p_{i+}p_{+j}$ are estimates of expected probabilities under the statistical model defining independence of the categories of the row from the column variable. In $p_{i+}p_{+j}\Sigma_k\mu_k r_{ik} c_{jk}$ the marginal proportions p_{i+} and p_{+j} correct for the fact that some rows and columns have higher marginal frequencies. Another property showing the relation of CA with the independence model is

$$\frac{\chi^2}{n} = \frac{(p_{ij} - p_{i+}p_{+j})^2}{p_{i+}p_{+j}} = \sum_{k=1}^{K} \lambda_k$$

where χ^2 is the Pearson chi-square statistic used for testing independence between the row and the column variable, and n is the sample size. This equation shows that the departure from independence, as measured by the Pearson chi-square, is decomposed over the K dimensions of the CA solution.

So there is a close connection between independence and CA: CA can be considered as a tool for residual analysis. By decomposing the residuals from the independence model a graphical representation is obtained of the interaction between the row and column variables. However, this tool is not very flexible, because the model for which the residuals are decomposed is always the independence model. In many applications this model is very often naïve: often we are interested in a model less naïve and less restrictive than independence.

The reader should notice that we are discussing a rather limited use of CA here: the above interpretation is only useful if the data stem from a (product)-

multinomial or a Poisson distribution. If this is not the case, then the interpretation of CA in terms of residual analysis will not be very fruitful (but see, for an application of the tools discussed here in the context of multiple correspondence analysis, van der Heijden and Meijerink 1989).

4.4.2 A generalization of correspondence analysis

When we want to study the residuals from less restricted models we can use a generalization of CA proposed by Escofier (1983, 1984). This generalization is

$$p_{ij} = q_{ij} + s_i t_j \sum_{k=1}^{K} \mu_k r_{ik} c_{jk}$$

Here the row scores r_{ik} and column scores c_{jk} are standardized such that

$$\sum_{i=1}^{I} s_i r_{ik} = 0 \qquad \sum_{i=1}^{I} s_i r_{ik}^2 = 1 \qquad \sum_{j=1}^{J} t_j c_{jk} = 0 \qquad \sum_{j=1}^{J} t_j c_{jk}^2 = 1$$

The terms s_i are used as weights for the row points, and as the metric defined by s_i^{-1} for the space of the column points, and the terms t_j are used as weights for the column points, and provide the metric defined by t_j^{-1} for the space of the row points (compare Chapter 1). These terms can be chosen by the user. For the elements q_{ij} we can choose estimates of expected probabilities under some contingency table model. If we choose $s_i = p_{i+}$, $t_j = p_{+j}$, and $q_{ij} = p_{i+} p_{+j}$, then we find 'ordinary' CA. Many different types of applications can be worked out using this generalization. We will discuss the two applications that have received most attention thus far.

4.4.3 Correspondence analysis of incomplete contingency tables

In the standard approach of CA an analysis is performed of a complete two-way contingency table. However, sometimes we would like to analyze only part of the cells in a contingency table. As a first example, consider a two-way table of regions by years, with the number of property crimes in the cells. It might be that in some of the regions no data are available for all years. So some of the frequencies are simply missing. A standard CA is not useful here, because it starts from the assumptions that there are no missing frequencies. Yet we would like to study the interaction in the available frequencies with a tool like CA.

As a second example, some of the cells can be structurally zero, i.e. they refer to combinations that cannot logically occur, such as diagonal cells in import–export tables. In such tables countries are the categories of the row variable as well as of the column variable. In the cells amounts of goods or money are provided that countries import from other countries (and that the latter countries export to the former countries). The values in the diagonal cells are undefined: it is not clear what should be filled in here, because countries do not export and/or import to themselves. This may cause a problem in a standard analysis of the data. A solution is to focus attention to the off-diagonal cells only.

Thirdly, for substantive reasons we may decide that we want to apply a model to some cells but not to others. For example, in a social mobility table it might be that we are only interested in the off-diagonal cells, that reflect changes. In the diagonal cells we find the number of father–son couples that have the same profession. Processes that lead to having the same profession are usually considered to be different from processes that lead to profession-differences for fathers and sons. Therefore it might be useful to model only the off-diagonal cells, so that only one process is studied. A similar example will be discussed in the next section.

In standard CA the independence model is the baseline model. In log-linear analysis the independence model is adjusted to a so-called quasi-independence model to deal with the three examples given above, where estimates under the independence model are $p_{i+}p_{+j}$, under the quasi-independence model they are $\hat{\alpha}_i\hat{\beta}_j$ for the cells for which we would like to fit (quasi-)independence. For the other cells the estimated probabilities are simply equal to the observed proportions.

We can analyze the residuals from quasi-independence by the generalization of CA in the following way: we choose $q_{ij} = \hat{\alpha}_i\hat{\beta}_j$. If we now choose $s_i = \hat{\alpha}_i$ and $t_j = \hat{\beta}_j$, then the generalization of CA has again the property that the Pearson chi-square (now for testing quasi-independence) is closely related to the sum of the squared singular values. This property was lost for the generalization in its most general form. So, like in ordinary CA, the information as measured by Pearson's chi-square (which is the criterion used to decide whether the residuals contain interesting information) is decomposed into a number of dimensions.

As de Leeuw and van der Heijden (1988) show, it turns out that this application is equivalent to a procedure to deal with missing data in CA for a long time (see, for example, Greenacre, 1984, Ch. 8). This procedure adjusts the table to be analyzed, *before* an actual analysis is performed. Specifically, for the cell values in which one is *not* interested, 'independent' values are filled in. These independent values are calculated iteratively as $p_{ij}^{(m+1)} = p_{i+}^{(m)} p_{+j}^{(m)}$ for iteration m, and iterating ends after convergence. As it turns out, for the converged values p_{ij}, $p_{ij} = \hat{\alpha}_i\hat{\beta}_j$,

This shows, first, that a well-known procedure in the CA tradition can be understood in terms of a residuals approach of log-linear models. It also shows that this application is easily carried out by ordinary CA programs.

De Leeuw and van der Heijden (1988) give more details about this procedure, and in van der Heijden *et al.* (1989) it is shown how this procedure can

Table 4.4

Number of students, subdivided by faculties and years. First six columns: first-year students; last six columns: graduate students.

Faculty	Year											
	1935	1946	1957	1968	1979	1983	1935	1946	1957	1968	1979	1983
ARTS	194	384	654	1129	4224	4101	144	46	188	527	1114	2075
THEOLOGY	167	236	123	317	246	248	124	37	56	53	152	100
MEDICINE	549	1504	868	2438	1940	2418	391	507	650	1320	2021	2174
MATHPHYS	257	560	796	2046	2537	2496	171	237	316	924	1218	1635
TECHNICS	296	1355	1251	2756	3140	3957	212	272	457	1050	1403	1727
AGRICULT	58	176	184	510	1155	959	71	106	71	138	343	649
LAW	464	768	365	2079	3417	4993	420	493	358	790	1385	2348
ECONOMIC	65	671	633	1584	2018	2513	81	314	226	571	807	868
SOC-CULT	0	0	313	1258	1194	1340	0	0	68	291	498	561
PSYCHOL	0	0	256	947	1247	1228	0	0	72	190	744	797
PEDAGOGY	0	0	59	542	1278	1112	0	0	10	71	670	867
GEOGRAPH	24	236	157	358	592	626	24	15	35	103	365	473
OTHER	0	0	0	0	224	719	0	0	0	0	112	432

be used fruitfully in the context of square contingency tables. They also show how it can be generalized to deal with the departure from other models in this context, such as the symmetry model and the quasi-symmetry model.

4.4.4 Correspondence analysis as a tool for the analysis of residuals from conditional independence in a higher-way table

In order not to make the discussion too abstract, we present this application by means of an example, omitting technical details (for these the references at the end of this section can be consulted).

Consider Table 4.4. It is a table with faculties in the rows, and types of student in specific years in the columns. In the cells we find the number of first-year or graduate students in different faculties in The Netherlands at successive time points. The faculties are the arts (including philosophy), theology, medicine (including veterinary and dentistry), mathematics and physics (together), technical science, agriculture, laws, economic sciences, socio-cultural sciences, psychology, pedagogy, geography, and others. The years chosen are 1935, 1946, 1957, 1968, 1979 and 1983, which are evenly spaced, except for the last year (1983). Some of the cells in the matrix have zero frequencies, due to the fact that some faculties did not exist at that time. So in fact, we could interpret Table 4.4 as an incomplete contingency table, and handle these zero cells appropriately (see above).

However, we do a standard CA first. The Pearson chi-square is 12 883 (df is 25). This chi-square reflects the fit of a (log-linear) model where the row variable is independent of the column variables, i.e. the faculties have no relation with the types of student and the years. The estimates of expected frequencies under this log-linear model have the following property: the relative proportions of first-year students and graduates are equal for each faculty, and both the number of first-year students and the number of graduate students develop for each faculty proportionally in the same way. In the terminology of CA we would say that under this log-linear model the row profiles are equal. But since this model is rejected (this is revealed by the Pearson chi-square, which is significant), it makes sense to study the departure from the log-linear model.

The standard CA shows how Table 4.4 departs from this situation. The first four singular values (with percentages of chi-square) are 0.245 (54.4%), 0.141 (18.0%), 0.104 (9.9%) and 0.083 (6.3%) so we can restrict attention to the first two dimensions (the singular values for dimension 3 and 4 do not differ enough to warrant attention to any of these two). These two dimensions display 72% of the total dependence between the row and the column variables (see Figure 4.2). Care should be taken that the points geography, economics and agriculture are not very well represented in the first two dimensions: 2.6%, 22.4% and 42.9%, respectively, are displayed in the first two dimensions;

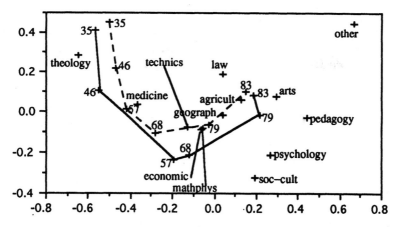

FIGURE 4.2   Correspondence analysis of Table 4.4. Joint plot, distances between rows and distances between column approximate chi-square distances. Solid line: first year students; dotted line: graduate students. See text for explanation.

similarly, only 8.6% and 50.5% are displayed from the graduate years 1979 and 1983.

The time points are roughly ordered in the first dimension from 1935 to 1983. We can conclude that, compared to the average faculty profile given by p_{+j}, the values p_{ij}/p_{i+} of many of the faculty profiles are either regularly growing or diminishing from 1935 until 1983, or have a peak in the middle period. Compared to the average faculty profile having elements p_{+j}, the faculties on the left of the origin have more students in the earlier years, whereas the faculties on the right of the origin have more students (than the average) in the later years. In the early years relatively many students were in theology and medicine. In recent years this has changed towards pedagogy, psychology, arts and 'other'. The peak for mathematics and physics (mathphys) and economy is found in the middle. All faculties are ordered along dimension 1 according to their peak compared with the average row profile.

Interpretation of the second dimension is more difficult. It seems that this dimension is a quadratic transformation of the first dimension. This effect is often found in CA when the first dimension is dominant. It is usually called the horseshoe effect, or Guttman effect. Different interpretations exist for this phenomenon (compare van Rijckevorsel 1986). For example, in Gifi (1990) an interpretation in terms of Hermite–Chebychev polynomials is given. Greenacre (1984, Ch. 8) presents an interpretation in terms of chi-square distances. Schriever (1983, 1986) presents an interpretation in terms of order dependence in categorical variables. It is clear from these interpretations that the second and higher dimensions need no substantive interpretations.

Let us now consider the solution in Figure 4.2 in terms of log-linear models again. In log-linear models the interaction in the three-way contingency table is attributed to different sources. There are three two-factor interactions, namely the two-factor interactions between faculties and years (do some faculties have more students in some years, relative to the average number of students over years?), between faculties and types of student (do some faculties have more graduate students, relative to the average proportion of graduate students over all the faculties?), and between years and types of student (are there more graduate students in some years, relative to the average proportion of graduate students?). Then there is also three-factor interaction: the way in which the relation between faculties and years differs for different types of student.

When we consider the standard CA of Table 4.4, it should first be noticed that the two-factor interaction between years and types of student is not displayed, because it is contained in the column margins of Table 4.4. This information could be used explicitly in choosing an appropriate way to code and analyze the three-way table: now the two-factor interaction between years and type of students is not displayed, but (two) different choices could have been made, namely either coding years and faculties interactively (then the two-factor interaction between years and faculties would not have been displayed) or coding faculties and students interactively (then the two-factor interaction between faculties and students would not have been displayed).

What *is* displayed in Figure 4.2 is the two-factor interaction between faculties and years, the two-factor interaction between faculties and types of student, and the three-factor interaction. So standard CA displays the departure from the log-linear model where faculty is independent from year and type of student jointly. The two-factor interaction between faculties and years is clearly overwhelming, and in fact it overshadows the other interactions. One reason is that this two-factor interaction consists for a large part of zeros in Table 4.4, namely that some faculties were founded in a later year. Therefore one might wonder whether it is possible to eliminate this two-factor interaction, in order to have a better view of the two-factor interaction between faculty and type of student, and the way in which this interaction differs in different years.

This can be accomplished by using the generalization of CA to study the departure from a different log-linear model. In this different log-linear model there is interaction between years and types of student (as before), but there is also interaction between faculties and years. This log-linear model is a conditional independence model: under this model faculties and types of student are independent for each year. So, if we study the *departure* from this model, we get a display of the two-factor interaction between faculties and types of student, and of the three factor interaction (i.e. how this interaction differs for different years). The Pearson chi-square for this new log-linear model is 2973, with 62 df. This model is still rejected, so in principle (generalized) CA can be useful to study the structure in the residuals. The generalized CA solution has first four singular values 0.118 (55.7%), 0.069 (19.3%), 0.056 (12.4%) and 0.050 (10.0%). Following the elbow criterion, we study only two dimensions. This solution in Figure 4.3 is closely related to the solution in

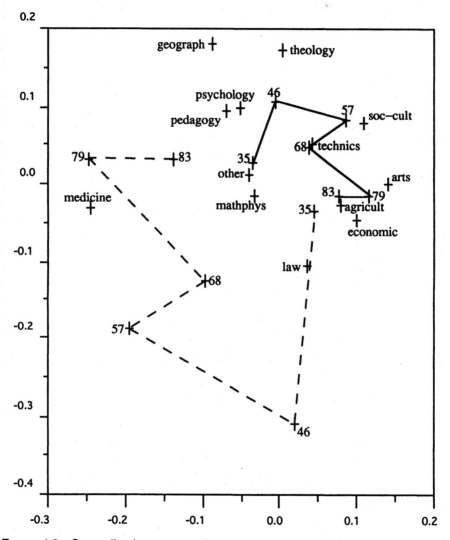

FIGURE 4.3 Generalized correspondence analysis of Table 4.4, where the two-factor interaction between faculties and years is eliminated. See text for explanation.

Figure 4.2: in the full-dimensional spaces corresponding to Figures 4.2 and 4.3 the distances between corresponding year points are identical. This holds, for example, for the distance between 1946 first-year students and 1946 graduates. However, whereas the (weighted) average of the two points in Figure 4.2 is clearly far from the origin (displaying the two-factor interaction between faculty and year), in Figure 4.3 it is located near the origin, because the interaction between faculties and years is eliminated from the (generalized) CA solution. Therefore we find in Figure 4.3 that each pair of year points has a weighted average of zero (when the marginal frequencies are used as weights). The first-order interaction between faculties and types of student is displayed on the first two dimensions: the faculties with relatively more first-year students (i.e. socio-cultural sciences (soc-cult), technical sciences (technics), arts and theology) are located in the direction of the first quadrant (solid line), whereas the faculties with relatively more graduates (medicine, law, mathematics and physics) are plotted in the direction of the third quadrant. Three-factor interaction can be derived from the plot as follows: medicine for example (but also pedagogics, psychology and geography) seems to have relatively more graduates in later years, and in these same years relatively less first-year students (as compared to the average). On the other hand, the law faculty (and also economy, agriculture) has relatively many first-year students in later years, and many graduates in early years. Inspection of the contributions shows that we have to be careful with the interpretation of the position of the points for mathematics/physics, technical sciences, psychology and pedagogy, because, respectively, only 29.9%, 47.2%, 51.0% and 44.7% of the total squared distance to the origin are projected on the first two dimensions.

We conclude that the approach to use log-linear models to eliminate information from standard CA solutions can be usefully applied when specific interactions in a higher-way table are so strong that they overwhelm the other interactions. For more applications, see van der Heijden *et al.* (1989). For more (technical) details concerning this application we refer to Escofier (1983, 1984), van der Heijden *et al.* (1989), and Takane *et al.* (1991b). Escofier (1983) calls the solution in Figure 4.3 a solution of an 'intra analysis', and she proposes to do this 'intra analysis' jointly with an 'inter analysis', which is the analysis of the marginal table of faculty by year. Van der Heijden and de Leeuw (1985) and van der Heijden *et al.* (1989) show the connection of 'intra analysis' with log-linear models, they show how this application can be extended to tables of more than three variables, and to more complicated log-linear models than 'simple' conditional independence models. They also propose to use CA and log-linear analysis complementary to each other: log-linear analysis could be used to answer the question of which variables are related by detecting the important interactions; in the second step CA can be used, for example, to explore the residuals when a model in which we are interested does not fit. Takane *et al.* (1991) show that this approach is closely related to other approaches with similar aims.

4.4.5 Conclusion

We think that by relating CA to log-linear models many insights are obtained. To mention a few:

- A new way to use CA is indicated, namely to see it as a tool for residual analysis. This might be particularly fruitful for large tables.
- It can also be emphasized that, given that the CA procedures study the residuals from log-linear models, these residuals should be meaningful, i.e. the models under study should not fit. This holds for the applications making use of the generalization of CA, but also for ordinary CA. In the applications that we see in the literature this is seldomly checked.
- For the analysis of higher-way tables by means of stacking categories of variables more insight is obtained into how to make an appropriate choice between different ways to stack variables.
- For incomplete contingency tables it is shown that a CA procedure already known can be understood in terms of the quasi-independence model.
- It is emphasized that for many applications and research questions the independence model is simply not the appropriate baseline model. Very often less restrictive models are needed to study these aspects of the data one is interested in. In the CA approach, the aspects we are *not* interested in are incorporated in the model, so that the interesting aspects are in the residuals.

For an extensive discussion and appreciation of this approach we refer to the discussion following the paper of van der Heijden *et al.* (1989).

4.5 CORRESPONDENCE ANALYSIS AND THE RC-ASSOCIATION MODELS

In this section we will discuss relations between CA and the RC-association model. This relation is mainly that, if certain conditions are fulfilled, parameter estimates found in CA are very similar to parameter estimates found in the RC-association model. Consider our example in Table 4.1. The results of the analysis with the so-called RC(2)-association model are displayed in panel C of Table 4.2. The similarity between the estimates in panel C and panel B is considerable. This is not a peculiarity of our example, but a systematic result, that will be worked out below.

In order to appreciate the form of the RC-association model, we first have to define the log-linear model for two-way tables. The saturated (i.e. unrestricted) log-linear model is

$$\log \pi_{ij} = u + a_i + b_j + c_{ij}$$

where the parameters add up to zero over each subscript. If $c_{ij} = 0$ we find the independence model.

The interaction parameters $u_{12(ij)}$ are often related to quantities called the 'log-odds ratio' $\theta_{ii'jj'}$, which is defined for four cells (i,j), (i,j'), (i',j) and (i',j'). For the saturated model the log-odds ratio is defined as

$$\theta_{ii'jj'} = \log \left[(\pi_{ij}\pi_{i'j'})/(\pi_{ij'}\pi_{i'j}) \right] = u_{12(ij)} + u_{12(i'j')} - u_{12(ij')} - u_{12(i'j)}$$

For the whole table $(I-1)(J-1)$ log-odds ratios describe the full pattern of associations. The definition of $\theta_{ii'jj'}$ shows that the log-odds ratio is independent of the margins, which is an attractive property for measures of interaction, because it allows one to compare the interaction strength in different subtables or tables. The definition also shows that under the independence model $\theta_{ii'jj'} = 0$.

There obviously is a need for models that are more restrictive than the saturated model, yet less restrictive than the independence model. Many models are proposed for this purpose, but recently models with bilinear terms (multiplicative terms) in the logarithm have received much attention.

4.5.1 The RC-association model

The basic model is the RC-association model (Goodman 1979, 1985, 1986, Andersen 1980), defined as

$$\log \pi_{ij} = u + a_i + b_j + \phi v_i w_j$$

where the v_i and w_j have to be constrained in order to identify the model. One way to identify them is by

$$\sum_{i=1}^{I} p_{i+}v_i = 0 \qquad \sum_{i=1}^{I} p_{i+}v_i^2 = 1 \qquad \sum_{j=1}^{J} p_{+j}w_j = 0 \qquad \sum_{j=1}^{J} p_{+j}w_j^2 = 1$$

These identifying constraints are chosen in such a way that they are very similar to those used for CA. The term $\phi v_i w_j$ is a bilinear term, and therefore the model is not log-linear any more. This term can be seen as a rank one constraint to the matrix of interaction parameters $u_{12(ij)}$. If I and J are large, the number of interaction parameters can be reduced considerably.

In terms of log-odds ratios, the model defines

$$\theta_{ii'jj'} = \log \left[(\pi_{ij}\pi_{i'j'})/(\pi_{ij'}\pi_{i'j}) \right] = \phi(v_i - v_{i'})(w_j - w_{j'})$$

This shows that $\theta_{ii'jj'}$ is a function of the association strength ϕ, of the difference between the scores for i and i', and of the difference between the scores for j and j'. The smaller any of these three quantities is, the more the quantity $\theta_{ii'jj'}$ approaches zero, which implies that in the subtable of cells (i,j), (i',j'), (i',j) and (i,j') the association approaches independence.

4.5.2 The RC(K)-association model

More recently Goodman (1985, 1986) proposed a natural extension of this model, namely the RC(K)-association model. This model is defined as

$$\log \pi_{ij} = u + a_i + b_j + \sum_{k=1}^{K} \phi_k v_{ik} w_{jk}$$

with constraints

$$\sum_{i=1}^{I} p_{i+} v_{ik} = 0 \qquad \sum_{i=1}^{I} p_{i+} v_{ik} v_{ik'} = \delta^{kk'} \qquad \sum_{j=1}^{J} p_{+j} w_{jk} = 0$$

$$\sum_{j=1}^{J} p_{+} w_{jk} w_{jk'} = \delta^{kk'}$$

where $\delta^{kk'} = 1$ if $k = k'$, and $\delta^{kk'} = 0$ if $k \neq k'$, that is, similar to CA, the scores are uncorrelated for different factors. If $K = \min (I-1, J-1)$, then $\Sigma_k \phi_k v_{ik} w_{jk} = u_{12(ij)}$, but if $K < \min (I-1, J-1)$, then the RC(K) association model is more restrictive than the saturated model.

In terms of log-odds ratios, the model is

$$\theta_{ii'jj'} = \log [(\pi_{ij} \pi_{i'j'})/(\pi_{ij'} \pi_{i'j})] = \Sigma_k \phi_k (v_{ik} - v_{i'k})(w_{jk} - w_{j'k})$$

The insight into the log-odds ratio $\theta_{ii'jj'}$ is getting more difficult now, because it is a function of parameters ϕ_k, differences of score i and i' on K dimensions, and differences of scores j and j' on K dimensions.

Consider again Table 4.2, panel C. In the first line the estimates for ϕ_k are found, then a 6×2 matrix with estimates for v_{ik} follows, and then a 6×2 matrix with estimates for w_{jk} follows. This model has $(I - K - 1)(J - K - 1)$ degrees of freedom, so for this example, where $K = 2$, the number of degrees of freedom is 9. The chi-square statistic is 5.828. The RC-association model, for which $K = 1$, the number of degrees of freedom is 16, and the chi-square is $G^2 = 90.2$.

Although the parameter estimates in Table 4.2, panel C are very close to these in panel A and B (reasons for this closeness are explained below), the interpretation of these parameters is different. Interpretation is simplified by means of a graphical display. A useful graphical display is obtained by using the coordinates $\hat{v}_{ik}^* = \hat{\phi}_k^{1/2} \hat{v}_{ik}$ for row point i on dimension k, and coordinates $\hat{w}_{jk}^* = \hat{\phi}_k^{1/2} \hat{w}_{jk}$ for column point j on dimension k (see Figure 4.4); thus

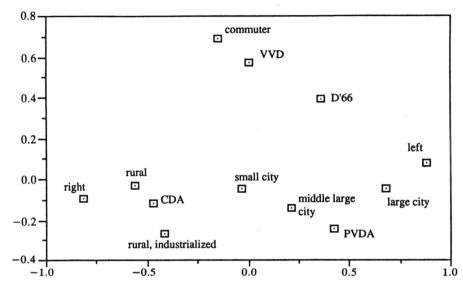

FIGURE 4.4   Graphical display of RC(2)-association model parameter estimates. See text for explanation.

$\Sigma_k \hat{v}_{ik}^* \hat{w}_{jk}^* = \Sigma_k \hat{\phi}_k \hat{v}_{ik} \hat{w}_{jk}$. Notice that, compared to Figure 4.1(a) and 4.1(b), the configuration of both the row points and the column is very similar, except for a stretching of the second dimension: on this dimension the distances are larger now. The reason is that in the CA representation distances between the row points, and distances between the column points, approximate chi-square distances, because $\hat{f}_{ik} \equiv \hat{\mu}_k \hat{r}_{ik}$ and $\hat{g}_{jk} \equiv \hat{\mu}_k \hat{c}_{jk}$ are used as coordinates. Here we do not use $\hat{\phi}_k$ but $\hat{\phi}_k^{1/2}$ to scale the row and column scores. Thus we may make a simultaneous display of row and column scores.

In Figure 4.4 estimates for $\Sigma_k \hat{v}_{ik}^* \hat{w}_{jk}^*$ could be derived to find the association for cell (i,j). This shows that the association for cell (i,j) is zero when row point i and column point j make a right angle with the origin. The association is positive when the angle is larger than $90°$, and negative when the angle is smaller than $90°$. The more the row point and the column point are away from the origin, and the smaller the angle between the row and column point, the more extreme becomes the association. Similarly, $\hat{\theta}_{ii'jj'} = \Sigma_k (\hat{v}_{ik}^* - \hat{v}_{i'k}^*)$ $(\hat{w}_{jk}^* - \hat{w}_{ik}^*)$ should be used to get the estimated log-odds ratio for the four cells (i,j), (i',j'), (i,j') and (i',j). So, if row points i and i' are close, and column points j and j' are close, then the log-odds ratio for the corresponding subtable approaches zero, and under the model the subtable of expected frequencies approaches independence. For many details concerning the interpretation of these graphical displays, and a discussion of some alternatives, we refer to Goodman (1991).

4.5.3 Estimation and restrictions

This model is usually fitted by ML. Apart from restrictions on the number of factors M, various additional restrictions can be imposed on the parameter estimates, such as fixed value constraints and equality constraints. If the row parameters are fixed, for example, to equidistant scores $v_i = \{-1, 0, 1\}$ if $I = 3$, the model becomes log-linear, and only w_j (and the scaling factor ϕ) has to be estimated. This model is called the C-association model. Similarly, if the column scores are fixed, and the row scores v_i have to be estimated, we find the R-association model. If both the row scores and the column scores are fixed to equidistant scores, then only ϕ has to be estimated, and this model is called the U-association model (see Goodman 1985 for details). Also, see Gilula and Haberman (1988). For square tables, it is possible to restrict $v_i = w_i$, and to eliminate the influence of the diagonal cells by including one parameter for each diagonal cell. The RC-association model is quite flexible in the sense that it can be adjusted to deal with many situations.

4.5.4 Relation to CA

For the example discussed above we found that the estimates for the RC(2)-association model were very similar to CA. This is not a coincidence. It is well known that CA, the RC-association model and the RC(M)-association model are closely related. We will now discuss the reasons for this.

First, it is shown by Goodman (1981) that if $k = 1$ and the proportions come from a discretized bivariate normal distribution (or a distribution that is bivariate normal after a suitable transformation of the rows and columns), CA in one dimension is closely related to the RC-association model with one dimension: it turns out that $\mu_1 \approx \phi$, $r_{i1} \approx v_i$ and $c_{j1} \approx w_j$. Second, the CA representation in k dimensions can be written in an adapted version of the reconstitution formula as

$$m_{ij} = p_{i+} p_{+j} \left(1 + \sum_{k=1}^{k^*} \mu_k r_{ik} c_{jk} \right)$$

where m_{ij} is the reconstituted value for cell (i, j). Escoufier (1982) noted that if

$$x = \sum_{k=1}^{k^*} \mu_k r_{ik} c_{jk}$$

is small compared to one (so that $\log (1 + x) \approx x$) we can rewrite the reconstitution formula as

$$\log m_{ij} \approx u + a_i + b_j + \sum_{k=1}^{K^*} \mu_k r_{ik} c_{jk}$$

where $u = 0$, $a_i = \log p_{i+}$ and $b_j = \log p_{+j}$. Since r_{ik} and c_{jk} are normalized in a similar way to v_i and w_j, Escoufier's condition roughly reduces to the situation that the departure from independence is not too large.

The conclusion is that in CA the interaction is decomposed approximately in a log–multiplicative way: the graphical displays show approximations of log–bilinear parameters. For empirical examples of this relation we refer, for example, to Goodman (1985). This close relation between CA and models with log–bilinear terms also holds for more complicated models. Examples are given in van der Heijden and Worsley (1988) and Green (1989) for a three-way table, and in van der Heijden (1992) for square tables where the CA approach to incomplete tables (see section 4.4.1) is compared with adjusted versions of the RC(M)-association model. Van der Heijden and Mooijaart (1991) show that the scores found in examples using the CA approach of section 4.4.1 where the departure from quasi-symmetry is studied, are very similar to the parameter estimates obtained in comparable models with log–bilinear terms.

Our experience is that the condition that $\log (1 + x) \approx x$ should be small, is not very restrictive. Even if for some of the cells x is rather large compared to one, then generally the interpretation will not change drastically. A reason might be that, if for some cells (i, j), x is large compared to one, this will not necessarily mean that the factorization of the matrix of cells (i, j) in the association model context will be very different from the factorization in the CA context. More research is needed in this area.

4.6 CORRESPONDENCE ANALYSIS AND IDEAL POINT DISCRIMINANT ANALYSIS

4.6.1 *Ideal point discriminant analysis*

Ideal point discriminant analysis (IPDA; Takane 1987, 1989, Takane *et al.* 1987) is a model that is inspired by the unfolding interpretation of CA. In unfolding (Coombs 1964) it is aimed to represent the rows and columns of a two-way table in one common spatial representation. Various specification can be given, but the general idea of unfolding is that, if row *i* has certain dissimilarities with the *J* columns, these dissimilarities are reflected by the distances of row *i* to the *J* columns. In correspondence analysis (approximated by least squares) a distance interpretation can be found if asymmetric scaling of rows and columns is used. Such an asymmetric scaling is obtained if for the row points of correspondence analysis coordinates r_{im} are used as coordinates, and for the column points g_{jk} (see section 2). Then the column points are in the centroids of the row points:

$$g_{jk} = \sum_{i=1}^{I} \frac{p_{ij}}{p_{+j}} r_{ik}$$

Thus column point j is in the weighted average of the row points, where the conditional proportions p_{ij}/p_{+j} are used as weights. It is allowed to calculate distances between row points i and column point j, and the order of these distances reflects the order of the dissimilarities. (Notice that this property does not hold for CA approximated by ML, unless instead of the observed conditional proportions p_{ij}/p_{+j} the conditional probabilities π_{ij}/π_{+j} are used as weight.) For more details concerning an unfolding interpretation of CA, see Takane (1980), Heiser (1981), Ihm and van Groenewoud (1984), ter Braak (1986) and Greenacre (1989).

Using this as a starting point, IPDA is defined as

$$\pi_{j|i} = \frac{w_j \exp(-d_{ij}^2)}{\sum_{j'}^{J} w_{j'} \exp(-d_{ij}^2)}$$

where d_{ij}^2 is defined as the Euclidean distance between row point i and column point j, column point j being derived as the weighted average of the row points. So, let x_{ik} be the IDPA parameter for the coordinate of row point i on dimension m, y_{jk} the IPDA column coordinate, then the column point j is derived from the row points i as

$$y_{jk} = \sum_{i=1}^{I} \frac{p_{ij}}{p_{+j}} x_{ik}$$

and the squared Euclidean distance d_{ij}^2 is

$$d_{ij}^2 = \sum_{k=1}^{K^*} (x_{ik} - y_{jk})^2$$

Notice that the coordinates y_{jk} are not parameters: the parameters that have to be estimated are the row parameters x_{ik} and the weights w_j (which will often approximate the marginal column proportions p_{+j}).

The denominator of the model for $\pi_{j|i}$ is only included in the model to ensure that $\Sigma_j \pi_{j|i} = 1$. It follows that $\pi_{j|i}$ is linearly related to w_j when $\exp(-d_{ij}^2)$ is fixed, or to $\exp(-d_{ij}^2)$ when w_j is fixed. This latter statement implies that, for fixed w_j, the conditional probability $\pi_{j|i}$ becomes larger if the distance d_{ij} becomes smaller, i.e. if row i is closer to column j.

Consider now the example in Table 4.1. The fit of the ideal point discriminant model with $K^* = 2$ dimensions is $G^2 = 7.02$ (df is 16). For $K^* = 1$ the fit is $G^2 = 90.48$ (df is 20), and for $K^* = 3$ the fit is $G^2 = 2.64$ (df = 13). The coordinates for $K^* = 2$ are presented in panel E of Table 4.2. A graphical display is found in Figure 4.5. In this display the distances between a row point and the column points are well defined. But now the distances between the rows, and the distances between the columns are not well defined. When we compare this with CA, we find that in a similar asymmetric display (where the column points are in the weighted average from the row points) the distances from each row to the columns are also well defined, but the distances between the column points are equal to chi-squared distances.

4.6.2 Relation to correspondence analysis and to RC-association models

It was indicated that IPDA was inspired by the unfolding interpretation of

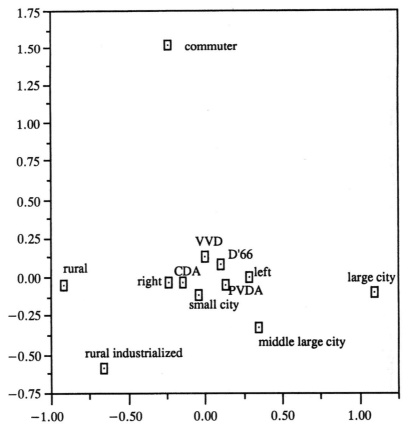

FIGURE 4.5 Graphical display of ideal point discriminant analysis parameter estimates. See text for explanation.

CA, because the IDPA column coordinates y_{jk} are derived as the weighted averages of the IPDA row parameters x_{ik}. This property holds for CA approximated by LS, but not by CA approximated by ML. Compared with CA approximated by ML and to the RC(M)-association model, in IPDA less parameters are estimated, because the column coordinates y_{jk} are derived from the row parameters x_{ik}. This usually does not lead to a great loss of fit, but, on the contrary, it does lead to a gain in number of degrees of freedom. It often leads to an increased stability of the parameters that *are* estimated in IPDA, in the sense that, if in some models one or more parameters are fixed, then the standard errors of the remaining parameters often become smaller.

Takane (1987) has shown that IPDA can be written as a constrained conditional version of the RC-association model:

$$\pi_{j|i} = \frac{b_j \exp\left(\sum_{k=1}^{K^*} \phi_k v_{ik} w_{jk}\right)}{\sum_{j'=1}^{J} b_{j'} \exp\left(\sum_{k'=1}^{K^*} \phi_{k'} v_{ik'} w_{jk'}\right)}$$

in the sense that w_j corresponds to w_j, $\phi_k v_{ik}$ corresponds to x_{ik}, and w_{jk} corresponds to the constrained y_{jk}. Takane (1987) calls this latter constraint critical in the comparison of IPDA and the RC-association model. The correspondence of IPDA with the RC-association model can be seen by working out $-d_{ij}^2$ as $-\Sigma_k(x_{ik} - y_{jk})^2 = -\Sigma_k x_{ik}^2 + 2\Sigma_k x_{ik} y_{jk} - \Sigma_k y_{jk}^2$, it follows that

$$\exp(-d_{ij}^2) = c_i c_j^* \exp\left(2 \sum_{k=1}^{K^*} (x_{ik} y_{jk})\right)$$

The parameter c_i cancels out in the numerator and denominator of the conditional model, and c_j^* is absorbed in w_j. This shows that, in IPDA, it is allowed to set $\Sigma_i p_{i+} x_{ik} = 0 = \Sigma_j p_{+j} y_{jk}$ without influencing the fit of the model. In order to simplify the comparison of IPDA with the RC-association model we now rescale x_{ik} to x_{ik}^* with the property that $\Sigma_i p_{i+} x_{ik}^* = 1$ and y_{jk} to y_{jk}^* with the property that $\Sigma_j p_{+j} y_{jk}^* = 1$. So we are rescaling $x_{ik} y_{jk} = \omega_k x_{ik}^* y_{jk}^*$. The corresponding parameter estimates are shown in panel F of Table 4.2. The similarity of the rescaled IPDA parameter estimates with the RC-association parameter estimates is striking, especially for x_{ik}^*. Notice that twice the parameter estimates for ω_k correspond to the parameter estimates for ϕ_k, since $-\Sigma_k(x_{ik} - y_{jk})^2$ corresponds to $2\Sigma_k x_{ik} y_{jk}$.

For more details concerning the relation of IPDA with CA and the RC(M)-association model, see Takane (1987) and Goodman (1991).

4.7 DISCUSSION

In this chapter we have compared the usual CA approach (which we called CA by least squares) with different sorts of models fitted by ML. As we see it, the most important advantage of CA by least squares is that the computations are straightforward, using the singular value decomposition, eigenvalue decomposition or reciprocal averaging algorithm. On the other hand, if we know that the assumptions of a (product)multinomial or Poisson distribution are fulfilled, then we should use this information in our model fitting, and thus ML estimation is a more natural candidate for the CA model.

When we compare CA estimated by ML with association models with log–bilinear terms (see section 4.4), a draw back of the correspondence analysis models is that the parameters for the interaction depend on the marginal distributions. This makes association models preferable. Another reason to prefer association models is that they are very easily extended to deal with higher-way tables, to deal with structural zeros, and to deal with specific patterns in the association. These are probably the more important reasons that there is a tendency in the Anglo-Saxon literature to prefer the RC association model over the CA model. In fact, we found only one exception to this, namely Wasserman and Faust (1989), who favour the use of the CA model over the RC association model for social network analysis. For other comparisons between the models we refer to Goodman (1985, 1986, with discussion; 1987, 1991, with discussion).

IPDA is closely related to both CA and to the RC-association model. CA estimated by least squares and IPDA have the unfolding interpretation in common (see section 4.4.2). The RC-association model and IPDA have the model formulation in common, where the IPDA can be seen as a version of the RC-association model parametrized differently and having additional constraints. These additional constraints lead to an increased stability of the model parameters, while they usually do not lead to a substantial reduction in fit.

APPENDIX: SOFTWARE

Most of the analyses were performed with special purpose software that we wrote ourselves. This special purpose software can be obtained from the authors upon request. However, there is much software available from other sources. In general, we refer to Andersen (1990), who wrote a program library called CATANA to support the analyses in his book, and Agresti (1990), who also discusses software.

CA estimated by least squares can be performed in software packages such as SPSS, BMDP and SAS. CA estimated by ML was fitted with a computer program written in APL by the second author (see also Siciliano *et al.* 1990). We have limited experience with a computer program sent to us by Haberman (see Gilula and Haberman 1986). Recently a program has been written in GLIM3.77 by de Falguerolles and Francis (in press).

The CA procedures to decompose residuals from log-linear models can usually be fitted with ordinary CA programs, if the input matrix is adjusted (see van der Heijden *et al.* 1989). These adjustments are done by us in special APL-programs.

There is more software available to fit association models with log-bilinear terms. See for references Agresti (1990). There are also programs written in GLIM3.77 by Dessens *et al.* (1985) and Becker (1989).

There is a computer program to perform IPDA that can be obtained from Takane.

Linear Constraints in Correspondence Analysis

Ulf Böckenholt and Yoshio Takane

5.1 INTRODUCTION

In any correspondence analysis (CA) study, external information is of paramount importance. External information may refer, for example, to prior knowledge about the row and column categories as well as expectations about their multidimensional representation. At a minimum, external information is necessary for an interpretation of the graphical representation, for instance, when labeling the dimensions. We may also include external information in the graphical display by fitting supplementary profiles (Greenacre 1984, p. 70) or by constraining the configuration. Both approaches are important tools in highlighting interesting features of the data that otherwise may be overlooked. However, in contrast to the use of supplementary points which do not affect the CA solution, the direct incorporation of external information in the form of constraints often leads to simplified multidimensional displays (Bentler and Weeks 1978, Carroll *et al.* 1980, Heiser 1981, p. 235, Ramsay 1982, Takane 1981). As a result, when analyzing the dependence between rows and columns, constraints are useful in the search for meaningful patterns, particularly in large data sets. The implementation and application of such constraints is the topic of this chapter.

Perhaps the simplest application of constrained CA is to explore whether row and column scores satisfy the order implied by the categories' labels (see Table 5.1), or, more parsimoniously, follow a linear order (Goodman 1991, Gilula and Haberman 1988). Although CA makes no assumptions about the spacing of the row and/or column scores, it is straightforward to estimate the scores under a linear order constraint and to compare the constrained solution

TABLE 5.1

Cross-classification of mental health status and parental socioeconomic status (Srole *et al.* 1962, p. 213).

Mental health category	Parental socioeconomic status stratum						Row totals
	A	B	C	D	E	F	
Well	64	57	57	72	36	21	307
Mild symptom formation	94	94	105	141	97	71	602
Moderate symptom formation	58	54	65	77	54	54	362
Impaired	46	40	60	94	78	71	389
Column totals	262	245	287	384	265	217	1,660

Note: A is the highest and F is the lowest socioeconomic status category

with its unconstrained counterpart. If the expectation about the spacing proves adequate, differences to the optimal solution are minimal and more than offset by the simplified and more parsimonious representation of the data. More generally, when the row or column categories form an (incomplete) factorial design, dummy or contrast variables used to code that design may be applied to constrain the graphical representation (Nishisato 1980). For example, Delbeke (1978) constructed different family compositions by factorially combining the number of sons and the number of daughters (which ranged from 0 to 3), and asked 82 students to rank order the 16 compositions according to their preference. In a constrained correspondence analysis of this data set, Takane *et al.* (1991a; see also Heiser 1981, p. 167) recoded the 16 family types as a factorial combination of the number of children and gender bias (defined as the difference between number of sons and daughters). Takane *et al*'s results showed that interactions between both factors can be ignored and that, in support of theoretical notions about family composition preferences (Coombs *et al.* 1973), subjects arrived at their preference judgments for the 16 family types by adding their separate utilities for the two factors gender bias and number of children. Many other applications with constrained configurations can be found in the multidimensional scaling literature. For example, Ekman's (1954) similarity ratings among pairs of 14 spectral hues are well described by a two-dimensional representation of the color circle. An analysis of Torgerson's (1958, p. 286) similarity data obtained for nine Munsell colors yields also a two-dimensional representation that corresponds closely to the colors' brightness and saturation (Takane 1978, Takane *et al.* 1991a). Clearly, the inclusion of known physical properties of the row and column categories in CA may not only reduce considerably the number of parameters to be estimated but may also lead to a much simplified interpretation of the data.

In this chapter we distinguish three general applications for imposing constraints in a CA. First, concomitant variables may be used to explain the association structure in the table. The equidistant spacing constraint is a simple example for this approach because it yields a readily interpretable representation of the results. Second, it may prove beneficial to partial out the effect of concomitant variables from a CA solution (Böckenholt and Böckenholt 1990, Gilula and Haberman 1986; van der Heijden *et al.* 1989). Third, in some applications it may be important to first partial out the effects of a subset of the concomitant variables and then to relate the residual information to the association structure in the table (ter Braak 1988). By incorporating external information through linear constraints on the row and/or column scores in these various ways, a representation of a contingency table is obtained that is not only more parsimonious but is also easier to understand. As a result, applications of constrained CA may prove especially useful in exploratory analyses of a contingency table (Escoufier and Junca 1986).

This chapter reviews and illustrates these three approaches for imposing linear constraints in a CA. Most of the theoretical results presented here can be found in Böckenholt and Böckenholt (1990), Golub and Underwood (1970), Rao (1964), Takane and Shibayama (1991), and Takane *et al.* (1991b). In particular, we refer to the last reference for rigorous derivations and comparisons between seemingly different approaches for incorporating linear constraints in the analysis of a contingency table.

5.2 CORRESPONDENCE ANALYSIS WITH LINEAR CONSTRAINTS

Correspondence analysis is a useful tool for obtaining a graphical display of the dependence between the rows and columns of a contingency table (e.g. Benzécri *et al.* 1980, Gifi 1990, Greenacre 1984, Lebart *et al.* 1984, Nishisato 1980). We first describe CA without constraints to introduce the notation used in this chapter. Consider an I by J contingency table $\mathbf{P}$ with proportions p_{ij} describing the joint distribution of two random categorical variables, X and Y, with I and J categories, respectively. Let $\mathbf{D_r}$ and $\mathbf{D_c}$ be diagonal matrices containing the row and column sums of $\mathbf{P}$, respectively. CA is the generalized singular value decomposition (GSVD) of

$$\mathbf{A} = \mathbf{D_r^{-1}}(\mathbf{P} - \mathbf{E})\mathbf{D_c^{-1}} = \mathbf{R}\mathbf{D_\lambda}\mathbf{C}' \qquad (5.1)$$

with $\mathbf{E} = \mathbf{D_r}\mathbf{1}\,\mathbf{1}'\mathbf{D_c}$ (where $\mathbf{1}$ is a unit vector), and $\mathbf{D_\lambda}$ is a diagonal matrix with $\min(I-1, J-1)$ singular values λ in descending order. The sum of the squared singular values is called the total inertia, and is equal to the χ^2-statistic for 'independence' divided by the sample size, n:

$$\sum \lambda_i^2 = \frac{\chi^2}{n}$$

The standard row and column coordinates (Greenacre 1984, p. 94), $\mathbf{R}$ and $\mathbf{C}$, satisfy the restrictions $\mathbf{R}'\mathbf{D_r}\mathbf{R} = \mathbf{I} = \mathbf{C}'\mathbf{D_c}\mathbf{C}$ and $\mathbf{1}'\mathbf{D_r}\mathbf{R} = \mathbf{0} = \mathbf{1}'\mathbf{D_c}\mathbf{C}$. In practice, the coordinates are computed by an ordinary SVD of the matrix $\mathbf{Z}$:

$$\mathbf{Z} = \mathbf{D_r}^{-1/2}(\mathbf{P} - \mathbf{E})\mathbf{D_c}^{-1/2} = \mathbf{U}\mathbf{D}_\lambda\mathbf{V}' \qquad (5.2)$$

with $\mathbf{U}'\mathbf{U} = \mathbf{I} = \mathbf{V}'\mathbf{V}$, and

$$\mathbf{R} = \mathbf{D_r}^{-1/2}\mathbf{U} \qquad \text{and} \qquad \mathbf{C} = \mathbf{D_c}^{-1/2}\mathbf{V}$$

The principal coordinates (Greenacre 1984, p. 90) are obtained by post-multiplying the standard scores by $\mathbf{D}_\lambda$. Usually, the interpretation of the data is based on a low-dimensional, graphical representation of the standard or the principal coordinates and is guided by the available background information about the row and column categories. In many applications, however, it may prove useful to explicitly take into account this background information when estimating the scores. As a result, the interpretation of a constrained representation is straightforward and differences between constrained and unconstrained solutions may point to unexplained features of the data. Linear constraints may be imposed by either the null-space or the reparametrization method (Böckenholt and Böckenholt 1990, Takane *et al.* 1991b). Both approaches can give identical results but because in some applications one method may be easier to use than the other, we review both procedures in the next two subsections.

5.2.1 The null-space method

According to the null-space method linear row and column constraints are defined by

$$\mathbf{G}'\mathbf{R}^* = \mathbf{0} \qquad \text{and } \mathbf{H}'\mathbf{C}^* = \mathbf{0}$$

where $\mathbf{G} = (\mathbf{D_r}\mathbf{1} \,|\, \mathbf{G}_*)$ is a known $I \times K$ matrix of rank K. Similarly, $\mathbf{H} = (\mathbf{D_c}\mathbf{1} \,|\, \mathbf{H}_*)$ is a known $J \times L$ matrix of rank L. The effects defined by the matrices $\mathbf{G}$ and $\mathbf{H}$ are partialed out from the standard row and column scores denoted by $\mathbf{R}^*$ and $\mathbf{C}^*$, respectively, by computing the complementary projection operators $\mathbf{Q_r}$ and $\mathbf{Q_c}$

$$\mathbf{Q_r} = \mathbf{I} - \mathbf{G}(\mathbf{G}'\mathbf{D_r}^{-1}\mathbf{G})^{-1}\mathbf{G}'\mathbf{D_r}^{-1}$$

and

$$\mathbf{Q_c} = \mathbf{I} - \mathbf{H}(\mathbf{H}'\mathbf{D_c}^{-1}\mathbf{H})^{-1}\mathbf{H}'\mathbf{D_c}^{-1}$$

The constrained standard scores are then obtained by the SVD of

$$\mathbf{Z}^* = \mathbf{D_r}^{-1/2}\mathbf{Q_r}(\mathbf{P} - \mathbf{E})\mathbf{Q_c'}\mathbf{D_c}^{-1/2} = \mathbf{U}^*\mathbf{D_\lambda^*}\mathbf{V}^{*\prime} \qquad (5.3)$$

with $\mathbf{U}^{*\prime}\mathbf{U}^* = \mathbf{I} = \mathbf{V}^{*\prime}\mathbf{V}^*$, yielding

$$\mathbf{R}^* = \mathbf{D_r}^{-1/2}\mathbf{U}^* \qquad \text{and} \qquad \mathbf{C}^* = \mathbf{D_c}^{-1/2}\mathbf{V}^*$$

and, consequently, $\mathbf{R}^{*\prime}\mathbf{D_r}\mathbf{R}^* = \mathbf{I} = \mathbf{C}^{*\prime}\mathbf{D_c}\mathbf{C}^*$ and $\mathbf{1}'\mathbf{D_r}\mathbf{R}^* = \mathbf{0} = \mathbf{1}'\mathbf{D_c}\mathbf{C}^*$. If constraints are imposed only on the row scores $\mathbf{H} = \mathbf{D_c}\mathbf{1}$ and, similarly, if constraints are imposed only on the column scores $\mathbf{G} = \mathbf{D_r}\mathbf{1}$. Thus, the matrices, $\mathbf{Z}^*$ in (5.3) and $\mathbf{Z}$ in (5.2), are identical when $\mathbf{G} = \mathbf{D_r}\mathbf{1}$ and $\mathbf{H} = \mathbf{D_c}\mathbf{1}$.

5.2.3 The reparametrization method

A second approach for imposing linear constraints on the standard row and column scores is given by

$$\mathbf{M}\mathbf{R_s} = \mathbf{R}^* \qquad \text{and} \qquad \mathbf{N}\mathbf{C_s} = \mathbf{C}^*$$

where $\mathbf{M} = (\mathbf{1} \mid \mathbf{M_*})$ is a known $I \times K$ matrix of rank K and $\mathbf{N} = (\mathbf{1} \mid \mathbf{N_*})$ is a known $J \times L$ matrix of rank L. The matrices $\mathbf{R_s}$ and $\mathbf{C_s}$ contain the reduced set of the scores. Thus, in contrast to the null-space method the constrained standard row and column scores are obtained by directly reparametrizing the unconstrained scores. The constrained standard scores are determined by computing the projection operators $\mathbf{O_r}$ and $\mathbf{O_c}$ as

$$\mathbf{O_r} = \mathbf{D_r}\mathbf{M}(\mathbf{M}'\mathbf{D_r}\mathbf{M})^{-1}\mathbf{M}'$$

and

$$\mathbf{O_c} = \mathbf{D_c}\mathbf{N}(\mathbf{N}'\mathbf{D_c}\mathbf{N})^{-1}\mathbf{N}'$$

and performing the SVD of

$$\mathbf{Z}^* = \mathbf{D_r}^{-1/2}\mathbf{O_r}(\mathbf{P} - \mathbf{E})\mathbf{O_c'}\mathbf{D_c}^{-1/2} = \mathbf{U}^*\mathbf{D_\lambda^*}\mathbf{V}^{*\prime} \qquad (5.4)$$

with $\mathbf{U}^{*\prime}\mathbf{U}^* = \mathbf{I} = \mathbf{V}^{*\prime}\mathbf{V}^*$, and

$$\mathbf{R}^* = \mathbf{D_r}^{-1/2}\mathbf{U}^* \qquad \text{and} \qquad \mathbf{C}^* = \mathbf{D_c}^{-1/2}\mathbf{V}^*$$

By setting $\mathbf{M}$ and $\mathbf{N}$ equal to an identity matrix we obtain the unconstrained CA solution.

5.2.3 Relationships between both methods

Because one can determine $\mathbf{N}(\mathbf{M})$ for a given $\mathbf{H}(\mathbf{G})$ and vice versa both the null-space and the reparametrization method can yield the same or ortho-complement results by appropriately defining the constraint matrices (Takane *et al.* 1991b). For example, the reparametrization and the null-space method give identical results in the case of row constraints when $\mathbf{O_r} = \mathbf{Q_r}$, and in the case of column constraints when $\mathbf{O_c} = \mathbf{Q_c}$. In contrast, if we impose constraints on the row scores, a residual analysis of the reparametrization method with $\mathbf{MR_s} = \mathbf{R^*}$ is equivalent to the null-space method with $\mathbf{G} = \mathbf{D_rM}$, and a residual analysis of the null-space method with $\mathbf{G'R^*} = \mathbf{0}$ is equivalent to the reparametrization method with $\mathbf{M} = \mathbf{D_r^{-1}G}$. In either case, $\mathbf{O_r} = \mathbf{I} - \mathbf{Q_r}$, or

$$\mathbf{D_rM(M'D_rM)^{-1}M'} = \mathbf{G(G'D_r^{-1}G)^{-1}G'D_r^{-1}}$$

Similarly, in the case of column constraints, the null-space and the reparametrization methods yield complementary results when $\mathbf{NC_s} = \mathbf{C^*}$ and $\mathbf{H} = \mathbf{D_cN}$, or when $\mathbf{H'C^*} = \mathbf{0}$ and $\mathbf{N} = \mathbf{D_c^{-1}H}$.

Clearly, the actual choice of the reparametrization or the null-space method depends only on the empirical application. The reparametrization method seems more natural when we want to directly constrain the coordinates, while the null-space method seems more natural when we want to exclude the effects of certain variables in interpreting a CA solution.

To summarize, we can decompose the $\mathbf{A}$ matrix in (5.1) into four components,

$$\mathbf{A} = \mathbf{D_r^{-1}(O_r(P - E)O_c'} + \mathbf{O_r(P - E)(I - O_c')}$$
$$+ (\mathbf{I} - \mathbf{O_r})(\mathbf{P} - \mathbf{E})\mathbf{O_c'} + (\mathbf{I} - \mathbf{O_r})(\mathbf{P} - \mathbf{E})(\mathbf{I} - \mathbf{O_c})')\mathbf{D_c^{-1}}$$

Each component refers to a particular effect of the constraints. To quantify the effects of these constraints the total inertia, $\Sigma\lambda_i^2$, may be decomposed into the corresponding four parts:

$$\Sigma\lambda_i^2 = \text{tr}(\mathbf{A'O_rD_rAO_cD_c}) + \text{tr}(\mathbf{A'O_rD_rA(I - O_c)D_c})$$
$$+ \text{tr}(\mathbf{A'(I - O_r)D_rAO_cD_c}) + \text{tr}(\mathbf{A'(I - O_r)D_rA(I - O_c)D_c})$$

The first component gives the part of the inertia obtained when both row and column scores are constrained, the sum of the first and second component equal the part of the inertia when only the row scores are constrained, and the sum of the first and third component equal the part of the inertia when only the column scores are constrained. Thus, the ratio of the first component and the total inertia gives the proportion of the χ^2-statistic that is accounted for when both row and column constraints are imposed.

In some applications it may prove useful to combine the ideas underlying the reparametrization and the null-space method. For example, a set of concomitant variables for the row scores may be divided into two subsets denoted by $\mathbf{X_1}$ and $\mathbf{X_2}$ and one may be interested in examining the effects of $\mathbf{X_2}$ while statistically controlling for the effects of $\mathbf{X_1}$. This is accomplished by first partialing out the effects of $\mathbf{X_1}$ from $\mathbf{X_2}$:

$$\mathbf{X_2^*} = (\mathbf{I} - \mathbf{X_1(X_1'D_rX_1)^{-1}X_1'D_r})\mathbf{X_2}$$

In the next step the residual information is related to the association structure in the table by setting M_* equal X_2^* in (5.4) (ter Braak 1988). Obviously, a similar procedure can be applied for the analysis of the column scores.

Only one set of constraints can be imposed by the reparametrization or the null-space method. Occasionally, it may be more appropriate to impose different sets of constraints on the scores corresponding to each singular value. For instance, it may be useful to impose uniform spacing on the scores of the first singular vector but equality constraints on the scores of the second singular vector (for an application see Gilula and Haberman 1986). Different constraints can be introduced by extracting the row and column scores corresponding to the first singular value λ_1^* and computing the rank-one reduced matrix Z_1^*:

$$Z_1^* = (I - u_1^* u_1^{*\prime})(P - E)(I - v_1^* v_1^{*\prime})$$

where u_1^* and v_1^* are the vectors corresponding to λ_1^*. In the next step, we substitute Z_1^* for $(P - E)$ in (5.3) or (5.4) and apply the different constraint matrices for the row and column scores corresponding to the second singular value. Although this approach is computationally straightforward it does not satisfy a global fitting criterion and different solutions may be obtained depending on the order by which the constraints are imposed. Consequently, it may be more appropriate to use an algorithm that allows for the simultaneous fitting of different constraint sets (Takane *et al.* 1991a).

5.3 APPLICATIONS

To illustrate the null-space and the reparametrization method, we report two examples in this section. For the sake of simplicity and reproducibility of the results, the selected data sets are rather small and do not represent typical applications of CA. Procedures for imposing the linear constraints are easily implemented, particularly when matrix commands (such as in SAS) can be used. For example, Blasius and Rohlinger (1989) provide a comprehensive documentation of a CA program written in SAS PROC MATRIX. The necessary modifications for constrained CA are straightforward and involve only the computation of the projection matrices.

5.3.1 Mental health status and parental socioeconomic status

The first data set (in Table 5.1) is taken from a study about the relationship between mental health status and parental socioeconomic status (Srole *et al.* 1962, p. 213). Subjects were assigned to one of four health categories and to one of six socioeconomic status strata (SES) according to composite scores derived from their fathers' schooling and occupational level. The SES were designated A through F to describe a sequence from highest to lowest position.

Previous analyses of this table can be found, for example, in Haberman (1979), Gilula (1986), Gilula and Haberman (1986), and Goodman (1985). This example illustrates in a simple way the equivalence between the null-space and the reparametrization method.

The independence model for the 4×6 table yields a Pearson χ^2-statistic of 46 with 15 degrees of freedom. To examine the relationship between the rows and columns of this table an unrestricted CA was computed which yielded three singular values $\lambda_1 = 0.161$, $\lambda_2 = 0.037$, and $\lambda_3 = 0.017$. The corresponding proportions of the χ^2-statistic are 0.94, 0.04, and 0.02. Clearly, a one-dimensional solution is sufficient for representing this data set. The first column of Table 5.2 contains the standard row and column scores obtained from (5.2). These scores reveal that the row and column category orders are natural. However, the scores corresponding to the second and third row are close together, indicating that the prevalence of 'mild' and 'moderate' symptoms is similar across socioeconomic statuses. The first and second as well as third and fourth column scores are also poorly distinguished. A simplified representation of this data set may be thus obtained by constraining

TABLE 5.2

Standard row and column scores for Table 5.1 by unconstrained and constrained CA.

	(1)	(2)	(3)	(4)	(5)
λ_1	0.161	0.156	0.157	0.150	0.158
%	94	88	89	81	91
No. of parameters	15	5	3	1	1
r_{11}	-1.609	-1.439	-1.617	-1.439	-1.625
r_{21}	-0.183	-0.481	-0.149	-0.481	-0.077
r_{31}	0.088	0.477	0.037	0.477	-0.077
r_{41}	1.472	1.436	1.472	1.436	1.472
c_{11}	-1.122	-1.067	-1.539	-1.539	-1.130
c_{21}	-1.147	-1.153	-0.918	-0.918	-1.130
c_{31}	-0.366	-0.343	-0.298	-0.298	-0.117
c_{41}	0.055	0.005	0.323	0.323	-0.117
c_{51}	1.025	0.952	0.944	0.944	0.896
c_{61}	1.783	1.874	1.565	1.565	1.909

Note: (1) Standard scores obtained from the first dimension of the unconstrained CA solution. (2) Row scores are restricted to follow a linear order. (3) Column scores are restricted to follow a linear order. (4) Both row and column scores are equidistant. (5) Row and column scores are equidistant and satisfy some equality constraints.

The number of parameters for unrestricted CA are equal to the degrees of freedom for the independence model. The remaining solutions are one-dimensional and the number of parameters is determined by $\{(I + J - 3) -$ number of linear restrictions$\}$. For example, in columns (4) and (5) λ_1 is the only parameter to be estimated.

the row scores, the column scores, or both to follow a linear order which takes into account that some of the categories are so similar that they can be combined.

To impose these constraints we make use of orthogonal polynomials which are convenient for subdividing the total variation of the scores into linear, quadratic, cubic, etc., components. Although higher-degree polynomials may be difficult to interpret, polynomials are quite useful in describing or approximating general forms of relationships within a limited value range. As discussed in the previous section, the reparametrization method for restricted CA is identical to simple CA when the constraint matrices $\mathbf{M}$ and $\mathbf{N}$ are set equal to an identity matrix. The basis vectors spanning the vector space of $\mathbf{I}$ may be changed without affecting the results of CA. For example, $\mathbf{M}_*$ may be equal to a matrix of orthogonal polynomials,

$$\mathbf{M}_* = \begin{bmatrix} -3 & 1 & -1 \\ -1 & -1 & 3 \\ 1 & -1 & -3 \\ 3 & 1 & 1 \end{bmatrix}$$

with the first, second, and third columns corresponding to the linear, quadratic, and cubic effects, respectively (see Bock 1975, p. 585). However, by considering the one-dimensional subspace spanned by the first column vector with

$$\mathbf{M}'_* = \begin{bmatrix} -3 & -1 & 1 & 3 \end{bmatrix} \tag{5.5}$$

we restrict the standard row scores obtained from Table 5.2 to be equally spaced. Thus, the constrained standard scores conform to the linear ordering

$$r^*_{11} - r^*_{21} = r^*_{21} - r^*_{31} = r^*_{31} - r^*_{41}$$

To satisfy the additional equality constraint between the scores of the second and the third mental health categories:

$$r^*_{21} = r^*_{31} \qquad (\text{and } r^*_{11} - r^*_{21} = r^*_{31} - r^*_{41})$$

we set

$$\mathbf{M}'_* = \begin{bmatrix} -1 & 0 & 0 & 1 \end{bmatrix} \tag{5.6}$$

Note that the equality constraint, $r^*_{21} = r^*_{31}$, is tantamount to combining the second and third categories. Thus, an equivalent approach for estimating the scores under the equality and linear spacing constraints is to group the second and third mental health category and to apply the linear constraint,

$$\mathbf{M}'_* = \begin{bmatrix} -1 & 0 & 1 \end{bmatrix}$$

to the collapsed table.

In a similar fashion we may constrain the column scores to follow a linear order. To facilitate the comparison between the unconstrained and constrained representation, we specify first N_* to be a matrix of orthogonal polynomials for the unconstrained estimation of the column scores

$$N_* = \begin{bmatrix} -5 & 5 & -5 & 1 & 1 \\ -3 & -1 & 7 & -3 & 5 \\ -1 & -4 & 4 & 2 & -10 \\ 1 & -4 & -4 & 2 & 10 \\ 3 & -1 & -7 & -3 & -5 \\ 5 & 5 & 5 & 1 & 1 \end{bmatrix}$$

(see Bock 1975, p. 585). The linear spacing of the column scores is obtained by using only the first column of N_*:

$$N_*^l = [-5 \quad -3 \quad -1 \quad 1 \quad 1 \quad 3 \quad 5] \tag{5.7}$$

Because the scores corresponding to the A and B and the C and D categories are poorly distinguished, we restrict the corresponding scores to be equal,

$$c_{11}^* - c_{21}^* = c_{31}^* - c_{41}^* = 0$$

This equality constraint in combination with the linear ordering constraint defined by (8),

$$c_{11}^* - c_{31}^* = c_{31}^* - c_{51}^* = c_{51}^* - c_{61}^*$$

is obtained by setting

$$N_*^l = [-7 \quad -7 \quad -1 \quad -1 \quad 5 \quad 11] \tag{5.8}$$

Because equality constraints are equivalent to collapsing the A and B as well as the C and D categories, we obtain the same results by applying the linear constraint

$$N_*^l = [-3 \quad -1 \quad 1 \quad 3] \tag{5.9}$$

to the collapsed data table.

The same constraints can be imposed by the null-space method. For example, we obtain a linear order for the standard row scores by partialing out the effects of the quadratic and the cubic trends. In this case G_* may be specified as

$$G_*^l = \begin{bmatrix} 1 & -1 & -1 & 1 \\ -1 & 3 & -3 & 1 \end{bmatrix} \tag{5.10}$$

and it is easy to see that G_* specifies the ortho-complement space of M_* in (5.5). Thus, by eliminating the effects of the quadratic and cubic trends, only the linear effect remains and the standard row scores conform to an equidistant spacing. The matrix $\mathbf{G}_*$ can be specified in different ways. For instance, an equivalent formulation of the linear spacing constraint given by (5.10) is

$$\mathbf{G}'_* = \begin{bmatrix} 1 & -2 & 1 & 0 \\ 0 & 1 & -2 & 1 \end{bmatrix}$$

The two rows of $\mathbf{G}'_*$ stipulate that there is no quadratic trend for any triple of consecutive categories. To satisfy the additional constraint for the data set that the scores corresponding to the second and third mental health categories are equal, we obtain

$$\mathbf{G}'_* = \begin{bmatrix} 1 & -1 & -1 & 1 \\ 0 & 1 & -1 & 0 \end{bmatrix} \tag{5.11}$$

or, equivalently,

$$\mathbf{G}'_* = \begin{bmatrix} 1 & -2 & 0 & 1 \\ 0 & 1 & -1 & 0 \end{bmatrix}$$

In both cases, the second column of $\mathbf{G}_*$ stipulates equality between the second and third row score.

A linear ordering of the column scores is obtained by defining the matrix $\mathbf{H}_*$ to contain the quadratic, cubic, quartic, and quintic contrasts,

$$\mathbf{H}'_* = \begin{bmatrix} 5 & -1 & -4 & -4 & -1 & 5 \\ -5 & 7 & 4 & -4 & -7 & 5 \\ 1 & -3 & 2 & 2 & -3 & 1 \\ -1 & 5 & -10 & 10 & -5 & 1 \end{bmatrix} \tag{5.12}$$

or, equivalently:

$$\mathbf{H}'_* = \begin{bmatrix} 1 & -2 & 1 & 0 & 0 & 0 \\ 0 & 1 & -2 & 1 & 0 & 0 \\ 0 & 0 & 1 & -2 & 1 & 0 \\ 0 & 0 & 0 & 1 & -2 & 1 \end{bmatrix}$$

To examine the linear spacing constraint with equated first and second and third and fourth column scores, one may set $\mathbf{H}_*$ to

$$\mathbf{H}'_* = \begin{bmatrix} 1 & 0 & -2 & 0 & 1 & 0 \\ 0 & 0 & 1 & 0 & -2 & 1 \\ 1 & -1 & 0 & 0 & 0 & 0 \\ 0 & 0 & 1 & -1 & 0 & 0 \end{bmatrix} \tag{5.13}$$

While the first two columns specify that there is no quadratic trend, the remaining columns imply the relevant equality constraints between the column scores.

Table 5.2 lists some of the results obtained when applying these constraints to the data in Table 5.1. The second column of Table 5.2 contains the standard row scores that satisfy a linear ordering defined by (5.5) or (5.10), the third column contains the constrained column scores defined by (5.7) or (5.12), and the fourth column contains the scores obtained by simultaneously constraining the row and column scores. The last column of Table 5.2 contains the constrained scores obtained by simultaneously imposing (5.6) and (5.8) (or (5.11) and (5.13)). In the second and third rows of Table 5.2 we also give the number of estimated parameters and the percentage of the χ^2-statistic accounted for by the first singular value, respectively.

Overall, the solution given in the fifth column is the most preferable. It requires the estimation of only one parameter and accounts for about 91% of the χ^2-statistic. From this representation, we conclude that the prevalence of well-being decreases with the socioeconomic status groups and that the opposite pattern is observed for the impaired mental health category. There is no appreciable difference between the mild and moderate symptom formation categories. Both categories' prevalences are similar across the status groups and can be combined. Moreover, the first and the second as well as the third and the fourth socioeconomic status categories relate to the mental health categories in a similar way and can also be combined with little loss in information. When combining these categories equally spaced row and column scores are obtained. As a result, the constrained CA yields a parsimonious and compact representation of the relationship between the mental health and socioeconomic status categories.

5.3.2 Magazine reading habits

In this study, reported by Böckenholt and Böckenholt (1991), 347 students were asked about their reading habits for the eight magazines: *People, Rolling Stone, Time, Sports Illustrated, Scientific American, National Geographic, Readers' Digest*, and *TV Guide*. These students were assigned to four groups and Table 5.3 lists how many members in each group read the magazines on a regular basis. For example, 31 out of a group of 91 students read regularly the magazine *People*. A subset of the respondents (53 students) was also asked to evaluate the magazines on several five-point rating scales. The total inertia of the data in Table 5.3 is 0.355 and the percentages of the inertia are 52.4, 41.1, and 6.4. A two-dimensional representation seems adequate and Figure 5.1 contains the principal coordinates obtained from the CA of these data. To obtain a simultaneous representation of the groups' selection and non-selection frequencies of the magazines we also analyzed the doubled data matrix but obtained virtually the same results.

We interpret the graphical display by inspecting groupings and contrasts in the configuration (Greenacre and Hastie 1987). Thus, differences between

TABLE 5.3
Tabulation of selection frequencies for eight magazines.

Group	PE	RS	TI	SI	SA	NG	RD	TV	Total	Size
1	31	55	1	55	24	16	6	47	235	91
2	32	20	0	3	4	1	15	14	89	57
3	71	59	66	28	11	23	79	39	376	150
4	8	6	30	10	23	32	12	5	126	49
Total	142	140	97	96	62	72	112	105	826	347

Note: PE = *People*, RS = *Rolling Stone*, TI = *Time*, SI = *Sports Illustrated*, SA = *Scientific American*, NG = *National Geographic*, RD = *Readers' Digest*, TV = *TV Guide*

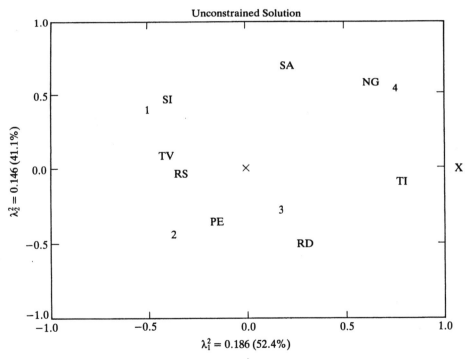

FIGURE 5.1   Two-dimensional display of the principal row and column coordinates obtained by unconstrained CA.
Note: PE = *People*, RS = *Rolling Stone*, TI = *Time*, SI = *Sports Illustrated*, SA = *Scientific American*, NG = *National Geographic*, RD = *Readers' Digest*, TV = *TV Guide*. The four groups are distinguished by the numbers 1 to 4.

magazines along an axis are small to the extent they were read on a similar recurrent basis. We note that the first axis separates *Scientific American, National Geographic, Time,* and *Readers' Digest* from the remaining four magazines, and that the second axis separates *Scientific American, National Geographic,* and *Sports Illustrated* from *People* and *Readers' Digest.* To guide the interpretation of these principal axes, the ratings of the magazines were averaged over the 53 respondents and included in the matrix N_* to constrain the column scores. This analysis showed that the two rating scales 'educational' and 'specialized' were particularly useful in distinguishing the magazines in the two-dimensional representation. Setting N_* equal to the (8×2) matrix of averaged ratings (for the eight magazines and the two attributes 'educational' and 'specialized') accounts for 87.6% of the total inertia. Figure 5.2 depicts the corresponding two-dimensional representation obtained from the column-constrained CA. This analysis indicates that *National Geographic* and *Scientific American* are perceived as more and *TV Guide* as less

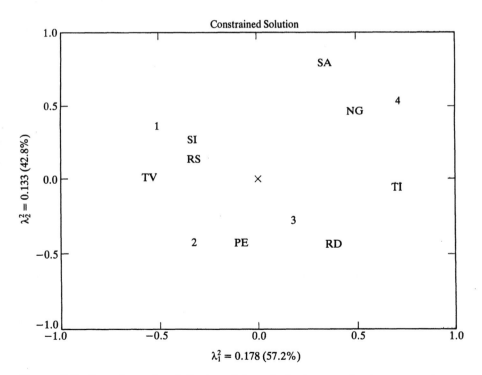

FIGURE 5.2   Two-dimensional display of the principal row and column coordinates obtained by reparametrization-method constrained CA.

educational magazines. *Readers' Digest* and *Scientific American* are at the two extreme positions on a continuum of increasing specialization.

A powerful comparison of the differences between the unconstrained and the constrained solution is provided by the null-space method with $\mathbf{H} = \mathbf{D_c N}$. This residual analysis examines the effects not accounted for by the two rating scales. Figure 5.3 depicts the first two dimensions which account for 10.3% of the total inertia. The first axis indicates that the constraints represent the relationships between *Scientific American* and *Sports Illustrated* less well than they represent the relationships among the other magazines. As a result, group 2, whose position is most affected by this result, is placed at the lower end of this axis. Overall, however, the residual analysis provides further support for the usefulness of the constraints. With the one minor exception, the mean ratings capture well the multidimensional structure of the magazines, and, consequently, facilitate a straightforward interpretation of the data.

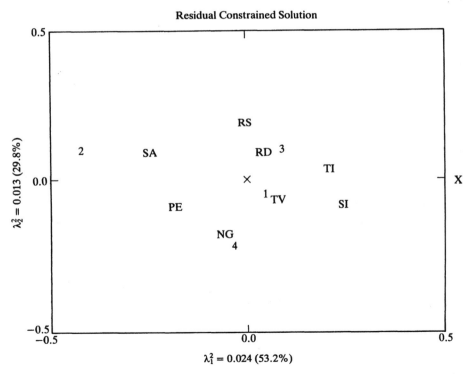

Residual Constrained Solution

FIGURE 5.3. Two-dimensional display of the principal row and column coordinates obtained by null-space-method constrained CA.

5.4 CONCLUDING REMARKS

This chapter presented two approaches for imposing linear constraints on the row and/or column scores by restricting the CA solution to lie either in a particular subspace (reparametrization method) or to be orthogonal to a subspace (null-space method). In both cases, we obtain a solution not in the full-space as in unconstrained CA but in a subspace that is directly or indirectly specified by external information. We see three major advantages in applying constraints in a CA. First, because of its simplicity and low computational cost for large problems, CA provides frequently an excellent starting point for further refinements and model building. Constraints that are formulated on the basis of some theoretical considerations can be examined for their empirical validity before more complicated models are applied. In particular, by incorporating hypothesized data structures in the singular value decomposition we gain much additional flexibility in developing a meaningful multidimensional representation of a data table. Similarly, the option to partial out the effect of certain variables in constrained CA should prove especially beneficial for the decomposition and graphical display of residuals obtained from models other than the log-linear independence model (de Leeuw and van der Heijden 1988, van der Heijden *et al*. 1989, Novak and Hoffman 1990). Second, constrained representations of a data matrix are more parsimonious and stable than unconstrained solutions and less sensitive to undesirable effects produced by, for example, outliers or coding errors. Third, by imposing constraints the search for patterns in the data may be considerably simplified. Although in some applications data structures may reveal themselves in an unconstrained analysis, in general, it is not trivial to separate 'noise' from 'signals' in a large data set. In these cases, constraints can prove helpful by either eliminating certain effects from the data or by directly imposing a certain structure. The resulting gains in interpretability may far exceed the loss in information as a result of the constraints.

Correspondence Analysis: A History and French Sociological Perspective

The BMS
(Karl M. van Meter, Marie-Ange Schiltz, Philippe Cibois and Lise Mounier)

6.1 A FRENCH HISTORICAL PERSPECTIVE

The most significant contribution that French researchers familiar with correspondence analysis can make to a study of this method is to present an 'insider's view' of the historical context in which correspondence analysis was developed and the initial uses made of it in French sociological research. Correspondence analysis found its first wide acceptance and use among French researchers and we shall describe the reactions in France to the relatively rapid and wide distribution of this new method of social science data analysis.

In the 1981 edition of *Année Sociologique*, Cibois presented an analysis both of the historical aspects, 'Jalons historiques' (Cibois 1981:335–337), of the development of the 'French Data Analysis' (FDA) approach (called '*Analyse des Données*' in French), and a sociological interpretation of this development – centred on Benzécri – as a religious, prophetic phenomenon, complete with its 'sacred texts' and its own language.

In his *Encyclopedia universalis* article 'La place de l'*a priori*' Benzécri (1973) states that the human sciences are threatened by idealism, a philosophy often implicit among researchers and which makes them substitute their own *a prioris* for observations of reality. To counter this tendency, Benzécri proposes his statistical method: 'This new tool which is the electronic computer can allow us to substitute in place of common sense qualitative notions, statistically defined quantities in such a manner that the final construction, founded upon an ample factual basis, will be independent from arbitrary constructions due to *a priori* ideas' (p. 21).

This text should be taken literally. The 'statistically defined quantities' are correspondence analysis factors. Thanks to these factors, Benzécri believes researchers can thus discover the essence of things created by God, a theme that Benzécri develops in an article published in a Catholic journal, *La Pensée Catholique* (1969).

This philosophical and religious thought has very important practical consequences. For Benzécri, what is important is not the data whose analysis gives an approximate image, but the factors which reflect the essence of reality. Thus one must eliminate any return to the data to interpret results, any use of post-factorial analysis which tries to find in the original data the results of the correspondence analysis which are only an approximation of the data themselves (Cibois 1983). This approach even has consequences for considering contributions to the factors. For Benzécri, one must consider the contribution as determining the quality of a point's representation on an axis (the squared correlations in the program results). This alone is a reliable measure and not the contribution to the factor which allows one to identify the points which contribute most to the construction of the factor since the data are less reliable than the axes. However, researchers tend to think that one should take factors more into consideration than the data itself for mathematical reasons. But for Benzécri it is a question of philosophical and even religious reasons which, of course, the researcher is not obliged to share.

6.2 FRENCH DATA ANALYSIS PERIODS OF DEVELOPMENT

6.2.1 *First period*

One of the best presentations of the recent history of French data analysis appears in 'Note on the historical background' in Chapter 2, 'Introduction of the French Contribution to the Joint Project', of the *'Comparative Study of Statistical Methods Applied to Social Science Data'*, by Rouanet (1988). He distinguishes three periods in the history of the development of modern French methods of data analysis: the emergence of the French data analysis or FDA approach in 1963–1973; the isolation of FDA in 1973–1981; and reluctant admittance since 1981.

According to Rouanet, during the first period, the theory and practice of correspondence analysis, taking the analysis of contingency tables as a leading case, began with Benzécri in 1963–1964, the first landmark being Cordier's doctoral dissertation in 1965 (Cordier 1965). The method was soon applied to a variety of fields, as reflected in Benzécri (1970) – some of these early applications, not all, were later reproduced in Benzécri *et al.* (1973). With the addition of complementary techniques such as disjunctive coding in MCA, the interpretation of contributions, etc., a comprehensive methodology for the analysis of multivariate data, especially those arising from surveys, became available and spread rapidly among French social scientists, including those working with Bourdieu. Toward 1973, the FDA approach, a new data analysis school, had emerged around Benzécri's laboratory. The first statistical textbook to include a presentation of correspondence analysis was written by Lebart and Fénelon (1971); followed by the monumental treatise of Benzécri *et al.* in 1973. During this first period, the main reaction to FDA from the Anglo–Saxon community – whether statisticians or social scientists – was polite ignorance.

According to our documentation, no review of Benzécri's treatise was made in any significant journal. The first published paper mentioning Benzécri's work and correspondence analysis as such was that of Hill (1974). Of course, one of the major obstacles to the dissemination of the method outside of France was that few Anglo–Saxon social scientists read French and the French did not make an effort to translate any basic presentations of correspondence analysis into English. Although this situation has changed over time, we have noticed a certain language barrier still exists and explains why no French texts are mentioned in the 'Factor Analysis' entry of the new *Encyclopedia of Sociology* (Borgatta and Borgatta 1992).

6.2.2 Second period

During the second period defined by Rouanet (1973 to 1981), FDA achieved an immense success in France. Numerous books about theory and applications were written (Bertier and Bouroche 1975, Cailliez and Pagès 1976, Lebart *et al.* 1977, Bouroche and Saporta 1980, Volle 1981, etc.), together with doctoral dissertations and innumerable papers in many fields: medical research, market research and econometric studies, social sciences, and others. In France, correspondence analysis was clearly becoming the most important, sophisticated method of data analysis. French data analysis began to be taught in many places as part of the standard training for statisticians. For instance, at the Université René Descartes (University of Paris V), it has been part of graduate statistical curriculum since 1971. It also started to enter certain social science curricula at the university or research level. For example, the first university credited class in FDA took place in 1971–1972 in the Social Science Department of the University of Paris V.

Meanwhile, there was a continued absence of FDA material in Anglo–Saxon literature and the language problem was not resolved. For instance, FDA was not mentioned in popular multivariate analysis handbooks such as Haberman (1978) or Bishop *et al.* (1975). However, as is typical in the confrontation between two schools of thought, with the increasing contribution of FDA to social science data analysis, silence tended to give way to condescension.

6.2.3 Third period

In the third period defined by Rouanet (from 1981 until now), the monopoly of FDA in France was shaken by the massive arrival on the French market of the large American computer program packages such as SPSS, SAS, BMDP, and others. Meanwhile, some interest in FDA was manifested outside France. A discussion of correspondence analysis was included in Nishisato's 1980 book on psychometrics. The Dutch began to incorporate the technique into the corpus of standard multivariate methods (Gifi 1981). In Agresti's 1984 handbook, there is a reluctant recognition of FDA, phrased in the following way: 'Finally, there are methods not considered in this book, such as the graphical method called correspondence analysis (see Benzécri 1976), that provide yet alternative views of the data.'

There have also been signs of deeper interest in FDA, such as the enthusiastic 1983 *Psychometrika* review, by Ramsay and de Leeuw, of Cailliez and Pagès' book; or the paper read at the Royal Statistical Society by Deville and Malinvaud in 1983; or the translation into English of the book of Lebart *et al.* (1984), and many other events. Conferences were organized in Great Britain where correspondence analysis was discussed: see for instance the contributions of Gower and Digby (1981), and, of course, Greenacre (1981). A major factor in this gradual change in attitude is most certainly the efforts made by statisticians of the FDA school to render their style more palatable for Anglo–Saxon statistical tastes. In this context, the valiant efforts of Greenacre – a former student of Benzécri – have certainly been decisive in bringing FDA to an increasingly international audience, with the publication of his book in 1984 and the entrance of correspondence analysis into JASA with the paper of Greenacre and Hastie (1987). In the same journal we also find maximum likelihood solutions to correspondence analysis discussed by Gilula and Haberman (1986, 1988).

6.3 A FRENCH PRESENTATION OF CORRESPONDENCE ANALYSIS

Perhaps the most recent and precise French presentation of correspondence analysis appears in a multi-method research report on 'A French Reanalysis of a British Survey: Comparative Study of Statistical Methods Applied to

Social Science Data' (Schiltz 1990)[1]. According to the presentation of this report, a sociologist studying a particular social topic usually tries to record and to measure all aspects of an observed reality considered relevant to the study in question. Because of cost, surveys are seldom repeated. Therefore, researchers often prefer large questionnaires with redundant or irrelevant questions rather than a parsimonious one.

Data collected in this context often consist of very large multidimensional tables of units (individuals, time periods, geographical units, etc.) cross-tabulated with categories of several discrete variables (sex, age, education, etc.). To explore these large multidimensional tables, statistical techniques are initially used to produce an intelligible, descriptive summary of data and to find 'possible' structures which verify or reject certain aspects of initial hypotheses.

Classical statistical tools offer little help in analyzing the 'messy' data often obtained in the context of nascent sociological theories. These tools were developed for the deductive or causal approaches of the natural sciences or the testing approaches of what Guttman (1984) called 'quality control' of a production process (calculating whether to throw out a certain produced object because of defects). Their purpose is largely to test hypotheses and statistical inference, and little attention is given to complementary problems such as the overall structure of a data set, the description of the data, and new ways of looking at the data set. Moreover, most of these classical techniques involve drastic assumptions concerning the data such as normality, independence, etc., which categorical survey data almost never satisfy.

The multidimensional tables we have mentioned can be analyzed with several different methods. A major advantage of correspondence analysis is that it deals with categorical data without postulates concerning the distributional characteristics of the initial variables. Since researchers often wish to study the nonlinear links that often exist in social phenomena between the different states of the variables, the relatively few quantitative variables present in typical French social science research surveys are split into categories to produce homogeneous multidimensional categorical tables. This is often done even if all the variables of the study are quantitative.

But once correspondence analysis is applied to the data, the interpretation of the graphic results is not straightforward, and can even be fallacious or

---

[1] Project jointly sponsored by the British Economic and Social Research Council (ESRC) and the French 'Centre National de Recherche Scientifique' (CNRS), and published as a research report (number P. 055) by the Centre d'Analyse et de Mathématique Sociales (CAMS) in Paris (Schiltz 1990). The report was written by Schiltz and the project was jointly directed by Everitt of the Biometric Unit, Institute of Psychiatry, University of London, and Rouanet, Groupe Mathématiques et Psychologie, Unité de Recherche Associée (URA) 1201 of the CNRS and Université René Descartes in Paris.

misleading. The results should be carefully interpreted by reading the table of coordinates and contributions (Le Roux and Rouanet 1984; see also Chapter 2 of this book).

Correspondence analysis became popular among French social science researchers in spite of statisticians' opposition to the method. Social science researchers found it to be a powerful method able to treat a wide range of data and furnish a robust image of data structure. The exploration of large data sets, such as those from sociological or epidemiological surveys, is the major use of the method in France. Results from these large data sets may be interpreted in two different but complementary ways. As with principal component analysis, the researcher can obtain a topology of the co-occurrences for all states of categorical variables. With this topology it is possible to determine the behavioural subgroups present in the graphical representations.

It is also possible, like in classical factor analysis, to classify the observed variables relative to the factors and thus explore new theoretical aspects. One of the most famous applications of MCA to sociological survey data is given by Bourdieu (1984).

The use of CA by Bourdieu in *La distinction* (1979) had a major impact due to the book's wide influence. In some cases, this use was quite clear and well presented where the actual two-dimensional graphical display was given. In other cases, it was rather rhetorical because it involved synthetic graphics which are presented as the results of several different correspondence analysis graphic displays. But these synthetic graphics were prepared by the author himself without direct reference to a specific correspondence analysis. Nonetheless, in all cases Bourdieu's intention is to demonstrate the association between certain types of taste (life-styles) and certain social situations. For example, a situation of high economic capital and low cultural capital corresponds with a 'beautiful blue Danube' taste. Inversely, low economic capital and high cultural capital corresponds with a taste for abstract painting, Picasso and *The well-tempered Clavier*.

It is this visualization of proximities between specific response levels of a questionnaire and characteristics of the respondents that has given such convincing force to Bourdieu's arguments. The visualization is quite convincing in spite of the lack of a detailed presentation of the analysis. However, the presence of cross-tabulations permits the interested reader to verify the results. Although readers rarely do examine the tables, they remain convinced by the visualizations of proximities. Due to Bourdieu's theory, correspondence analysis was more and more accepted and was applied to an increasing extent in social science research (see for example, Chapters 14 and 15).

6.4 TWO CONCEPTUAL PERSPECTIVES – TWO METHODOLOGICAL APPROACHES

6.4.1 French data analysis versus statistical methods

During the 1960s, important advances were made in the application of statistical methods to qualitative data, including developments in linear modelling (Goodman 1969) and in correspondence analysis. However, two opposing approaches developed (Schiltz 1991).

One approach was based on the principles of classical statistical inference developed at the beginning of the century for the analysis of data from experiments. This largely Anglo–Saxon approach required researchers first to state a model, then try to fit the data to it. Log-linear models, logistic regression and latent variable models are products of this school. Such procedures can only model a limited number of variables, and their use is restricted to local aspects of the data. The problem with such methods is to move the focus from wide-ranging questions down to narrower questions concerning particular relationships.

On the other hand, the description-oriented FDA approach, exemplified by correspondence analysis and Benzécri, followed the principle of its founder: 'the model must follow the data and not the reverse'. Other descriptive techniques – cluster analysis, or classification analysis, and multidimensional scaling – also cover wide areas of data simultaneously and are used widely in France.

The way in which a researcher formulates questions to be asked of research data will clearly depend on the methodology the researcher intends to use. Anglo–Saxon researchers often attempt to match their theoretical hypotheses with empirical observations, but with large multidimensional tables they encounter problems in selecting a limited number of effective variables that can be fitted to the model. The use of log-linear modelling techniques involves almost formal methods of statistical inference: the researcher has to generalize conclusions from the observed sample to the whole population. On the other hand, French social scientists appear to be interested in finding patterns in their data that are representative of particular subpopulations of the data set and can be represented graphically. However, French social scientists are rarely interested in formal methods of inference or in testing models based on hypotheses concerning relationships between variables.

6.4.2 Pros and cons of both approaches

The application of the rigorous statistical tools of quantitative statistical analysis, such as measures, levels of significance, tests, etc., to 'messy' social science data can give the results the false appearance of scientific fact.

Moreover, the literal acceptance of the results may lead to fallacious conclusions.

However, correspondence analysis can deal with this type of data and can also treat individual responses favouring different states of a variable and not whole variables, as modelling techniques do. For example, it is perhaps more meaningful to know if persons who are *young* (a particular state of the variable *age*) establish distinctive relationships between different states of specific variables in a survey, than to know whether *age*, without any further specification, is related to a set of variables taken in its entirety.

The most common criticism of the use of widely-applicable descriptive techniques is that little attention is therefore paid to developing good models; anyone can then produce results without constructing hypotheses prior to data collection. Such descriptive techniques can produce a 'black-box effect', and analysis is done without any consideration of what type of data is involved. Furthermore, the researcher is not obliged to be familiar with and does not necessarily handle his or her own data personally. Thus the use of complex, descriptive statistical techniques is often suspected of covering up poor data quality and theoretical weaknesses.

On the other hand, modelling methods are more to the liking of statisticians. While these methods are popular in biometrics and the behavioural sciences, their use is less universal in the social sciences because of the restrictions they place on data collection and data structure. In addition, because the social scientist is limited to the collection of non-experimental data, the researcher may have difficulty reducing observed reality to a restricted set of variables.

6.4.3 Complementarity and possible reconciliation

In order to reconcile the two approaches, it is not uncommon among statisticians to recommend a complementary use of the two techniques. For example, correspondence analysis can be used first of all to describe a large multidimensional set of collected data, yielding a simplified picture of the data and suggesting a model and relevant variables. Then a log-linear model is used to fit these hypotheses. Another example is proposed by a Dutch statistics team which suggests the opposite, complementary use of the two techniques: using correspondence analysis to explore the residuals from the log-linear model (van der Heijden and De Leeuw 1985; see also Chapter 4 of this book).

Another recent and effective junction between the quantitative and qualitative approaches also exists. It is now formally possible to base the interpretation of correspondence analysis results on a direct interpretation of the graphics using the 'data interrogation language' (LID for *langage d'interrogation des données*) computer program (Bernard *et al.* 1989). This program introduces measure into qualitative analyses. The basic idea is to apply the analysis of variance technique to results of the FDA approach such

as correspondence analysis or clustering method results. It permits the visualization of different subsets of individuals and the selection of relevant variable
states. In general, it brings together quantitative measures and the quality of
the selected representation.

Although it lacks a theoretical basis and can serve only as a starting point
for such future developments, a final and interesting combination of these two
approaches is the method for the visualization of the decomposition of
variance developed by Le Guen and Jaffre (1988). With this method a data set
is first analyzed by correspondence analysis and by classification analysis. The
classes of the classification analysis are projected onto the graphical display
determined by the first and second axes both to visualize the classes and to
guarantee their coherence. The variance for each variable is calculated within
each coherent class and plotted as a point on the same graphical display. For
each class, these points are joined together and generate nearly identical
figures. The similarity of these figures reflects the coherence between the
decomposition of the data into classes and the analysis of their variance.

Despite these interesting and encouraging developments in France, there
currently remains a dominant use of correspondence analysis, a significant
minority of social scientists working with more Anglo–Saxon methods, and a
small but growing minority of researchers practising a combined approach.

It was once commonplace in French statistical circles to describe FDA as
'descriptively-minded', in accordance with Benzécri's famous principle
mentioned above: 'The model must follow the data, not the reverse'; as
opposed to the Anglo–Saxon more 'sample-minded' doctrine: 'First state a
model, then try to fit it to the data'. But over the last few years, the dissemination in France of Anglo–Saxon techniques such as log-linear models is
encouraging complementary use along with FDA methods.

The point has also been made that there is nothing 'intrinsically descriptive'
in correspondence analysis as such, which after all, may be regarded as the
fitting of a sequence of models (see Chapter 4). This was suggested previously
by Gower and Aitkin in their reactions to the article by Deville and Malinvaud
(1983) in the *Journal of the Royal Statistical Society*. Conversely, one may
make a meaningful use of log–linear methods in a descriptive spirit, as Daudin
and Trécourt (1980) have shown.

Such issues have been discussed in France over the past few years,
particularly in connection with a French–British project similar to Rouanet's
conducted by Caussinus (Toulouse) and Aitkin (Lancaster) from 1984 to 1986.
The special issue of *Revue de Statistique Appliquée* in 1987 with contributions
from Daudi, de Falguerolles, van der Heijden, de Leeuw, Escoufier, and
others was largely due to this project. It began with the same observation
mentioned above: to analyze complex data sets French statisticians often use
data analysis methods adapted to qualitative (categorical) data while their
British colleagues usually resort to modelling methods. But a finely tuned

interaction between the two approaches is possible where correspondence analysis can simplify the data set and modelling methods can eliminate factors when tests show their lack of contribution. This project also examined the Dutch team's idea of visualizing model residuals with correspondence analysis.

Most of these procedures are clear examples of complementary uses of different methods of data analysis. They point in the direction of increased use in the future of multimethod analysis, no matter if the methods are called descriptive and confirmatory, or qualitative and quantitative, or ascending and descending (Van Meter, 1994, Smelser, in press).

APPENDIX

Besides the well-known classical statistical packages such as SAS and SPSS where correspondence analysis is available, there are some special software packages which deal with French data analysis methods. For further information, please contact the following addresses:

ADDAD, French package of multivariate descriptive data analysis, ADDAD, 25 avenue Charcot, 75013 Paris, France.

EYELID, for graphical inspection of multivariate data using the data inter-rogation language (LID) and introduces measure into French qualitative statistical analysis, Infodis, Centre d'affaires Paris-Nord, Immeuble 'Le Continental', 93153 Le Blanc Mesnil cedex, France.

SPAD, includes all the essential techniques for data analysis and classification analysis, CESIA, 25 avenue de l'Europe, 92310 Sèvres, France.

TRI-DEUX, a free program focusing on correspondence analysis, Philippe Cibois, Université René Descartes, 12 rue Cujas, 75005 Paris, France.

STATlab, complete collection of data analysis techniques using Windows as an interface, France Telecom, CNET, 51–59 rue Ledru-Rollin, 94853 Ivy-sur-Seine, France.

Part 2

Generalization to Multivariate Data

Multiple and Joint Correspondence Analysis

Michael Greenacre

7.1 INTRODUCTION

In Chapter 1 a geometric framework for simple correspondence analysis (CA) was introduced. CA is primarily applicable to a contingency table in which two categorical variables are cross-tabulated. The basic concepts of profile, mass and (chi-squared) distance all have a clearly justified interpretation in this context, as does the derived concept of inertia and its decomposition. These concepts form the backbone of the interpretation of the numerical and graphical results.

In this chapter a more general situation is considered where the associations among more than two categorical variables are of interest. These inter-associations are often referred to as *interactions* in this context. This generalization to the 'multi-way' case is often referred to as multiple correspondence analysis (MCA), although we shall also discuss some different generalizations which have some advantages over MCA.

Consider, for example, the case of three variables, labelled I, J and K, with categories indexed as follows: $i = 1, ..., I, j = 1, ..., J, k = 1, ..., K$. Cross-tabulation of these three variables leads to a three-way contingency table containing IJK cells (Figure 7.1(a)). This 'cubic' table is usually sliced up for purposes of reporting, for example for each category (or 'level') k of variable K the two-way table cross-tabulating J and K is given (Figure 7.1(b)). All the three-way interactions are reported in these table slices. The two-way interactions can be imagined to be the faces of the data cube onto which the frequencies are collapsed. For example in Figure 7.1, adding up all the frequencies over the levels of variable K gives a table cross-tabulating variables

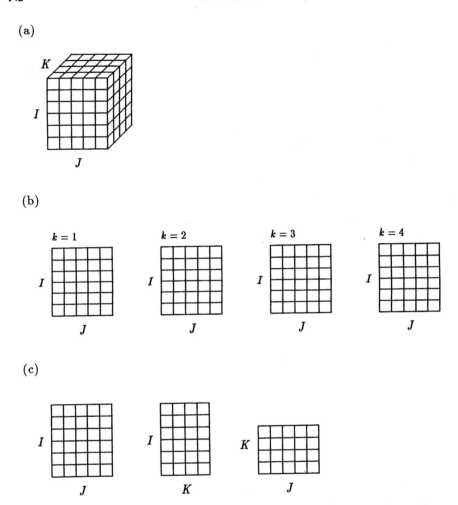

FIGURE 7.1 (a) The three-way 'cubic' table of frequencies of the three variables I (6 categories), J (5 categories) and K (4 categories). (b) The four 'slices' of the table showing the cross-tabulation of variables I and J for each level of variable K. (c) The two-way margins of the three-way table, i.e. the table collapsed onto each of the three faces in (a), e.g. the I × J margin is the sum of the four slices in (b).

I and J (the front face of the table). The three two-way tables which can be derived from the pairs of the three variables, (I, J), (I, K) and (J, K) are the two-way margins (Figure 7.1(c)).

If we go to four variables, I, J, K and L, then the four-way table is a more complex structure involving a 'hypercube' of *IJKL* frequencies. As the number of variables increases, so the number of cells in the hypercube increases and the frequencies in each cell get smaller. For an increasing number of variables and categories it becomes difficult to analyze the full multi-way table. We thus have to limit our investigation to simpler aspects of the table where the data are more concentrated, for example, the two-way margins or some set of appropriately chosen slices. In this chapter we restrict our attention to sets of two-way margins of the multi-way table.

In the general case of Q categorical variables, there are $Q(Q-1)/2$ possible two-way marginal cross-tabulations of pairs of variables. To start with, we consider 'concatenated' tables, where cross-tabulations of a number of variables with a single variable are analyzed jointly (section 7.2). This case is of interest when one variable is considered as a variable to be described by the other variables (see Preface and Chapter 2). For example, this situation is common in the social sciences where the responses to several biographical questions, e.g. age group, income group and educational group, are cross-tabulated with the responses to a question of central interest in a study such as political affiliation. We then consider the case where all $Q(Q-1)/2$ cross-tabulations are analyzed jointly (section 7.3), which we call 'joint corre-spondence analysis' (JCA). This case is of interest when all the variables are considered to have the same role in the study, so that it is the joint association structure between the variables that we want to visualize. For example, a set of statements might be presented to each respondent, who then has to express agreement or disagreement with each statement on a five-point scale. The objective would be similar to that of factor analysis, namely to uncover latent dimensions in the responses. Finally, we review the classical definition of MCA and describe its relationship with JCA, principal components analysis and exploratory factor analysis.

7.2 CA OF CONCATENATED TABLES

We use data given by Lebart *et al.* (1984, pp. 100–108) involving $n = 1000$ respondents and $Q = 7$ variables: sex, education level, lodging status, owner-ship of shares, ownership of real estate, age group and size of town. These data, which are also used by Rovan in Chapter 10, originate from a socio-economic survey relating to the living conditions and aspirations of the French (Lebart and Houzel van Effenterre 1980). The variables have 2, 5, 4, 2, 2, 5 and 5 levels respectively, which we denote by $J_1, J_2, ..., J_7$. Let us suppose that one of the seven variables, say 'lodging status', can be considered a variable to be described, or dependent variable, whereas the others are describing variables, or independent variables. Then each cross-tabulation of the third variable with the others is of interest (see Table 7.1, which also gives the names of the categories of each variable).

TABLE 7.1
Cross-tabulations of variables with 'lodging status'.

	LOD1	LOD2	LOD3	LOD4
SEXM	62	151	224	32
SEXF	58	142	302	29
EDU1	17	59	105	15
EDU2	45	116	151	15
EDU3	16	37	98	9
EDU4	27	39	87	14
EDU5	15	42	85	8
STO1	11	60	45	5
STO2	109	233	481	56
HOU1	7	48	23	4
HOU2	113	245	503	57
AGE1	3	9	22	6
AGE2	23	24	180	20
AGE3	68	91	182	15
AGE4	20	80	79	9
AGE5	6	89	63	11
SIZ1	7	59	11	6
SIZ2	20	33	29	5
SIZ3	27	62	80	6
SIZ4	48	72	191	18
SIZ5	18	67	215	26

Description of variables:
Sex: SEXM – male; SEXF – female
Level of education: EDU1 – none; EDU2 – grade school; EDU3 – some high school; EDU4 – high school; EDU5 – some college
Lodging status: LOD1 – mortgage; LOD2 – owner; LOD3 – tenant; LOD4 – rent free
Ownership of stocks and shares: STO1 – yes; STO2 – no
Ownership of real estate: HOU1 – yes; HOU2 – no
Age: AGE1 – under 25 years; AGE2 – 25–34 years; AGE3 – 35–49 years; AGE4 – 50–64 years; AGE5 – 65 years and over
Size of town: SIZ1 – less than 2000 inhabitants; SIZ2 – 2000–50 000; SIZ3 – 50 000–100 000; SIZ4 – 100 000–500 000; SIZ5 – more than 500 000.

We can now perform a CA on each of these tables to display the separate relationship between each of the six variables and lodging status (Figure 7.2). Each of the six maps in Figure 7.2 represents the maximum percentage of inertia possible in a display of at most two dimensions. Notice that when one variable has only two categories the display is unidimensional, i.e. the total inertia is represented by a single line. Thus in the first map where 'sex' is cross-tabulated with 'lodging status', we see all the points lying on a straight line, with 'females' more associated with 'tenant' (LOD3). The association is, however, very low (inertia = 0.008 31). The highest inertia is between 'age group' and 'lodging status' (the fifth map, inertia = 0.145 38), showing 'owners' (LOD2) associated with the higher age groups (AGE4 and AGE5). (Notice that in order to compare tables of different sizes, the magnitude of the inertia should be related to the number of categories of the two variables, as discussed in Chapter 2, section 2.2 – a coefficient such as Cramer's V could be used.) Apart from the first map, LOD2 is also separated from the other lodging categories, associated with the lower educational groups, owners of stocks and shares, owners of property, high age groups and small towns. Apart from the first map again, categories 'tenant' (LOD3) and 'rent-free' (LOD4) are consistently close together, opposing LOD2. In the two-dimensional maps (i.e. the second, fifth and sixth maps) lodging category 'mortgage' (LOD1) consistently separates out along the second dimension, associated with 'high school' (EDU4), '35–49 years' (AGE3) and '2000–50 000 inhabitants' (SIZ2).

In each map in Figure 7.2 the categories LOD1 to LOD4 of 'lodging status' appear in different positions. The question now arises whether we can reduce the six maps to just one map in which the pairwise relationships can be represented. In such a map the categories of 'lodging status' should be represented only once. To achieve this combined map, it can be shown that we need to perform a CA on the set of tables as they are given in Table 7.1, concatenated into one table by stacking one on top of each other. In other words the complete 21×4 matrix of tables $\mathbf{N}$ is analyzed as if it were a single cross-tabulation:

$$\mathbf{N} = \begin{pmatrix} \mathbf{N}_1 \\ \mathbf{N}_2 \\ \vdots \\ \mathbf{N}_Q \end{pmatrix}$$

where $Q = 6$ since one of the seven variables is used as the column variable.

Assuming no missing data, the inertia $\phi^2(\mathbf{N})$ of this stacked table can be shown to be the average of the inertias of the individual tables:

$$\phi^2(\mathbf{N}) = \frac{1}{Q} \sum_{q=1}^{Q} \phi^2(N_q)$$

This result holds only approximately if there are missing data (see Chapter 2,

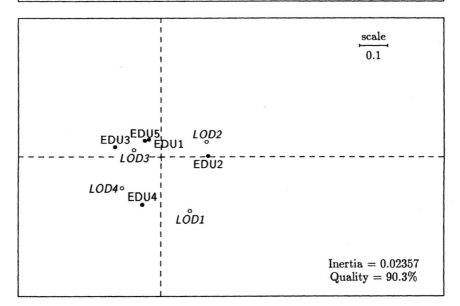

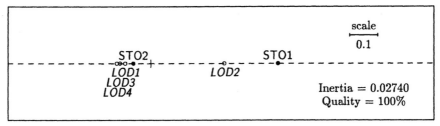

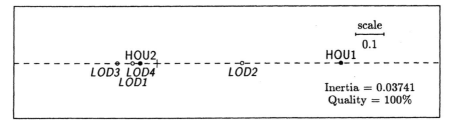

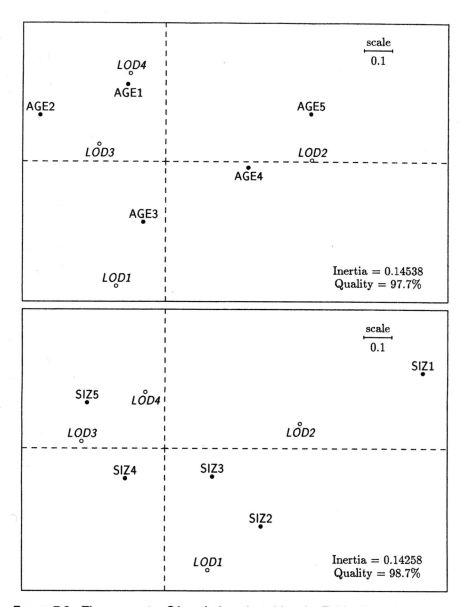

FIGURE 7.2 The separate CAs of the six tables in Table 7.1, showing the symmetric map in each case. The total inertia and percentage of inertia (quality) explained in the map are indicated. In the first, third and fourth maps the displays are exactly one-dimensional because one of the variables has only two levels.

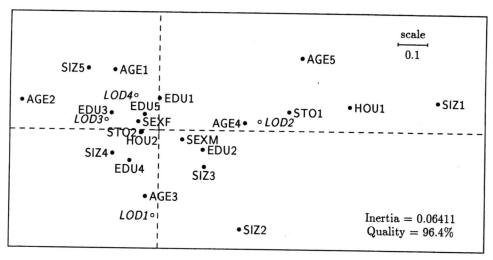

FIGURE 7.3   The joint analysis of the six tables in Table 7.2, showing the symmetric map.

section 2.3). Figure 7.3 shows the 'joint' analysis of all six tables. The average inertia is 0.064 11, of which 96.4% is displayed in the map. It is clear that this map condenses the results of the six maps of Figure 7.2. LOD2 on the right is associated with the same categories as before, such as SIZ1, HOU1, STO1, AGE5, AGE4 and EDU2, opposing LOD3 and LOD4 on the left. Similarly, LOD1 lies along the second axis, associated with AGE3, EDU4 and SIZ2 as before.

7.3   JOINT CORRESPONDENCE ANALYSIS OF ALL TWO-WAY TABLES

Instead of jointly analyzing the cross-tabulations of a single variable with all the others, we can consider the analysis of all pairs of cross-tabulations. In this example this is the joint analysis of $Q(Q-1)/2$ cross-tabulations, where $Q = 7$. These 21 cross-tabulations are outlined in the upper triangle of Table 7.2. The complete table is called the 'Burt matrix', which is the cross-tabulation of all seven variables with themselves, giving 49 tables altogether. Of these, seven square tables cross-tabulating each variable with itself lie down the diagonal. These are diagonal matrices of marginal frequencies of each variable. In the upper and lower triangles of the matrix we find the 21 pairwise cross-tabulations in identical but transposed pairs: for example, in the upper triangle we find the cross-tabulation of sex (rows) and education (columns) and in the

TABLE 7.2

Burt matrix of the seven variables, showing all the two-way cross-tabulations of pairs of variables as well as the cross-tabulations of each variable with itself.

| | Sex | | Education | | | | | Lodging | | | | Stocks | | Property | | Age | | | | | Size of town | | | | |
	SEXM	SEXF	EDU1	EDU2	EDU3	EDU4	EDU5	LOD1	LOD2	LOD3	LOD4	STO1	STO2	HOU1	HOU2	AGE1	AGE2	AGE3	AGE4	AGE5	SIZ1	SIZ2	SIZ3	SIZ4	SIZ5
SEXM	469	0	102	164	69	65	69	62	151	224	32	54	415	35	434	18	111	169	84	87	42	40	81	161	145
SEXF	0	531	94	163	91	102	81	58	142	302	29	67	464	47	484	22	136	187	104	82	41	47	94	168	181
EDU1	102	94	196	0	0	0	0	17	59	105	15	12	184	11	185	9	27	54	47	59	19	17	35	70	55
EDU2	164	163	0	327	0	0	0	45	116	151	15	26	301	21	306	0	55	125	84	63	43	34	67	110	73
EDU3	69	91	0	0	160	0	0	16	37	98	9	19	141	14	146	14	47	60	27	12	12	13	34	52	49
EDU4	65	102	0	0	0	167	0	27	39	87	14	27	140	19	148	16	62	63	16	10	8	15	21	57	66
EDU5	69	81	0	0	0	0	150	15	42	85	8	37	113	17	133	1	56	54	14	25	1	8	18	40	83
LOD1	62	58	17	45	16	27	15	120	0	0	0	11	109	7	113	3	23	68	20	6	7	20	27	48	18
LOD2	151	142	59	116	37	39	42	0	293	0	0	60	233	48	245	9	24	91	80	89	59	33	62	72	67
LOD3	224	302	105	151	98	87	85	0	0	526	0	45	481	23	503	22	180	182	79	63	11	29	80	191	215
LOD4	32	29	15	15	9	14	8	0	0	0	61	5	56	4	57	6	20	15	9	11	6	5	6	18	26
STO1	54	67	12	26	19	27	37	11	60	45	5	121	0	39	82	2	18	35	27	39	4	9	22	36	50
STO2	415	464	184	301	141	140	113	109	233	481	56	0	879	43	836	38	229	321	161	130	79	78	153	293	276
HOU1	35	47	11	21	14	19	17	7	48	23	4	39	43	82	0	3	11	27	20	21	7	12	12	25	26
HOU2	434	484	185	306	146	148	133	113	245	503	57	82	836	0	918	37	236	329	168	148	76	75	163	304	300
AGE1	18	22	9	0	14	16	1	3	9	22	6	2	38	3	37	40	0	0	0	0	2	4	8	15	11
AGE2	111	136	27	55	47	62	56	23	24	180	20	18	229	11	236	0	247	0	0	0	13	16	34	102	82
AGE3	169	187	54	125	60	63	54	68	91	182	15	35	321	27	329	0	0	356	0	0	31	34	71	100	120
AGE4	84	104	47	84	27	16	14	20	80	79	9	27	161	20	168	0	0	0	188	0	25	20	26	60	57
AGE5	87	82	59	63	12	10	25	6	89	63	11	39	130	21	148	0	0	0	0	169	12	13	36	52	56
SIZ1	42	41	19	43	12	8	1	7	59	11	6	9	79	7	76	2	13	31	25	12	83	0	0	0	0
SIZ2	40	47	17	34	13	15	8	20	33	29	5	9	78	12	75	4	16	34	20	13	0	87	0	0	0
SIZ3	81	94	35	67	34	21	18	27	62	80	6	22	153	12	163	8	34	71	26	36	0	0	175	0	0
SIZ4	161	168	70	110	52	57	40	48	72	191	18	36	293	25	304	15	102	100	60	52	0	0	0	329	0
SIZ5	145	181	55	73	49	66	83	18	67	215	26	50	276	26	300	11	82	120	57	56	0	0	0	0	326

lower triangle the cross-tabulation of education (rows) and sex (columns). We
are concerned here with the joint analysis of the 21 tables in either format, say
those indicated in the upper triangle. Six of these tables (Table 7.1) have
already been analysed in section 7.2 – these tables are in the columns headed
'Lodging' in Table 7.2.

As in section 7.2, we can imagine 21 separate CAs of each two-way table,
leading to 21 maps in which each variable appears six separate times. To

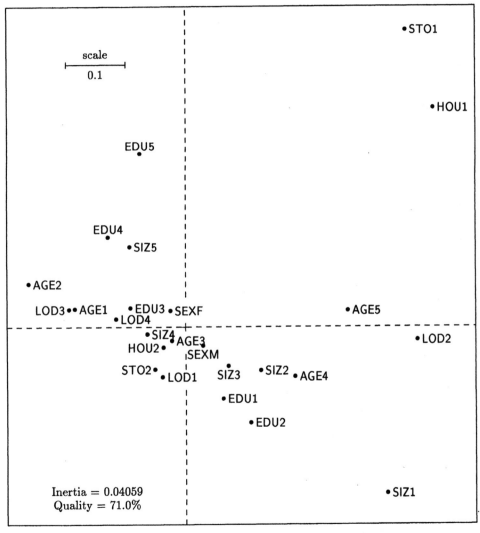

FIGURE 7.4   Joint correspondence analysis (JCA) of all 21 cross-tabulations of
distinct pairs of variables in Table 7.2.

simplify these maps to a single map where each variable appears once is not as simple a task as before, since it is not possible to express this problem as the CA of a particular two-way table known in advance. This problem, called 'joint correspondence analysis' (JCA), has been described in detail by Greenacre (1988, 1993). An iterative alternating least-squares algorithm has been proposed to obtain a solution (Greenacre 1988) and is described further in section 7.5 below. The objective is, as before, to account for as much of the inertia in all 21 tables in one joint map.

The JCA of the seven-variable application is given in Figure 7.4. The inertia is now 0.040 59, which is the average of all the 21 two-way tables. The quality of the map is 71.0%, i.e. 71% of the inertia in the 21 pairwise contingency tables is accounted for in the map. There is a diagonal gradient of points from top left to bottom right, stretching from the lower age groups (AGE1, AGE2), higher education levels (EDU4, EDU5), big towns (SIZ5), tenants (LOD3) across to the higher age groups (AGE4, AGE5), owners (LOD2), small towns (SIZ1), and lower education levels (EDU1, EDU2). Perpendicular to this gradient we have the separation of two points which are strongly associated, owning stocks and shares (STO1) and owning property (HOU1).

7.4 MULTIPLE CORRESPONDENCE ANALYSIS

The classical way of generalizing CA to the multivariate case is to apply the usual CA algorithm to a matrix known as the 'super-indicator matrix', denoted by Z. This matrix has as many rows as there are respondents and as many columns as the levels of the categorical variables. For the data set given in Tables 7.1 and 7.2 there are 1000 cases and a total of 25 categories, so Z is a 1000×25 matrix. Each row of Z consists mainly of zeros, but there is a one in each column corresponding to the options chosen by the particular respondent. For example, the following row corresponds to a respondent who is male, has high school education, owns a home, owns shares and real estate, is in the age group 35–49 years and lives in a town of between 2000 and 50 000 inhabitants, in other words the series of responses $(1, 4, 2, 1, 1, 3, 2)$:

$$10\ 00010\ 0100\ 10\ 10\ 00100\ 01000$$

(in Z this is a continuous set of 25 digits, blanks are inserted between every set of digits above for legibility). The first two columns refer to the two categories of sex, SEXM and SEXF. Since the respondent is a male, this is indicated by the two digits 10. The next five columns refer to the five categories of education, and the information that the respondent is in the fourth education category is reflected in the series of digits 00010, and so on for the other five categorical variables. The responses are all coded in this way, leading to a large matrix consisting mostly of zeros, with seven ones in each row indicating a particular set of responses. Just as we have separated the $Q = 7$ groups of columns in the above example, so we can denote $\mathbf{Z}$ in general as the concatenation of Q matrices, one for each of the Q variables:

$$\mathbf{Z} = [\mathbf{Z}_1 \ \mathbf{Z}_2 \ ... \ \mathbf{Z}_Q]$$

Notice in the above that we have implicitly assumed that there are no missing data. There are various strategies for handling missing data. If there are very few missing data which occur fairly randomly in the data (i.e. they are not concentrated into a small subset of respondents or variables), then one can simply code the corresponding set of columns by zeros. This has the effect of changing the profiles slightly for some cases and variables. A large amount of missing data for a respondent should lead to elimination of the respondent from the study, whereas in the case of a substantial amount of missing data, say more than 10%, for a variable, an additional category labelled 'missing' can be defined for that variable. Various approaches to the missing data problem are discussed by Meulman (1982) and Greenacre (1984, section 5.3).

As we said at the start of this section, MCA is the CA of this matrix $\mathbf{Z}$. Since the row totals of $\mathbf{Z}$ are all the same, the masses of the rows are all equal to $1/n$, where n is the number of cases. The column sums of $\mathbf{Z}$ are equal to the frequencies of response of the categories, so that the column masses are proportional to these marginal frequencies, as in simple CA. Chi-squared distances between cases are a variation of the so-called 'matching coefficient' for categorical variables. The matching coefficient is a simple count of how many different responses there are between two respondents and, since the responses are coded as zeros and ones only, this coefficient can be defined as the sum of squared differences between the two respective rows. In the chi-squared formulation we still count only the differences between rows but each difference is multiplied by the inverse of the respective category mass. This means that differences coinciding with a rare response count higher in the distance calculation than differences coinciding with a frequently occurring response, which is the general idea underlying the chi-squared distance. The justification of this distance is the same here as for simple CA (see Chapter 1, section 1.2.3).

The geometric concepts of correspondence analysis become more difficult to justify when applied to the columns of $\mathbf{Z}$, as described in detail by Greenacre (1989). For example, consider the following simple case of two columns of $\mathbf{Z}$, corresponding to categories of different variables, given on the left. On the right we give the corresponding column profiles:

$$
\begin{pmatrix} 1 \\ 0 \\ 1 \\ 1 \\ 0 \\ 1 \\ 1 \\ 0 \\ 1 \\ 0 \end{pmatrix}
\begin{pmatrix} 0 \\ 0 \\ 1 \\ 0 \\ 0 \\ 0 \\ 0 \\ 0 \\ 1 \\ 0 \end{pmatrix}
\begin{pmatrix} \frac{1}{6} \\ 0 \\ \frac{1}{6} \\ \frac{1}{6} \\ 0 \\ \frac{1}{6} \\ \frac{1}{6} \\ 0 \\ \frac{1}{6} \\ 0 \end{pmatrix}
\begin{pmatrix} 0 \\ 0 \\ \frac{1}{2} \\ 0 \\ 0 \\ 0 \\ 0 \\ 0 \\ \frac{1}{2} \\ 0 \end{pmatrix}
$$

It is difficult to apply the same geometric approach as in simple CA to justify the use of such profile points to display the categories.

The rationale underlying the analysis of Z becomes clearer, however, when it is realized that there is a connection between Z and the pairwise contingency tables described in section 7.3. If we pre-multiply Z by its transpose, we obtain the $J \times J$ Burt matrix B in Table 7.2. B contains all the pairwise cross-tabulations of the Q variables, including the cross-tabulations of each variable with itself:

$$
B = Z^T Z = \begin{pmatrix}
D_1 & N_{12} & \cdots & N_{1Q} \\
N_{21} & D_2 & \cdots & N_{2Q} \\
\vdots & \vdots & \ddots & \vdots \\
N_{Q1} & N_{Q2} & \cdots & D_Q
\end{pmatrix}
$$

where $N_{qs} = Z_q^T Z_s$. MCA can be alternatively defined as the CA of the Burt matrix B, since this is closely related to the CA of Z, thanks to the connection between the singular-value decompositions implicit in the analysis of B and of Z (see Chapter 3, sections 3.2.10 and 3.2.11). Both decompositions have the same right singular vectors, and the singular values in the analysis of B are the squares of those of Z. The geometric relationship between the two analyses is thus as follows. First, the standard coordinates of the categories (i.e. columns of Z or either the rows or columns of B since B is symmetric) are the same. Second, the principal inertias of B are the squares of those of Z.

Figure 7.5 shows the MCA of Table 7.2. The differences between this map and the map of Figure 7.4 obtained by JCA are fairly slight. The main difference is in the scale of the maps (see the scale interval of 0.1 indicated in the top right-hand corner of each map), with the points in Figure 7.5 being more spread out. From a qualititative point of view, the two maps are similar, but in the case of the MCA the impression is given that the points are poorly fitted since the quality of display is only 26.4%, compared to 71.0% for JCA.

There are several important implications of these results. In theory, it means that MCA can be performed by applying correspondence analysis to either the

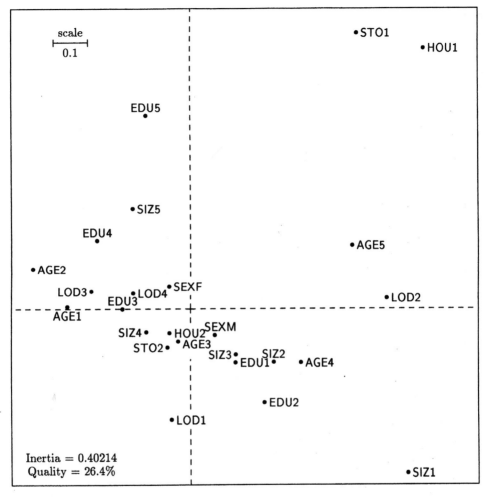

FIGURE 7.5   Multiple correspondence analysis (MCA) of the complete Burt matrix
in Table 7.2.

indicator matrix or the Burt matrix. If the Burt matrix is analyzed and if the
positions of individual respondents, or respondent groups, are of interest,
these can be added as supplementary points (for the use of supplementary
points see Chapters 12 and 13). Using the Burt matrix as initial data makes
more sense from a geometric point of view and can be interpreted in the same
way as JCA. Thus MCA can be explained as the joint analysis of all pairwise
cross-tabulations, including the cross-tabulations of each variable with itself.
It is the inclusion of the latter tables, however, that is the complicating factor
in the assessment of the map's quality. Each cross-tabulation of a variable

with itself is a diagonal matrix and thus contains the most extreme form of association in a two-way table: in fact, the inertia ϕ^2_{qq} of the $J_q \times J_q$ diagonal matrix is equal to $(J_q - 1)$. These values can be seen down the diagonal of Table 7.3, which shows the inertias of all cross-tabulations (for example, the inertia between education and lodging is $\phi^2_{23} = 0.0236$). Notice how small the off-diagonal inertias ϕ^2_{qs} are compared to the diagonal inertias. The presence of the high inertias on the diagonal grossly inflates the inertia of **B**. In fact, if the inertia of **B** is denoted by $\phi^2(\mathbf{B})$ and the average of the off-diagonal inertias by $\bar{\phi}^2$, i.e.

$$\bar{\phi}^2 = \frac{1}{Q(Q-1)} \Sigma\Sigma_{q \neq s}\phi^2_{qs}$$

then it can be shown (Greenacre 1993, formula (17.3)) that:

$$\bar{\phi}^2 = \frac{Q}{(Q-1)} \left(\phi^2(\mathbf{B}) - \frac{(J-Q)}{Q^2} \right)$$

Inverting this formula we obtain:

$$\phi^2(\mathbf{B}) = \frac{Q-1}{Q} \bar{\phi}^2 + \frac{J-Q}{Q^2}$$

with the second term due to the diagonal being generally much larger than the first term. To illustrate this in our example, $\bar{\phi}^2 = 0.040\ 59$, $J = 25$ and $Q = 7$, so that

$$\phi^2(\mathbf{B}) = \frac{6}{7} \times 0.040\ 59 + \frac{18}{49} = 0.034\ 79 + 0.367\ 35 = 0.402\ 13$$

which demonstrates that in this case there is more than ten times the inertia in the seven tables on the diagonal than in the 21 tables on either side of the diagonal. These results can be checked by directly computing the average inertia of the individual inertias given in Table 7.3. Geometrically the high inertias on the diagonal tend to push all the category points outwards, as seen

TABLE 7.3
Inertias for each two-way cross-tabulation.

	Sex	Education	Lodging	Stocks	Property	Age	Size of town
Sex	1	0.0087	0.0083	0.0003	0.0006	0.0023	0.0018
Education	0.0087	4	0.0236	0.0368	0.0074	0.1508	0.0745
Lodging	0.0083	0.0236	3	0.0274	0.0374	0.1454	0.1426
Stocks	0.0003	0.0368	0.0274	1	0.1056	0.0029	0.0081
Property	0.0006	0.0074	0.0374	0.1056	1	0.0103	0.0042
Age	0.0023	0.1508	0.1454	0.0290	0.0103	4	0.0274
Size of town	0.0018	0.0745	0.1426	0.0081	0.0042	0.0274	4

in Figure 7.5 compared to Figure 7.4, and the percentages of inertia recovered in the map are much lower than in JCA. Notice that if we had analyzed **Z** instead of **B**, then category points would be in the same relative positions along the principal axes as in Figure 7.5, the only difference being an even larger scale and a poorer quality of display (18.2%).

There is a way of remedying the apparently poor quality of fit in an MCA. Greenacre (1988, 1991) proposed that the scale of the MCA solution be changed to allow the map to fit the off-diagonal tables better at the expense of fitting the diagonal tables worse. In other words, we can imagine the scale along the principal axes in Figure 7.5 being changed until this map is as close as possible to the map in Figure 7.4. This is a compromise between MCA, which is easier to calculate than the more computationally-expensive JCA. The simplest type of rescaling which can be performed is to use the following principal inertias (eigenvalues) rather than those (λ_k) obtained in the analysis of **B** (see Greenacre 1993, formula (17.4)):

$$\left(\frac{Q}{Q-1}\right)^2 \times \left(\sqrt{\lambda_k} - \frac{1}{Q}\right)^2 \qquad k = 1, 2, \dots$$

with the condition that these be calculated for $\sqrt{\lambda_k} > 1/Q$ only. Using these adjusted inertias leads to the quality of the MCA map improving from 26.4% in this example to 68.4%. These are the same adjusted principal inertias used by Rovan in his analysis of these data in Chapter 10. Rovan, however, calculates his percentages of inertia in a different way according to a proposal by Benzécri (1979), and obtains a quality of 83.3% in the adjusted MCA map. This percentage is, by contrast, an over-estimate of the quality of the map (Greenacre 1988).

7.5 DECOMPOSITION OF INERTIA AND CONTRIBUTIONS

If JCA is performed on part of the Burt matrix, the question arises how the contributions are to be calculated. At each iteration of the JCA algorithm, estimates are obtained for the 'missing' blocks on the diagonal of the Burt matrix, using the reconstitution formula in two dimensions (Chapter 3, sections 3.2.38 and 3.2.39). At the optimal JCA solution, these estimates are perfectly fitted, and the contributions can be computed in the usual way on the Burt matrix with its diagonal replaced by these optimal estimates. This 'modified' Burt matrix is shown in Table 7.4, where the replacement estimates on the diagonal have been rounded to the nearest integers. We can thus apply CA to this modified matrix to obtain the contributions, given in Table 7.5. But to obtain the estimates which replace the diagonal of the Burt matrix requires an iterative algorithm, described by Greenacre (1988). The listing in Table 7.5,

TABLE 7.4
Modified Burt matrix with frequencies estimated by a two-dimensional JCA solution in the diagonal blocks.

	Sex		Education					Lodging				Stocks		Property		Age					Size of town				
	SEXM	SEXF	EDU1	EDU2	EDU3	EDU4	EDU5	LOD1	LOD2	LOD3	LOD4	STO1	STO2	HOU1	HOU2	AGE1	AGE2	AGE3	AGE4	AGE5	SIZ1	SIZ2	SIZ3	SIZ4	SIZ5
SEXM	223	246	102	164	69	65	69	62	151	224	32	54	415	35	434	18	111	169	84	87	42	40	81	161	145
SEXF	246	285	94	163	91	102	81	58	142	302	29	67	464	47	484	22	136	187	104	82	41	47	94	168	181
EDU1	102	94	44	77	29	26	20	17	59	105	15	12	184	11	185	9	27	54	47	59	19	17	35	70	55
EDU2	164	163	77	138	47	38	26	45	116	151	15	26	301	21	306	0	55	125	84	63	43	34	67	110	73
EDU3	69	91	29	47	27	30	27	16	37	98	9	19	141	14	146	14	47	60	27	12	12	13	34	52	49
EDU4	65	102	26	38	30	37	36	27	39	87	14	27	140	19	148	16	62	63	16	10	8	15	21	57	66
EDU5	69	81	20	26	27	36	40	15	42	85	8	37	113	17	133	1	56	54	14	25	1	8	18	40	83
LOD1	62	58	17	45	16	27	15	32	65	178	7	11	109	7	113	3	23	68	20	6	7	20	27	48	18
LOD2	151	142	59	116	37	39	42	65	76	348	13	60	233	48	245	9	90	91	80	23	20	33	62	110	68
LOD3	224	302	105	151	98	87	85	178	348	503	37	45	481	23	503	24	128	182	79	113	34	21	76	162	233
LOD4	32	29	15	15	9	14	8	7	13	37	4	5	56	4	57	4	6	15	9	27	22	13	10	9	7
STO1	54	67	12	26	19	27	37	11	60	45	5	60	61	39	82	2	18	35	27	39	4	9	22	36	50
STO2	415	464	184	301	141	140	113	109	233	481	56	61	818	43	836	38	229	321	161	130	79	78	153	293	276
HOU1	35	47	11	21	14	19	17	7	48	23	4	39	43	23	59	3	11	27	20	21	2	13	31	25	11
HOU2	434	484	185	306	146	148	133	113	245	503	57	82	836	59	859	37	236	329	168	148	81	74	144	304	315
AGE1	18	22	9	0	14	16	1	3	9	24	4	2	38	3	37	5	13	15	6	5	2	4	8	15	11
AGE2	111	136	27	55	47	62	56	23	90	128	6	18	229	11	236	13	91	94	29	20	13	16	34	102	82
AGE3	169	187	54	125	60	63	54	68	91	182	15	35	321	27	329	15	94	128	60	59	31	34	71	100	120
AGE4	84	104	47	84	27	16	14	20	80	79	9	27	161	20	168	6	29	60	51	42	25	20	26	60	57
AGE5	87	82	59	63	12	10	25	6	23	113	27	39	130	21	148	5	20	59	42	43	12	13	36	52	56
SIZ1	42	41	19	43	12	8	1	7	20	34	22	4	79	2	81	2	13	31	25	12	13	10	19	24	13
SIZ2	40	47	17	34	13	15	8	20	33	21	13	9	78	13	74	4	16	34	20	13	10	9	17	27	24
SIZ3	81	94	35	67	34	21	18	27	62	76	10	22	153	31	144	8	34	71	26	36	19	17	33	56	50
SIZ4	161	168	70	110	52	57	40	48	110	162	9	36	293	25	304	15	102	100	60	52	24	27	56	111	110
SIZ5	145	181	55	73	49	66	83	18	68	233	7	50	276	11	315	11	82	120	57	56	13	24	50	110	129

TABLE 7.5
Numerical output for the CA of Table 7.4.

J	Name	QLT	MAS	INR	$k=1$	COR	CTR	$k=2$	COR	CTR
1	SEXM	500	67	5	−29	230	2	−31	270	5
2	SEXF	505	76	5	26	239	2	27	267	4
3	EDU1	409	28	26	−62	88	5	−118	321	29
4	EDU2	859	47	43	−108	268	23	−160	591	88
5	EDU3	375	23	12	92	335	8	32	40	2
6	EDU4	589	24	34	129	243	17	154	346	41
7	EDU5	819	21	52	76	50	5	296	769	138
8	LOD1	114	17	26	40	22	1	−82	92	8
9	LOD2	954	42	147	−397	950	279	−24	3	2
10	LOD3	898	75	69	196	877	122	30	21	5
11	LOD4	259	9	9	113	255	5	15	4	0
12	STO1	980	17	147	−372	342	101	507	638	326
13	STO2	981	126	20	52	346	14	−70	635	45
14	HOU1	944	12	84	−421	523	88	378	422	123
15	HOU2	945	131	8	38	527	8	−34	418	11
16	AGE1	182	6	22	179	177	8	30	5	0
17	AGE2	878	35	63	263	814	103	73	63	14
18	AGE3	93	51	13	24	49	1	−23	44	2
19	AGE4	813	27	28	−184	676	38	−83	137	13
20	AGE5	653	24	59	−273	647	76	26	6	1
21	SIZ1	821	12	58	−337	491	57	−276	329	66
22	SIZ2	426	12	13	−124	322	8	−70	103	5
23	SIZ3	464	25	11	−71	252	5	−65	211	8
24	SIZ4	440	47	10	65	427	8	−11	13	0
25	SIZ5	693	47	39	93	221	17	136	472	63

from the program SimCA 2 (Greenacre 1986), is in a similar summary format as in section 3.2.40 except the results have all been multiplied by 1000 to avoid printing decimal points.

Because of the inclusion of high inertias on the diagonal of the Burt matrix and their high influence on the map, it follows that the contributions, which are based on the decomposition of inertia, will be strongly affected as well. Table 7.6 shows the corresponding numerical output for the CA of **B**, i.e. the MCA. Notice that the qualities of display (column QLT, or SQCOR in section 3.2.40) are all much lower than the equivalent quantities in Table 7.5, showing that the points are displayed more poorly in the MCA. The same holds for the squared correlations (columns labelled COR, or QCOR in section 3.2.40) which indicate the quality on individual axes.

TABLE 7.6
Numerical output for the CA of Table 7.2.

J	Name	QLT	MAS	INR	k = 1	COR	CTR	k = 2	COR	CTR
1	SEXM	66	67	28	71	30	5	−76	35	8
2	SEXF	66	76	24	−63	30	5	67	35	7
3	EDU1	69	28	43	132	28	8	−159	41	15
4	EDU2	386	47	38	217	145	35	−280	242	77
5	EDU3	51	23	44	−199	51	14	−1	0	0
6	EDU4	156	24	46	−273	97	28	213	59	23
7	EDU5	399	21	48	−130	19	6	582	380	152
8	LOD1	100	17	48	−55	3	1	−331	98	39
9	LOD2	756	42	46	578	754	224	31	2	1
10	LOD3	555	75	29	−290	538	101	53	18	4
11	LOD4	14	9	49	−168	13	4	47	1	0
12	STO1	739	17	54	490	192	66	828	547	248
13	STO2	739	126	7	−67	192	9	−114	547	34
14	HOU1	582	12	54	688	254	89	782	328	150
15	HOU2	582	131	5	−61	254	8	−70	328	13
16	AGE1	36	6	51	−360	36	12	7	0	0
17	AGE2	454	35	44	−461	424	120	123	30	11
18	AGE3	39	51	34	−37	5	1	−95	34	10
19	AGE4	199	27	44	325	161	45	−160	39	14
20	AGE5	331	24	48	476	282	87	197	48	20
21	SIZ1	365	12	52	640	231	78	−486	134	59
22	SIZ2	55	12	48	244	39	12	−158	16	7
23	SIZ3	52	25	43	133	25	7	−136	27	10
24	SIZ4	71	47	35	−130	56	13	−68	15	5
25	SIZ5	373	47	38	−168	86	21	307	287	92

7.6 DISCUSSION

The relationship between MCA and JCA is analogous to that between principal components analysis (PCA) and exploratory factor analysis (FA). In PCA the model for the correlation matrix is:

$$\mathbf{R} = \mathbf{L}\mathbf{L}^{\mathrm{T}} \tag{7.1}$$

where $\mathbf{L}$ is the complete matrix of factor loadings. In Chapter 3, formula (3.2), we called (7.1) the complete solution. We can write this as a reduced solution (in K^* dimensions, say $K^* = 2$):

$$\mathbf{R} \approx \hat{\mathbf{L}}\hat{\mathbf{L}}^{\mathrm{T}} \tag{7.2}$$

where $\hat{\mathbf{L}}$ is the matrix consisting of the first K^* columns of $\mathbf{L}$, and $\approx$ means 'approximated by least squares'. In FA the model is modified by introducing additional parameters called 'specificities', denoted by the diagonal matrix $\mathbf{D}_\psi$, corresponding to the diagonal elements of $\mathbf{R}$:

$$\mathbf{R} \approx \hat{\mathbf{L}}\hat{\mathbf{L}}^\mathrm{T} + \mathbf{D}_\psi \qquad (7.3)$$

This allows attention to be taken off the fitting of the unit correlations down the diagonal of $\mathbf{R}$ to the benefit of fitting the off-diagonal correlations themselves. In MCA the centered and standardized Burt matrix:

$$\mathbf{B}^* = \mathbf{D}_r^{-1/2}(\mathbf{B} - \mathbf{r}\mathbf{r}^\mathrm{T})\mathbf{D}_r^{-1/2}$$

(cf.the formula (3.12) for the standardized residuals in Chapter 3) is analogous to the correlation matrix $\mathbf{R}$, with the difference that each variable defines a set of rows and columns of $\mathbf{B}^*$ and each 'correlation' is now a submatrix. As a consequence of the reconstitution formula in section 3.2.28, the reduced model for the MCA of $\mathbf{B}^*$ can be written as:

$$\mathbf{B}^* \approx \hat{\mathbf{M}}\hat{\mathbf{M}}^\mathrm{T}$$

which is analogous to the PCA reduced model in (7.2). In JCA new parameters for the blocks down the diagonal of $\mathbf{B}^*$ are introduced in a similar way:

$$\mathbf{B}^* \approx \hat{\mathbf{M}}\hat{\mathbf{M}}^\mathrm{T} + \mathbf{C}$$

where $\mathbf{C}$ is a block diagonal matrix of the following form:

$$\mathbf{C} = \begin{pmatrix} \mathbf{C}_{11} & \mathbf{0} & \cdots & \mathbf{0} \\ \mathbf{0} & \mathbf{C}_{22} & \cdots & \mathbf{0} \\ \vdots & \vdots & \ddots & \vdots \\ \mathbf{0} & \mathbf{0} & \cdots & \mathbf{C}_{QQ} \end{pmatrix}$$

The specificities are thus matrices which play the same role as in (7.3), namely to take the attention off the diagonal blocks and to allow better fitting of the off-diagonal block 'correlations'. It is an interesting problem for the future to clarify the interpretation of the estimated diagonal blocks, the 'block communalities', and the 'block specificities' in the JCA model. These estimated square tables lie somewhere 'between' the independence model (analogy of zero correlation) and the complete association model (analogy of unit correlation).

Finally, we also need to pay attention to the following problem. The inertia in MCA of each variable is equal to the number of categories minus 1 (see the diagonal of the matrix of inertias in Table 7.3). This suggests that a further standardization step is needed in MCA and JCA, namely to equalize the contributions to inertia of each variable. This can be achieved by dividing each subtable N_{qs} of B by $J_q - 1$ or $J_s - 1$, whichever is the smaller. Thus the inertias ϕ^2_{qs} would now become $\phi^2_{qs}/(\min\{J_q, J_s\} - 1)$, which is the square of Cramer's V association coefficient (see Chapter 2, section 2.2). The inertias ϕ^2_{qq} on the diagonal of the Burt matrix would now be one and the analogy with a (squared) correlation matrix would be complete.

Complementary use of Correspondence Analysis and Cluster Analysis

Ludovic Lebart

8.1 INTRODUCTION

Correspondence analysis rarely provides an exhaustive insight of a set of data. When processing the large data arrays issued from sample surveys through multiple correspondence analysis, it often occurs that the results are still too complex to be easily read: the configurations of points need further summarizing. In many cases, this can be achieved satisfactorily with the help of clustering techniques. The contribution of these techniques, however, is not limited to this practical phase of the processing. The complementarity between correspondence analysis and classification concerns both the basic comprehension of the data structure and the facilities provided in the final steps of interpretation of the results. The viewpoints of the two approaches as well as their outputs are fundamentally different. Consequently, a combined use of both sets of techniques is highly recommended for a thorough description of any complex data set.

After a first section dealing with the various aspects of the complementary use of correspondence analysis and classification, an artificial numerical example will help the reader to explore the boundaries between the two techniques. The third section presents a real example of survey data processing to emphasize the practical usefulness and the heuristic power of these methods in combination.

8.2 THE VARIOUS FACETS OF THE COMPLEMENTARITY

Correspondence analysis describes the main features of the data as they appear in the space spanned by the first principal dimensions. This could involve a substantial shrinkage (as a consequence of a projection onto a subspace) and/or some distortions due to the sensitivity to outliers of the principal axes. A remote profile point can notably influence, for example, the first principal axis, and therefore all the subsequent dimensions, since these dimensions are related to the first axis through the constraints of orthogonality of the axes. By contrast, most of the classification algorithms, and particularly the agglomerative algorithms, are locally robust in the sense that the lower parts of the produced dendrograms are largely independent of possible outliers. For all these reasons, it is highly advisable to complement the correspondence analysis with a classification performed on the whole space, or at least in the high-dimensional space spanned by all the significant axes. Classification could also be used to pinpoint some high-order interaction, to go beyond the classical limitations of eigen-analyses which are basically limited to second-order moment matrices.

In the case of large data sets (the only circumstances actually justifying the use of these techniques, according to the experience of the author of this chapter) a purely practical issue reinforces this need for both approaches: it is much easier to describe a set of clusters than a continuous space, even if it is two-dimensional. The most significant categories or variables characterizing each cluster are automatically selected and sorted, therefore producing a computer-aided description of the classes, and, hence, of the whole multi-dimensional space (Lebart *et al.* 1984).

It is often more efficient to perform a classification using a limited number of factors issued from correspondence analysis (case of large data sets, such as survey data files). A technique of hierarchical clustering such as the reciprocal neighbour algorithm (see MacQuitty 1966), and particularly the chain search algorithm (Benzécri 1982) can be performed without storing the array of distances in the central memory. The distances between pairs of points are recomputed when needed within the space spanned by the q first principal axes. The storage of the $n \times q$ array consisting of the q principal coordinates of the n observations is usually much smaller than the array of the $n(n-1)/2$ distances. Conversely, it is sometimes useful to perform a preliminary clustering of the observations (using an algorithm adapted to large data sets, such as the hybrid method described in section 8.5), in order to reduce the complexity of the analysis. Classification is used in this case as a prior condensation of the data.

8.3 THEORETICAL LINKS BETWEEN THE TWO APPROACHES: A BRIEF REVIEW

Besides these practical incentives to use correspondence analysis and classification simultaneously, one can find a series of more theoretical bridges between these two approaches, exemplified by some particular models. Two characteristics of correspondence analysis are in favour of a reconciliation with classification: the symmetry of the roles of rows and columns in the analysis, and the property of distributional equivalence, assuring stability of the results when agglomerating elements with similar profiles. To aggregate the rows or the columns of a contingency table is natural in the sense that it is merely replacing classes by classes (instead of replacing individuals by groups, or variables by groups of variables...).

The questions of clustering in contingency tables based on grouping of homogeneous items are discussed in Cazes (1986), Escoufier (1988), Greenacre (1988), Gilula (1986), Goodman (1981a), Govaert (1984), Jambu (1978). In the case of many categorical variables, van Buuren and Heiser (1989) propose an algorithm which simultaneously achieves a coding of the variables and a clustering of the individuals. Mirkin (1990, 1992) proposes several sub-optimal algorithms to achieve the simultaneous clustering of rows and columns of a contingency table through the optimization of a global criterion.

To sketch the approach of Mirkin, let us consider for each row i and column j the value

$$q_{ij} = (n \times n_{ij})/(n_{i+}n_{+j}) - 1 = p_{ij}/(p_{i+}p_{+j}) - 1$$

which expresses the relative increment (or decrement) of the probability of row i due to the knowledge of column j of the contingency table N. Note that we have the two following relationships expressing the classical chi-square χ^2 as a function of the q_{ij} coefficients:

$$\chi^2 = \sum_i \sum_j n_{ij}q_{ij} = (1/n) \sum_i \sum_j n_{i+}n_{+j}q_{ij}^2 = n \sum_i \sum_j p_{i+}p_{+j}q_{ij}^2$$

The elementary q_{ij} concept underlies the basic reconstitution formula:

$$q_{ij} = \sum_k \sqrt{\lambda_k} x_{ik} y_{jk} \tag{8.1}$$

where x_{ik} and y_{jk} are the standard coordinates corresponding to eigenvalue λ_k.

This allows us to formulate the partitioning problem of a contingency table $N = [n_{ij}]$ as an approximation problem: to find a pair of partitions, $\{V_s\}$ on the rows and $\{W_t\}$ on the columns, and values corresponding to $\sqrt{\lambda_k}$ for $k = (s, t)$ to approximate the matrix $Q = [q_{ij}]$, by the weighted least-squares criterion L^2:

$$L^2 = \sum_i \sum_j p_{i+}p_{+j} \left(q_{ij} - \sum_k \sqrt{\lambda_k} x_{ik} y_{jk} \right)^2 \tag{8.2}$$

where x_{ik} and y_{jk} are restricted to the values 0 and 1. Thus, minimizing the same criterion (8.2) can lead to correspondence analysis of the matrix N (with classical constraints of normalization and orthogonality upon the coordinates) or to a simultaneous clustering of rows and columns of **N** (with Boolean constraints upon the coordinates.

8.4 AN ARTIFICIAL EXAMPLE OF COINCIDENCE BETWEEN CLUSTERING AND SIMPLE CORRESPONDENCE ANALYSIS

The following symmetric 6×6 contingency table has the property of providing an exact coincidence between correspondence analysis and hierarchical clustering using a weighted version of Ward's criterion (see Ward 1963, Jambu 1978, Greenacre 1988) in the following sense: each eigenvalue of the correspondence analysis corresponds exactly to a node of the classification. The associated axis of the CA separates the two sets of elements constituting this node. These results are proved in a much more general framework by Benzécri *et al.* (1973, T II B, no. 11). These authors have shown that any binary hierarchy can be summarized by a set of orthogonal functions representing the different nodes. Through the reconstitution formula (8.1), where $\sqrt{\lambda_k}$ designates the square root of the value corresponding to node k, these specific functions lead to a symmetric contingency table similar to Table 8.1. Correspondence analysis of this symmetric table necessarily gives back the coordinates x_{ik} and y_{jk}, and the values $\sqrt{\lambda_k}$ associated with the nodes. By inspecting eigenvectors given in Table 8.3, one can see their link with the hierarchical structure.

Such a data matrix is similar to a 'confusion matrix', whose rows often represent some emitted stimuli (colours for example) and columns represent the recognized stimuli. Such matrices are most often almost symmetric. As it is the case here, they usually have a heavily loaded diagonal when the stimuli are distinct.

The symmetry of the matrix is immaterial: the main features of the following results hold if a random whole number (within the range 1–8, for

TABLE 8.1
A symmetric 'confusion' matrix.

	COL1	COL2	COL3	COL4	COL5	COL6
ROW1	30	18	6	2	2	2
ROW2	18	30	6	2	2	2
ROW3	6	6	42	2	2	2
ROW4	2	2	2	25	19	10
ROW5	2	2	2	19	25	10
ROW6	2	2	2	10	10	34

example) is added to each cell of the matrix of Table 1. Correspondence analysis of this Table leads to five eigenvalues (see Table 8.2). The first plane (represented in Figure 8.1) accounts for 82.6% of the total variance.

The sequence of patterns that can be observed in the columns of Table 8.3 is typical of a hierarchical structure: the non-zero coordinates on each

TABLE 8.2
Correspondence analysis of Table 8.1: eigenvalues.

	Eigenvalues	%	% Cumul	Histogram
1	0.64	52.89	52.89	************************************
2	0.36	29.75	82.64	*******************
3	0.16	13.22	95.87	********
4	0.04	3.31	99.17	**
5	0.01	0.83	100.00	*

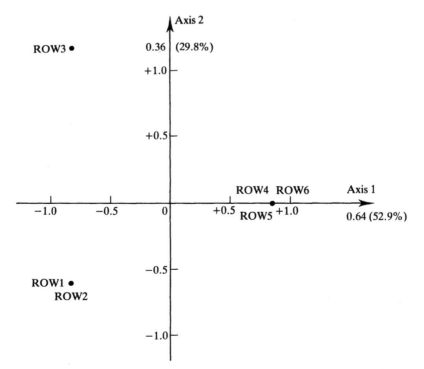

FIGURE 8.1 Plane spanned up by axes 1 and 2 (correspondence analysis of Table 8.1).

TABLE 8.3
Correspondence analysis of Table 8.1: coordinates.

| Ident. | Masses | Coordinates | | | | |
		F1	F2	F3	F4	F5
ROW1	0.167	−0.80	−0.60	0.00	0.35	0.00
ROW2	0.167	−0.80	−0.60	0.00	−0.35	0.00
ROW3	0.167	−0.80	1.20	0.00	0.00	0.00
ROW4	0.167	0.80	0.00	0.40	0.00	−0.17
ROW5	0.167	0.80	0.00	0.40	0.00	0.17
ROW6	0.167	0.80	0.00	−0.80	0.00	0.00

principal axis can take only two distinct values, opposing two groups of elements.

The first axis, for instance, opposes (ROW1, ROW2, ROW3) to (ROW4, ROW5, ROW6). The second axis, within the first group isolated by axis 1, opposes (ROW1, ROW2) to ROW3. The third axis, within the second group isolated by axis 1, opposes (ROW4, ROW5) to ROW6. The last two axes oppose, within the previous subgroups, ROW1 to ROW2 and ROW4 to ROW5. Correspondence analysis functions in this case like a divisive algorithm, working iteratively from the upper to the lower level of a hierarchy.

Hierarchical clustering using the chi-squared distances (Table 8.4), and Ward's criterion, implies a minimum loss of inertia at each step. This loss is given by the index (Table 8.5). It has been shown that the use of Ward's criterion ensures a good compatibility between CA and classification (see Benzécri 1973, Greenacre 1988). We note in this case that the eigenvalues and the indices of the classification coincide. The first eigenvalue corresponds to the upper part of the dendrogram, whereas the second eigenvalue represents the upper part of the main branch (i.e. with the higher index). The last eigenvalue corresponds to the first index in Table 8.5, in other words, the most

TABLE 8.4
Hierarchical cluster analysis of Table 8.1: triangular table of distances.

	ROW1	ROW2	ROW3	ROW4	ROW5
ROW2	0.48				
ROW3	3.36	3.36			
ROW4	3.23	3.23	4.19		
ROW5	3.23	3.23	4.19	0.12	
ROW6	3.68	3.68	4.64	1.47	1.47

TABLE 8.5
Hierarchical cluster analysis of Table 8.1: the agglomerative process.

Node	Index	Selected pair		Weight	Content of clusters
7	0.010	4	5	2	ROW4 ROW5
8	0.040	1	2	2	ROW1 ROW2
9	0.160	6	7	3	ROW6 ROW4 ROW5
10	0.360	3	8	3	ROW3 ROW1 ROW2
11	0.640	9	10	6	ROW6 ROW4 ROW5 ROW3 ROW1 ROW2

similar rows (ROW4 and ROW5) which give the first cluster are the last distinction in CA.

Figure 8.1 is the display of the first principal plane issued from the CA, whereas Figure 8.2 represents the dendrogram issued from the hierarchical clustering. We notice that the configuration of points represented in Figure 8.1 highlights only a limited part of the underlying structure, by comparison with the tree (also a planar representation) given by Figure 8.2. Figure 8.1 gives no detailed information about the distances between ROW1 and ROW2 (the corresponding points are superimposed on the plane, which suggests a null distance). Similarly, ROW4, ROW5 and ROW6 are also superimposed on the graphical display. This shrinkage of distances is easily explained by the geometrical properties of the initial cloud of points.

Although this artificial example has been built on purpose to pinpoint some of the drawbacks of correspondence analysis, the user must keep in mind, in a real situation, the limitation of the graphical displays. The combined use of clustering and correspondence analysis can only improve the understanding of the data.

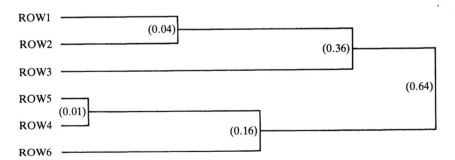

FIGURE 8.2   Dendrogram associated with Table 8.1.

8.5 COMPLEMENTARITY IN REAL LIFE

The example presented in this section aims at presenting an assessment of the complementary use of correspondence analysis and classification in the case of full scale socio-economic sample surveys. It consists of the processing of the data set provided by the *National Survey about the Aspiration and the Living Conditions of the French*, from 1978 to 1984 (see Lebart 1986). We will deal successively with the scope of the problem, the data set, the descriptive tools, some methodological remarks about the original contribution of the methods and finally an outline of the main results.

8.5.1   The initial purpose of the survey

The demand of the users and clients of this research and the nature of the corresponding questions implied a descriptive and exploratory approach:

- What kind of organization or affinities can be detected among the opinions of the French in relation to socioeconomic problems?
- Is it possible to build a meaningful typology of the respondents with respect to their opinions?
- How are the opinions about socioeconomic problems conditicned by the situations of the respondents (i.e: their sociodemographic characteristics)?
- Is the structure of opinion stable? Can we detect or describe a shift in this structure?

8.5.2   Main features of the data set

The data set consists of an array with approximately 14 000 rows and 200 columns stacking row-wise seven independent samples of 2000 in size, each one being representative of the French population over the age of 18, for each year from 1978 to 1984 (for more details, see Lebart 1986). The questionnaires comprised about 200 questions, half of which describes the socioeconomic characteristics of the respondents. The other half relates to opinions and attitudes towards general topics such as economic situations, family, physical and technological environment, social change and justice. The responses are mainly of a nominal (or categorical) nature, but also lead to textual data (answers to open-ended questions) which have been processed separately by using specialized software such as SPAD (see Lebart *et al.* 1987, 1991).

8.5.3   Presentation of the exploratory tools: the hybrid clustering

The two complementary techniques used to describe this large set of categorical data are on the one hand multiple correspondence analysis, and on the other hand a composite technique designated here as hybrid clustering (see, for instance, Wong 1982). Since MCA has already been presented in this book, we limit ourselves to giving a basic idea about the clustering technique.

The technique of hybrid clustering comprises three steps

Step 1. Preliminary clustering, using agglomeration around variable centers (*k*-means method, or dynamic cluster method). The search for stable groups (Diday 1972) enables improvement and validation of the clusters. This search consists of obtaining several partitions, starting with different sets of provisional centers. The stable groups are the groups of individuals always clustered together (i.e. always belonging to a same cluster). In our example, the 14 000 individuals were assigned to about 100 clusters.

Step 2. Hierarchical clustering of the previously obtained clusters (using Ward's criterion), and determination of the cutting of the dendrogram, which determines simultaneously the number of final clusters and a provisional set of centers for the final partition. The dendrogram provides invaluable assistance in determining the number of classes, a number which is unknown beforehand. Inspection of the sequence of indexes suggests the most suitable partition which should correspond to a significant jump of the index.

Step 3. Reallocation (using an iterative procedure similar to the *k*-means method) of the individuals or objects in order to improve the quality of the partitioning (even a relatively clear-cut partition according to the shape of the dendrogram does not always produce a locally optimal partition).

Details about these methods and the corresponding software can be found in Morineau and Lebart (1986).

8.5.4 Methodological prerequisites

The fundamental notion of active variable

The results of a typological study (be it obtained by MCA, from a clustering procedure, or from a combination of both these techniques) make sense only if the list of 'active variables' is clearly specified. These active variables (see Chapter 1) are used to compute the distances or similarities between the respondents. If this list is sufficiently large, we can reasonably expect that the results obtained will be somewhat independent of the presence or the absence of a particular variable within this set. The set of active variables must satisfy a criterion of homogeneity whose rationale is intuitive: the computations of distances must make sense, so that the patterns or groupings formed out of these distances also make sense. Whereas the homogeneous set of *active variables* allows for the definition of a specific point of view, the set of *supplementary* (or *illustrative*) *variables* (see Chapter 1) will allow for the *a posteriori* characterization or identification of the structural features produced. This set is not necessarily homogeneous, since its elements are neither used to determine the basic MCA nor the basic partition.

In the analysis presented in this paper, there will be 14 active questions (and about 60 modalities of responses). These questions are aimed at giving an overall description of attitudes, opinions and perceptions *vis-à-vis* the living conditions of the French. The remaining variables have the status of illustrative variables.

Choice of the active variables

Let us recall that our purpose is to *explore* the organization of the opinions of the French in relation to socioeconomic problems, to build a meaningful typology of the respondents with respect to their opinions, and to investigate to what extent these opinions are conditioned by the situations of the respondents (i.e. their sociodemographic characteristics). To achieve this exploratory phase, a subset of questions has been selected. It aims at representing, somewhat arbitrarily, the main dimensions of the phenomenon under study. These questions have been used since 1978 in the longitudinal survey about living conditions of the French (see Lebart and Houzel 1980, Babeau *et al.* 1984, Lebart 1987). They have also been included in the questionnaires of several international surveys (see Hayashi *et al.* 1986, 1992).

The active questions can be summarized as follows:

- Two questions relating to the perception of the evolution in living conditions of the interviewed person and of the French in general.
- Three questions about the image of the family, about the meaning of the marriage, the activity of women.
- Three questions about the physical and technological environment (satisfaction towards the aspects of immediate surroundings of the house, care for the protection of the environment, belief in the benefit of scientific progress).
- Three questions about health (satisfaction towards personal state of health, role of the health care system).
- Three more general questions about society in general: opinions on justice, social change, collective facilities and services.

More details about the wording of some of these questions can be found in Figures 8.4 and 8.5, whereas a full list is in the above references.

8.5.5   Outline of the main results

MCA produces planar maps where the points represent the response-items and the proximities between points represent the affinities between these responses (i.e. two close responses have been given by almost the same individuals). The clustering will highlight the main grouping of individuals with respect to their most significant profiles. These groupings will be described in a systematic way, using all the objective characteristics of the individuals (sex, age, educational level, occupation of the interviewed person, of the head of household, size of town, region, etc.). The centroids of these groupings can be plotted onto the maps issued from MCA, thus providing assistance in deciphering and interpreting these rather complex outputs.

Based on the 14 responses to the active questions, the respondents are positioned in a spatially continuous 'cloud of points'. There are no clear-cut groupings in this continuum, but it is possible to divide it into major clusters by using the hybrid clustering algorithm. Note that the percentages of inertia corresponding to the first two axes (respectively 10% and 8%) are highly significant (as shown for instance by a simulation study, see Lebart *et al.* 1984) although representing a modest part of the total trace. This pessimistic order of magnitude is usual in classical MCA. It would be erroneous to interpret these percentages as 'percentages of information' (see Chapter 7).

Eight clusters are revealed through the classification process. The dendrogram (Figure 8.3) describe the proximities between these clusters. The

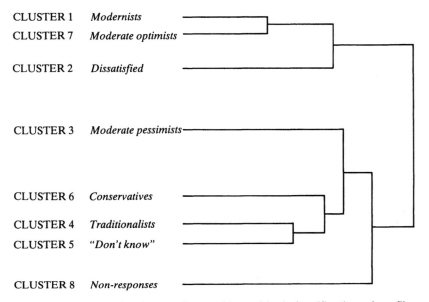

FIGURE 8.3 Eight main clusters from a hierarchical classification of profiles.

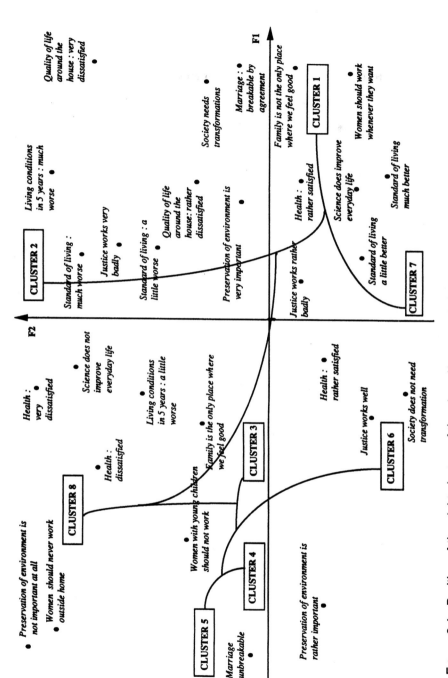

FIGURE 8.4 Position of the eight clusters of the plane (Axes 1 and 2) issued from the MCA of the active variables.

centroids of the clusters are plotted as supplementary elements in the planar map of Figure 8.4, where some of the most characteristic response-points are represented. The dendrogram of Figure 8.3 has also been sketched on Figures 8.4 and 8.5, to recall the distances between clusters in the whole space.

All the items characterizing the clusters are selected by the computer according to their value (or their percentage) within the cluster, as compared to their value (or percentage) in the global population (see Morineau and Lebart 1986). For example, within cluster 1, 87% of the respondents think that 'Family is not the only place where one feels good', whereas the average percentage of this response in the whole population is 35%. This difference between 87% and 35% is highly significant. The responses can be sorted according to their level of significance in characterizing cluster 1.

Each cluster, defined according to significant groupings of responses, is identified *a posteriori* by the objective characteristics of the individuals involved. Each cluster is provided with a mnemonic label, and a fictitious *modal element* is described, possessing the most salient sociodemographic features of the cluster. Notice that the characteristics of the clusters and the modal elements are not necessarily depicted in Figures 8.4 and 8.5, respectively. The following is a description of the eight most stable clusters which appear each year in independent samples.

Cluster 1 'Modernists' (about 18%)
This cluster, with its very marked and significant characteristics, includes people with modernist ideas on the family (family is not the only place where one feels good; marriage can be broken upon agreement). They are also in favour of environmental protection as well as technological progress (preservation of environment is very important; science improves everyday life). They have an active social life, are mobile and not very worried. *Modal element*: A young Parisian who does not have children, and who has a high level of education but an average standard of living.

Cluster 2 'Dissatisfied/Isolated' (about 11%)
This second cluster includes pessimistic, dissatisfied, critical and socially isolated people (standard of living worsening; dissatisfied with health; dissatisfied with the environment of the house; science does not improve everyday life; society needs deep changes). *Modal element*: A worker or jobless person, with an increasing number of serious problems of different types (housing, family, health) in fairly unpleasant surroundings.

Cluster 3 'Moderate pessimists' (about 16%)
This is an intermediate cluster: satisfaction *vis-à-vis* lifestyle, health; divided opinions about family; slightly pessimistic about everything else. *Modal*

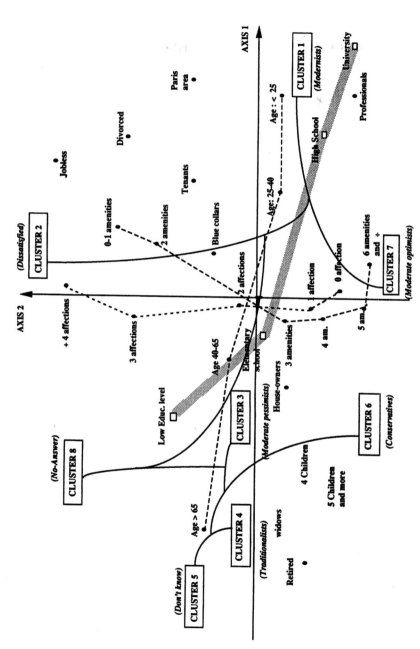

FIGURE 8.5 Illustrative (or supplementary) socioeconomic categories projected onto the plane of Figure 8.4.

element: A person from a rural area, 50 years old, married, with children, with average level of education and standard of living.

Cluster 4 (left of Figure 8.4) 'Traditionalists' (about 15%)
In some years, individuals in this cluster were integrated into cluster 3 or 6. Characteristics include traditionalist opinions on the family (marriage unbreakable; family is the only place where one feels good; women should not work outside home) but quite critical in relation to justice and society. *Modal element*: a retired person who has had children, a low level of education, a house in a rural area.

Cluster 5 'Don't know' (about 8%)
This cluster of 'non-expression' includes those with lack of interest in the survey and those who refuse to become involved. *Modal element*: An old woman (often a widow) with a low level of education in a rural area.

Cluster 6 'Conservatives' (about 13%)
Characteristics include: no change needed in a society where justice works well; satisfaction *vis-à-vis* living conditions; optimism in relation to science and technological progress. *Modal element*: A person approaching fifty, a high level of income and amenities, medium-high level of education, house owner.

Cluster 7 'Moderate optimists' (about 12%)
This is an intermediate cluster, as in cluster 3. Moderation also means hesitation and not very marked opinions (intermediate response-items are generally chosen), but also indicating a fairly general feeling of satisfaction (health, daily life style, living conditions). *Modal element*: A person about thirty years old, outside the Paris area, active, with children and an average level of qualification.

Cluster 8 'No answer' (about 5%)
This is a second cluster with responses similar to the 'don't know' type: these responses relate mainly to the issue of family. *Modal element*: A single man of any age, with low education and with a very low income.

In Figure 8.4, cluster 2 ('dissatisfied people') is closer to cluster 8 ('no answer') than to cluster 7 ('moderate optimists'). We can see on the dendrogram (Figure 8.3) that clusters 2 and 7 merge early, while cluster 8 comes later. Thus, the dendrogram is successfully used here to remedy the distortion of distances due to the planar approximation of Figure 8.4, as illustrated in the artificial example in section 8.4.

Figure 8.5 shows the dispersion of the supplementary socioeconomic variables in the plane of the first two principal axes. Variables *age* and *educational level* are spread along the horizontal axis. The age decreases from the left to the right, while the level of education increases along a similar trajectory.

Low standard of living categories are positioned in the upper part of the display. A dotted line describing the number of amenities of the household (number of items in a list of 10 comprising dish-washer, colour TV set, etc.) runs vertically along axis 2. Another battery of questions relating to various current ailments or affections (headache, insomnia, etc.), also line up in the vertical direction but in the opposite sense (second vertical dotted line). People with a small number of affections are positioned in the lower part of the display.

8.6   DISCUSSION

Multivariate descriptive techniques are generally intended to discover something, and not to prove anything. However, there are some favourable situations in which graphical displays and typologies are worth reporting: as is the case here, when the stability (and hence the relevance) of the structures highlighted is tested in independent samples. Affinities between expressed opinions are the result of a compromise; to make matters simple, we will draw a distinction between three levels of affinities or similarities:

(a) The *semantic level* that one can study directly from an analysis of the questionnaire itself, before gathering the responses. This level will be responsible for the 'created' part of the structure, due to the affinities of meaning between questions.

(b) The *psychological level*, or more generally the personal level, responsible for structural features through the effect of latent factors depending on the personality or the socioeconomic status of the respondent (a depressive state, for instance, could lead to negative opinions about justice as well as to living conditions).

(c) The *basic structural level*, designating here the complex network of relationships between objective situations, physical and economic constraints (location, income, age, etc.).

As a first approximation, one may consider the observed structure as a compromise between the structures induced by these three levels. One of the main points in the present example of application is the evidence of the prominent role played by the basic structural level, in other words, the strong influence of situations over perceptions. As shown in Figure 8.5, the objective situations intervene in a rather complex system of interactions making the interpretations all the more difficult.

Concerning the combined use of correspondence analysis and clustering, four points need to be stressed, which will serve as a conclusion of this chapter:

(1) *The descriptive power of the clusters* (as opposed to configurations in a continuous space).

As mentioned previously, it is easy to describe a group. It suffices to compute for each numerical variable the average within the group and the global average, or for each category of nominal variable the percentage of respondents within the group and the global percentage. Two series of statistical tests allow us to select and to sort (according to the computed levels of significance) the most characteristic items. A detailed description of the clusters is now readily available as computer output (see for instance Morineau and Lebart 1986, Lebart *et al.* 1987).

(2) *The supplementary information produced by the clusters.*

The clusters are not only used to describe regions of the factorial plane. They are established in a high-dimensional space, supplying many more elements of information, and in particular information that could have been hidden by the projection onto a two- or three-dimensional subspace (remember the distortion of the distances between clusters 2, 8, 7).

(3) *The descriptive power of the axes, the importance of continuous latent factors and of tendencies, and consequently the importance of the spatial configuration of points.*

This contribution of the MCA is apparent in Figures 8.4 and 8.5, for example, the regular decrease of age from left to right coincides with the concomitant increase of 'modernism' along the same axis and the opposition of cluster 5 to cluster 1. Figure 8.3 and the description of the corresponding clusters do not suffice to provide this kind of result.

(4) *The discovery of latent phenomena.*

Independently of the 'dissection effect' of the clustering (working as multi-dimensional histograms, used to divide a huge space in smaller portions easier to analyze), this technique can evidently help to discover possible existing groups. Similarly, correspondence analysis can put forward some unexpected latent factors. To uncover such hidden patterns or dimensions is indeed the original goal of both families of methods, and the most ambitious one. Their complementary use is necessary to achieve this quest, as it is necessary to undertake the humbler task of description.

Homogeneity Analysis: Exploring the Distribution of Variables and their Nonlinear Relationships

Willem J. Heiser and Jacqueline J. Meulman

9.1 INTRODUCTION

When several variables are measured on the same group of individuals, one of the first questions that needs an answer is: how does one distinguish the reliable variability from the unreliable variability? As will become evident shortly, this question can be approached by phrasing it as a problem of finding out what the distribution of the variables is, a matter that is of concern independently from considerations about the distribution of individuals. The discussion starts with a key attribute of empirical distributions, their homogeneity.

9.1.1 Homogeneity of individuals and variables

The term homogeneity is predominantly used in statistics in connection with samples from different populations, which may – or may not – exhibit identical behaviour, or display similar characteristics. If the populations are the same for the purposes of the study, they are collectively called *homogeneous*, meaning *of one kind*, and different random samples from homogeneous populations will be called homogeneous as well. A first thing to note is that populations and subpopulations can be homogeneous in one respect and heterogeneous in another, as for example in the idealized circumstances of the familiar *t*-test, where it is assumed that the observations are drawn from two populations that are homogeneous in their variance and heterogeneous in their means.

Populations of individuals that are heterogeneous in their means are so common (or perhaps, believed to be common) in the social and behavioural sciences that they form the almost exclusive object of study, both in theoretical and in empirical research. To fully appreciate the truth of this remark, we have to make a slight digression to look at the distinct *aggregation levels* at which any variable can be studied.

In an operational sense, a 'variable' refers to some specified *rule* that assigns values to individuals like when one calls a variable 'school achievement' while in fact grade point averages of a number of students are being studied. But for the purposes of analysis, individuals can be grouped into school classes, or into schools, into school districts, or into other higher-order aggregates, and the grouping defines the analysis variable. By determining the mean and the variance, and perhaps other distributional properties like the skewness or an extra bump in the tail of the distribution, we express our knowledge at the group level, and thereby characterize a certain population of students. From such a choice of aggregation level it follows that not much more can be said about any *particular* student than that it is a member of a population with such-and-such characteristics. However, often a social or behavioural scientist is not just interested in the distribution properties at the highest aggregation level, or in a single, *a priori* chosen, grouping, because the concept of a variable is also linked with the idea of a *measuring instrument*, or a *scale*, on which the individuals have definite mixed scores, known up to some *measurement error*.

One tends to think about measurement error as an unpredictable influence that is unavoidable, albeit hopefully small, or at least not so large as to be forced to confuse, for instance, the ability of a person at one end of the scale with the ability of a person at the other end of the scale. It is one of the central tenets in social science methodology that it should be possible to cut back on the aggregation level by distinguishing subgroups of individuals − or subgroups of individual measurements − along the measurement scale. Such differently located subgroups constitute a translation family of populations, and that is the reason why we can say that the assumption of populations being heterogeneous in their means is so common.

The availability of a reliable measurement scale is prerequisite if we try to answer questions such as whether females are brighter than males, whether different teaching instructions are really effective or not, whether a distinction in social environment is noticeable on the scale or not, and so on. Subgroups may be defined by different experimental treatments or interventions, such as teaching instructions, or by an observable background characteristic, such as sex or age. In other situations it may be desirable to find the *right* characteristic from a number of plausible candidates. Since the distinction in background variable and response variable (similar contrasting terms are: exogenous–endogenous, or independent–dependent) is a distinction in role, not in substance, one cannot say once and for all which variable is of one type and which of the other. Even sex and age can be response variables, as for instance in a demographic study, where regional characteristics are the background variables. Conversely, a typical response variable like endorsing a statement of opinion might serve as a background characteristic in a study of political voting.

Is it possible to reduce the uncertainty that follows from the measurement error on the response scale, or in the background characteristic, and thereby to improve the chances of convincingly demonstrating the existence of subgroups that are heterogeneous in their means? An old answer to this recurring question is to *average* over a number of *parallel* measurements, the idea being that errors on different versions of the variable will tend to cancel one another out (Spearman 1904). When things really matter, we do not evaluate a student on a single grade, but on a grade point average. The remarkable concept here is that we consider replication across 'variables', not across individuals! Ideally, the parallel measurements are equal in the location of each individual (or of subgroups of them), and the actual data are expected to vary randomly around these equal values. The average location will tend to have smaller variance than any of the single measurements on which it is based, provided that the average is taken over a homogeneous set of variables. Because the psychometric notion of parallel measurement involves a distribution of scores across variables (in sociology the term 'indicator' denotes the same idea of using a variety of observables to pin down some theoretical construct), we can talk about homogeneous samples and populations of variables. By implication, samples of variables can be *heterogeneous* too, in which case it would not seem to be very sensible to expect improvement of reliability by averaging.

How to find out whether a set of variables is homogeneous or heterogeneous? There are a number of answers to this question, and one of them is – unsurprisingly – to do a homogeneity analysis. Although the primary aim of this type of analysis is to characterize a distribution of variables, it always *also* leads to a characterization of individuals, because homogeneity analysis relies on a common basis of comparison in which samples of individuals are maximally heterogeneous.

We have seen that the idea of a statistical distribution and related concepts can be applied to sets of individuals and sets of variables alike. It is now time to define more precisely what is meant by 'parallel measurement' or 'homogeneous sample of variables'. Indeed, various definitions are possible, and it is helpful to order them from simple to more complex, which then yields a natural ordering of the methods of homogeneity analysis.

9.1.2 Forms of homogeneity

Since homogeneity is a concept that describes the relationship between the elements of a distribution, our task is to specify in what sense variables can be different while still being of the same kind. A distribution is a multidimensional concept, because it has various aspects, like a centrality, a dispersion, and perhaps irregularities of shape, as noted before. A variable is a multidimensional concept too, because it specifies the different locations of (groups of) individuals. To keep the complexity within bounds, the following concepts and definitions are very useful:

(a) a sample of variables is called homogeneous if its elements (i.e. the variables) are equal, up to some prespecified class of information-preserving operations, and up to random deviations;

(b) information-preserving operations are operations like changing the mean, rescaling, taking the logarithm, and so on, that are considered to leave the information on the individuals, carried by the variable, unchanged;

(c) random deviations among the variables are measured through a loss function, in which they are compared with one another, or with a latent variable, which is the centre of the distribution.

The simplest, and most restrictive type of information-preserving operation is the identity, which keeps any variable the same, and in this pure form the loss function can be expressed as follows:

$$\sigma^2(\mathbf{x}) = (Nm)^{-1}\Sigma_j \| \mathbf{h}_j - \mathbf{x} \|^2. \tag{9.1}$$

Here $\mathbf{h}_j$ is an N-vector containing the observed scores of N individuals (also called *objects*) on variable j, with $j = 1, ..., m$, and the function $\| . \|$ is the Euclidean norm, which is the square root of the sum of squares of its elements. The common basis of comparison is the unknown variable $\mathbf{x}$, a vector of the same length, for which the scores of the N individuals, called *object scores*, are to be determined. In classical psychometrics, the object scores are called *true scores*, corresponding to the conceptualization of the random deviations $(\mathbf{h}_j - \mathbf{x})$ as *errors*. Geometrically, the value of $\sigma^2(\mathbf{x})$ — called *loss of homogeneity* — is the mean squared Euclidean distance between the vectors $\mathbf{h}_1, ..., \mathbf{h}_j, ..., \mathbf{h}_m$ and $\mathbf{x}$. Loss of homogeneity is the multidimensional analogue of the variance; it measures the departure from $\mathbf{x}$, the multidimensional analogue of the mean.

Indeed, the minimum loss of homogeneity, denoted as $\sigma^2(*)$, is attained for $\mathbf{x} = \bar{\mathbf{h}} = m^{-1}\Sigma_j\mathbf{h}_j$, the mean across variables for each object (individual). Here, as elsewhere in the paper, rigorous proofs are omitted; the interested reader is referred to Nishisato (1980) or Gifi (1990) for more details. The minimal value turns out to be

$$\sigma^2(*) = (Nm)^{-1}\Sigma_j \| \mathbf{h}_j - \bar{\mathbf{h}} \|^2 = N^{-1}[m^{-1}\Sigma_j \| \mathbf{h}_j \|^2 - \| \bar{\mathbf{h}} \|^2] \qquad (9.2)$$

up to a factor N the mean squared length of the $\mathbf{h}_j$ minus the squared length of their mean – analogous again to the familiar pocket calculator formula for the variance. It is possible to simplify $\sigma^2(*)$ still further, if it is assumed – as will be done in the remainder of the chapter – that all variables are in deviations from their own mean, and standardized as $s^2(\mathbf{h}_j) = N^{-1} \| \mathbf{h}_j \|^2 = 1$; in that case, $\sigma^2(*) = 1 - r_{..}$, where $r_{..}$ is the average correlation between all $\mathbf{h}_j$, including the self-correlations equal to 1, and all correlations r_{ij} are counted twice. Thus minimal loss of homogeneity attains its lowest value if all variables are perfectly correlated ($\sigma^2(*) = 0$), and it is maximal if the variables are altogether uncorrelated ($\sigma^2(*) = 1 - 1/m$).

Now suppose that we allow for rescaling of the variables. There can be a variety of reasons for taking this step, for instance because it is expected that the variables are different in their power to discriminate the objects and are therefore to be assigned a different weight in the analysis. In this case it is natural to modify the definition of loss of homogeneity into

$$\sigma^2(\mathbf{a}, \mathbf{x}) = (Nm)^{-1}\Sigma_j \| a_j\mathbf{h}_j - \mathbf{x} \|^2, \qquad (9.3)$$

where each variable is rescaled by the scaling factor a_j. The coefficients a_j are the elements of the m-vector $\mathbf{a}$ that is included in the list of unknowns in $\sigma^2(\mathbf{a}, \mathbf{x})$. Note that we do not restrict a_j in any way; if $a_j < 0$, then the scores in $\mathbf{h}_j$ change sign, a phenomenon called *reflection*; if $a_j = 0$, the jth variable drops out of the analysis. The only type of restriction that ought to be considered here consists of normalization to identify $\mathbf{a}$ and $\mathbf{x}$, e.g. $\| \mathbf{a} \|^2 = m$. Before continuing the discussion of rescaling, a number of introductory remarks on computation are in order.

9.1.3 Alternating least squares

Although computation is of no great concern in this chapter, the question of determining the optimal value of $\sigma^2(\mathbf{a}, \mathbf{x})$ provides an excellent opportunity to introduce alternating least squares (ALS), which is a very flexible computational strategy for many methods in multidimensional data analysis. Briefly, suppose we fix the a_j at some particular value a_j^* (for $j = 1, ..., m$), obtained in a previous round of computation. Then actually rescaling the variables as $\mathbf{q}_j^* = a_j^*\mathbf{h}_j$ shows that our best current guess $\mathbf{x}^*$ of $\mathbf{x}$, which still is the average across variables, becomes equal to

$$\bar{\mathbf{q}}^* = m^{-1}\Sigma_j\mathbf{q}_j^* = m^{-1}\Sigma_j a_j^*\mathbf{h}_j, \qquad (9.4)$$

i.e. the *weighted mean* of the original variables. Because of their role in (9.4), the scaling factors a_j are also called *weights*. Fixing $x = \bar{q}^*$ in turn, the ALS principle assures us that the conditional minimum of $\sigma^2(a, \bar{q}^*)$ over a always improves (decreases) loss of homogeneity. The conditional minimum is attained by choosing the new scaling factors $\hat{a}_j$ as

$$\hat{a}_j = c(h_j, \bar{q}^*)/s^2(h_j), \tag{9.5}$$

where $c(h_j, \bar{q}^*)$ denotes the covariance between h_j and $\bar{q}^*$, and $s^2(h_j) = N^{-1} \| h_j \|^2$ the variance of h_j (which was assumed to be equal to one, but which is included in (9.5) to be fully explicit). Alternating between (9.4) and (9.5) yields a convergent process. Note that $\hat{a}_j$ in (9.5) is simply the regression coefficient for the linear regression of $\bar{q}^*$ on h_j. In order to avoid a solution with $a_j = 0$ for all variables and, correspondingly, $x = 0$, which trivially minimizes $\sigma^2(a, x)$, some convention on *normalization* is required; usually, one fixes either the sum of squares of the object scores or the sum of squares of the scaling factors at some prechosen value. Application of alternating least squares generally requires that the loss function can be split into independent components, and several such decompositions will be demonstrated in the sequel.

9.2 PRINCIPAL COMPONENTS ANALYSIS AS A METHOD OF HOMOGENEITY ANALYSIS

Optimally rescaling the variables before averaging determines another centre of the distribution, different to that obtained by directly averaging them, but what is so special about this for the study of homogeneity? This question will be answered by discussing the loss of homogeneity function in more detail.

9.2.1 The mean squared correlation and the eigenvalue

Frequently, the elements of x in (9.3) are scaled as z-scores. Since the latter normalization satisfies $s^2(x) = 1$, substitution of the optimal weights (9.5) in (9.3) then allows us to write the loss function as

$$\sigma^2(*, x) = 1 - m^{-1}\Sigma_j r^2(h_j, x), \tag{9.6}$$

where $r(h_j, x)$ denotes the correlation between h_j and x. So under the rescaling definition of loss (9.3), the centre of the distribution of variables, x, is chosen in such a way that it has maximal mean squared correlation with the original variables; this centre − generally different from the unweighted mean − is called the (first) *principal component*. The correlation between h_j and x is often called the *loading* of variable j on the principal component. The squared loading then is the *variance* of h_i that can be *accounted for* by x in linear regression terms, and the mean squared correlation in (6) is the average variance accounted for.

Just as one can go from (9.3) to (9.6) by substituting the conditionally optimal weights, it is possible to re-express (9.3) in a form that does not involve the principal component. First, we switch from normalization of the component to normalization of the weights, so that $\| \mathbf{a} \|^2 = m$, while the conditionally optimal value of $\mathbf{x}$ remains exactly as indicated in (9.4). Next, it is not difficult to verify a fundamental identity in Euclidean space (Gower 1975), written in the present notation as

$$(Nm)^{-1}\Sigma_j \| a_j\mathbf{h}_j - \mathbf{x} \|^2 = (2m)^{-1}(Nm)^{-1}\Sigma_j\Sigma_l \| a_j\mathbf{h}_j - a_l\mathbf{h}_l \|^2, \quad (9.7)$$

which expresses the fact that the dispersion of the rescaled variables may be formulated either with respect to the centre $\mathbf{x}$, or in terms of the sum of the squared distances between all of them taken in pairs; the two formulations differ only by a factor $2m$. As a final step, using the fact that $c(a_j\mathbf{h}_j, a_l\mathbf{h}_l) = a_j a_l r(\mathbf{h}_j, \mathbf{h}_l)$, the right-hand side of (9.7) may be rewritten to obtain

$$\sigma^2(\mathbf{a}, *) = 1 - m^{-2}\Sigma_j\Sigma_l a_j a_l r(\mathbf{h}_j, \mathbf{h}_l). \quad (9.8)$$

So weights that minimize loss of homogeneity will maximize the weighted mean correlation among all pairs of variables. From (9.8) it becomes clear how the value $1 - r_{..}$ was obtained earlier for the case of all weights equal to one. When optimal weights $\hat{a}_j$ and $\hat{a}_l$ are inserted in the weighted mean correlation, the quantity

$$\lambda = m^{-1}\Sigma_j\Sigma_l \hat{a}_j\hat{a}_l r(\mathbf{h}_j, \mathbf{h}_l) \quad (9.9)$$

is called the (maximum, or first) *eigenvalue* of the correlation matrix, i.e. the $m \times m$ matrix that consists of elements $r(\mathbf{h}_j, \mathbf{h}_l)$. The eigenvalue expresses the homogeneity of the variables in terms of their mutual correlations. Inserting (9.9) into (9.8), and comparing the result with (9.6), we reach the important conclusion that $m^{-1}\lambda$ is *also* equal to the mean squared correlation of the variables with their principal component.

9.2.2 Cronbach's α

Loss of homogeneity was expressed in (9.8) without reference to a centre, but directly as a weighted mean of the correlations between the variables, for any set of weights satisfying the normalization constraint $\| \mathbf{a} \|^2 = m$. If all correlations are one, this weighted mean is bounded by

$$m^{-2}\Sigma_j\Sigma_l a_j a_l = (m^{-1}\Sigma_j a_j)(m^{-1}\Sigma_l a_l) = \bar{a}^2 \leqslant m^{-1}\| \mathbf{a} \|^2 = 1, \quad (9.10)$$

where the inequality follows from the elementary fact that $\Sigma_j(a_j - \bar{a})^2 \geqslant 0$. The average weight, $\bar{a}$, can never be larger than one, and it can only become equal to one if all weights become equal to one. In that case, $\sigma^2(\mathbf{a}, *)$ attains its natural minimum of zero. When all correlations are zero, it follows from (9.8) that $\sigma^2(\mathbf{a}, *)$ becomes $1 - m^{-1}$, for any choice of $\mathbf{a}$, where the presence of m^{-1} reflects the inclusion of $j = l$ in the summation. Because $\sigma^2(\mathbf{a}, *)$ involves correlations, which are bounded by -1.0 and 1.0, and weights with a fixed sum of squares, it is natural to consider a related measure that itself has the properties of a correlation. Multiplying $\sigma^2(\mathbf{a}, *)$ by the factor $m/(m-1)$ corrects for the undesired maximum, and after reflection the quantity

$$\alpha(\mathbf{a}) = [1 - m/(m-1)\sigma^2(\mathbf{a}, *)]/[1 - \sigma^2(\mathbf{a}, *)] \qquad (9.11)$$

has the desired properties of a correlation coefficient. By substitution of (9.8) and some algebra on the sum of the correlations, (9.11) becomes

$$\alpha(\mathbf{a}) = [(m(m-1))^{-1}\Sigma_j\Sigma_{l \neq j}c(a_j\mathbf{h}_j, a_l\mathbf{h}_l)]/s^2(m^{-1}\Sigma_j a_j\mathbf{h}_j), \qquad (9.12)$$

i.e., the mean covariance among the weighted variables, *excluding* the variances, divided by the variance of the mean scores. In this form or in a number of similar forms, the coefficient in (9.12) is best known as Cronbach's α. Originally, it was defined for binary variables only, but it soon became obvious that this limitation was unnecessary. Rather than deriving α from (strong) assumptions, Cronbach (1951) settled upon a definition and studied its properties. The coefficient is written here as a function $\alpha(\mathbf{a})$ to emphasize its dependence on the weights. In those days, weights were determined by a variety of methods; they were not frequently optimized, as in (9.9). Therefore, coefficient α was welcomed as a convenient measure to evaluate different weighting schemes or different selections of variables, most frequently test items, for computing *total scores*.

Although Cronbach's α is probably the most frequently used omnibus coefficient in applied psychometrics, it has a number of interpretations that are less known than they should be. The primary interpretation follows from (12): it is a measure of homogeneity or *internal consistency*. If one includes uncorrelated variables, the value of α drops; if one adds correlated variables, the value of α increases. Cronbach (1951) gave a number of additional properties that refer more specifically to Spearman's (1910) *split-half* approach for determining the reliability of a series of measurements, in which the battery is rescored, half the variables at a time, to get *two* estimates of the mean score. Split-half reliability is then defined as the correlation between these two sets of scores, corrected with the so-called Spearman–Brown formula, which accounts for the fact that only half of the information is used in estimating the mean score. The conventional split-half approach had been criticized because of its lack of uniqueness, due to the many different ways in which one can split m variables into two sets. However, Cronbach (1951) showed that α is the mean of all possible split-half coefficients if all splits are weighted equally. Moreover, he showed that α is the expected value of the correlation between the object scores in two independent random samples of length m, under fairly general regularity conditions on the distribution of variables (e.g. unimodality), and with the same weighting. These remarkable properties of α (and similar coefficients) relate the early psychometricians' concern with reliability to present-day statistical practice of estimating the variance of a statistic by cross-validation.

Considering α is relevant for homogeneity analysis, because homogeneity is determined by minimizing $\sigma^2(\mathbf{a}, *)$ over $\mathbf{a}$, and it may be seen from (9.11) that $\alpha(\mathbf{a})$ increases when $\sigma^2(\mathbf{a}, *)$ decreases. According to Nishisato (1980, p. 100), it was Lord (1958) who first demonstrated that maximization of $1 - \sigma^2(\mathbf{a}, *)$ leads to maximization of $\alpha(\mathbf{a})$. Simple calculations with (9.8) and (9.9) show that, for optimal $\hat{\mathbf{a}}$, the value of $\alpha_{max} = \alpha(\hat{\mathbf{a}})$ can be expressed in terms of the eigenvalue as

$$\alpha_{max} = m(\lambda - 1)/(m - 1)\lambda \tag{9.13}$$

The relationship between α_{max} and the eigenvalue is nonlinear and monotonically increasing. Two examples are plotted in Figure 9.1, for $m = 6$ and $m = 11$, to give an impression of the rather severe nonlinearity when m grows. Note that α_{max} would become negative if $m^{-1}\lambda < m^{-1}$. However, it can be shown that this condition cannot occur for the largest eigenvalue, which is the reason that each curve in Figure 9.1 starts with a value of m^{-1} on the

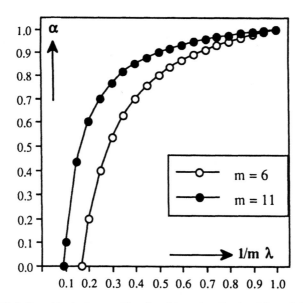

FIGURE 9.1   Relationship between Cronbach's α (vertical axis) and the eigenvalue (horizontal axis).

horizontal axis. Clearly, especially for large m the interpretation of α_{max} and $m^{-1}\lambda$ must be carefully adjusted to their different ranges.

Summarizing, the result of a homogeneity analysis – as defined so far – can be given in terms of the weights or in terms of the object scores, and the measure of homogeneity can be given either as a variance, $\sigma^2(*, *)$, or as a property of the correlation matrix of the rescaled variables, $m^{-1}\lambda$, which is a mean, or as an upper bound to Cronbach's α, which is a correlation.

9.2.3 Irregularities of shape: more principal components

A uniform distribution of variables corresponds to correlations of about equal size, so that $r_{..}$ is sufficient to describe it. Non-uniform distributions can be classified by the number of regions of high density. With only one such region, the distribution is called *unimodal*. There will be a characteristic pattern of higher and lower correlations, called the *Spearman hierarchy*, and the eigenvalue will satisfactorily describe relative homogeneity, in so far as $m^{-1}\lambda$ is close to one.

When $m^{-1}\lambda$ is not close to one, it may be rewarding to look for multiple solutions. Apart from possible multimodality, there is also the possibility that some of the variables may have got small – maybe even close to zero – weights in the first principal component, so that initially they do not really enter into the analysis. In the latter case, interest may be in finding out whether a second principal component yields larger weights for the neglected variables – an indication that the total set is distributed around a plane, rather than around a single central direction in m-space, or around a number of scattered directions. Deviations from uniformity and unimodality are instances of what was called earlier *irregularities of shape* of a distribution, and in practice they turn out to be the rule, rather than the exception.

In one way or another, the characteristics of the first weighting scheme should not reappear in the second weighting scheme (and perhaps in further ones), or at least such re-occurrences are to be avoided as much as possible. This objective can be translated into the formal analysis by requiring that the components are uncorrelated, i.e. $c(\mathbf{x}_s, \mathbf{x}_t) = 0$, where $\mathbf{x}_s$ is a particular principal component and $\mathbf{x}_t$ another one. Now, consider the minimization of

$$\sigma^2(\mathbf{A}, \mathbf{X}) = (Nmp)^{-1}\Sigma_j\Sigma_s \| a_{js}\mathbf{h}_j - \mathbf{x}_s \|^2, \qquad (9.14)$$

in which $s = 1, ..., p$, with p the number of components sought, where $\mathbf{X}$ is an $N \times p$ matrix of object scores, with columns $\mathbf{x}_s$, and $\mathbf{A}$ is an $m \times p$ matrix of weights $\{a_{js}\}$ for variable j with respect to principal component s (see (9.3), where $p = 1$). Various algorithms exist for finding multiple principal components, but they all lead to the same result, up to a scaling factor. As before, the scale of $\mathbf{x}_s$ is chosen by convention to identify a solution – a choice that also determines the range of $\sigma^2(\mathbf{A}, \mathbf{X})$, while leaving the solution essentially the same, because the coefficients in **A** get uniformly adjusted when $\mathbf{x}_s$ is scaled differently.

The optimal choice of a_{js} still is the covariance between $\mathbf{h}_j$ and $\mathbf{x}_s$ divided by the variance of $\mathbf{h}_j$, analogous to (9.5), and substitution in (9.14) with $s^2(\mathbf{h}_j) = 1$ and $s^2(\mathbf{x}_s) = 1$ shows that multiple principal components are uncorrelated variables $\mathbf{x}_s$ that optimize the function

$$\sigma^2(*, \mathbf{X}) = 1 - (mp)^{-1}\Sigma_j\Sigma_s r^2(\mathbf{h}_j, \mathbf{x}_s), \qquad (9.15)$$

again one minus the mean squared correlation, but now also averaged across components. The components can be ordered by the size of their eigenvalues, which are (cf. equations (9.8) and (9.9)) equal to the mean squared $r(\mathbf{h}_j, \mathbf{x}_s)$ per component. When the squared correlation is averaged across components, a measure called the *fit per variable* is obtained, which indicates how much each variable contributes to the analysis. This remark on the alternative decomposition of the total variance concludes the discussion of *linear* principal components analysis (PCA) as a method for studying the homogeneity of variables.

9.3 NONLINEAR TRANSFORMATIONS OF THE VARIABLES

Linear PCA allows for linear transformations of the variables to optimize their homogeneity. All lack of homogeneity necessarily is identified as random deviation. One way to split off further systematic components from the total variability is to allow for nonlinear transformations of the variables, which can often *also* be justified on theoretical grounds, or on characteristics of the observational setting.

9.3.1 Scaling and nonlinearity

A lot of measurements in the social and behavioural sciences are recorded on a scale with uncertain units of measurement. In the case of a five-category Likert item, for example, it is usually quite arbitrary whether we should assign scores {1, 2, 3, 4, 5} or {−3, −1, 0, 1, 3} to the categories {'strongly disagree', 'disagree', 'neutral', 'agree', 'strongly agree'}. Therefore, both the origin of the scale and the distance between consecutive values are uncertain. Another example is time as a response variable: in studies of attention and memory one may use *reaction time* as the behavioural response, and an important aspect of emotional processes is their *duration*. Of course, one measures time in milliseconds or in minutes, but the psychological calibration of time may be different from the series of unit intervals on the dial of a clock. Therefore, it is often appropriate to use a log scale, or some other nonlinear transformation of physical time. A frequently used background variable is *age*, measured in years; here, nonlinearities arise because a lot of developmental processes have typical patterns of acceleration and deceleration. Many achievement variables first improve, then level off, and eventually deteriorate with age.

When such nonlinearities are a possibility, the uncertainty in the unit of measurement is not just a matter of measurement error, because its variability may have a systematic component. On a log scale, for example, a distance at the lower end is longer than a distance at the upper end, compared to unit distances on the original scale. Now suppose that all observed variables are in fact different transformations of the same basic variable; then there must exist *inverse* transformations $\phi_j(\mathbf{h}_j)$ that make them equal again. Instead of a simple rescaling $a_j\mathbf{h}_j$ we thus consider a one-to-one mapping $\phi_j(\mathbf{h}_j) = \mathbf{q}_j$ of the original observations $\{h_{ij}\}$ to new *quantifications* $\{q_{ij}\}$. The mapping ϕ_j allocates to each different value of variable $\mathbf{h}_j$ a new value that can be chosen as to minimize

$$\sigma^2(\phi_1, \ldots, \phi_m, \mathbf{x}) = (Nm)^{-1}\Sigma_j \| \phi_j(\mathbf{h}_j) - \mathbf{x} \|^2. \tag{9.16}$$

In this extended, nonlinear definition of homogeneity, the class of information preserving operations is determined by the specification of ϕ_j, which in the general case is a nonlinear mapping, and which can be chosen independently for each variable. Note that explicitly taking the logarithm is not included in formulation (9.16); that would simply involve the preliminary recoding $\mathbf{h}_j \Leftarrow \ln \mathbf{h}_j$ and does not form a family of transformations. However, if the homogeneity of variables can be improved by increasing the distance between smaller values of a particular variable $\mathbf{h}_j$ and by decreasing distances between larger values of the same variable, and if the class of transformations is broad enough to allow for such changes, then a plot of the transformed values $\phi_j(\mathbf{h}_j)$ against the original values $\mathbf{h}_j$ may well reveal a logarithmic type of function for ϕ_j. Thus homogeneity analysis suggests useful transformations on *a posteriori* grounds, by considering equivalent forms of the variables, and then selecting precisely those that yield a distribution with minimal dispersion.

9.3.2 *Finding the best transformation by regression*

It will be clear that the centre $\mathbf{x}$ of the distribution equals the mean of the transformed variables $m^{-1}\Sigma_j\phi_j(\mathbf{h}_j)$, just as the optimal location under simple rescaling was the mean of the weighted variables (see (9.4)). But it is not known beforehand what transformation would be best to take. Using the ALS principle, a conditionally optimal transformation can be found for each variable separately, given the best current guess $\mathbf{x}^*$ of $\mathbf{x}$, and keeping the other variables constant at their current values $\mathbf{q}_l^*$, because then (9.16) can be decomposed into a constant and a variable term:

$$\sigma^2(\phi_j) = \text{constant} + (Nm)^{-1} \| \phi_j(\mathbf{h}_j) - \mathbf{x}^* \|^2. \tag{9.17}$$

Minimizing $\sigma^2(\phi_j)$ over all possible choices of ϕ_j is a regression problem, in which the *estimate* $\mathbf{x}^*$ is regressed on the space of transformations of the *data* $\mathbf{h}_j$.

In the previous section, the case $\phi_j(\mathbf{h}_j) = a_j\mathbf{h}_j$ was already considered, which corresponds to linear regression without an intercept, and which is appropriate if − but not only if − the measurements are recorded on a ratio scale. Next, the case $\phi_j(\mathbf{h}_j) = a_j + b_j\mathbf{h}_j$, appropriate if − but not only if − one deals with an interval scale, can be solved by linear regression with an intercept, and when $\phi_j(\mathbf{h}_j) = a_j + b_j\mathbf{h}_j + c_j\mathbf{h}_j^2$, formulation (9.17) leads to polynomial regression. Yet another possibility would be to choose piecewise polynomials, also called *splines*, which very naturally adjust themselves to nonlinear relationships, while requiring only a modest number of parameters (Winsberg and Ramsay 1983). After all variables have been processed according to their prespecified class of transformations, yielding current values $\mathbf{q}_j^*$, the ALS principle tells us to continue with an updated guess $\mathbf{x}^*$ equal to the standardized version of $m^{-1}\Sigma_j\mathbf{q}_j^*$, unless this set of scores is close enough to the previous one to stop the process.

Whatever type of regression is chosen, in the majority of cases it will be possible to isolate a scaling factor from the transformed variable, so that we can write $\mathbf{q}_j = a_j\phi_j(\mathbf{h}_j)$ with $s^2(\phi_j(\mathbf{h}_j)) = 1$. Therefore, the interpretation of loss of homogeneity in (9.6) is still possible: the transformed variables will maximize the mean squared correlation with their principal component. Likewise, the interpretation in terms of the eigenvalue and of Cronbach's α remains valid. The next question is: what type of regression is used in the treatment of variables − like the five-category Likert item mentioned earlier − that are *non-numerical*, such as nominal and ordinal variables?

9.3.3 Nominal variables: intra-class regression

A nominal variable is a rule for identifying classes of individuals that are equivalent in some aspect or another. For example, the variable 'religion' groups individuals through their connection with a system of religious belief, and the variable 'nationality' through their connection with a nation. Typically, the individuals in one class of a nominal variable − called *category* − share at least one attribute, making them similar in a specific sense, but there is no preferred way of comparing individuals from different categories. Sometimes there might be an obvious suggestion to order or even to scale the classes, yet the analysis is supposed to ignore this information − an example being the nominal treatment of age when nonlinear relationships between age groups and other variables are anticipated. Since so few constraints are taken into consideration, nominal treatment of a variable is the most general of a range of possibilities.

The values of a nominal variable are just labels to identify the categories, and therefore they can be replaced by any other set of values. Let the number of categories of variable j be denoted by k_j. Thus there are k_j different values in $\mathbf{h}_j$, and we are looking for k_j new values, called *category quantifications*, that minimize the sum of squared deviations in (9.17). This sum of squares can be decomposed into k_j separate parts, corresponding to the given grouping – by equal h_{ij}-value – of individuals into the k_j classes of variable j, and the category quantifications must be the mean $\mathbf{x}^*$-value (object score) of the individuals in each group. Together, the category quantifications form the *intra-class regression* of $\mathbf{x}^*$ onto $\mathbf{h}_j$, and in as much as they are different, they measure the extent to which the subgroups defined by variable j are heterogeneous in terms of the estimated object scores. The mean of the object scores can be freely chosen to be zero, and the sum of squares of the category quantifications, weighted by the proportion of individuals in each group, then becomes a between-group variance. When the k_j classes are well-separated in terms of the object scores, variable j is a good discriminator, and therefore the between-group variance is also called *discrimination measure* (Gifi 1990).

As a historical remark, it was in this form – all variables nominal – that Guttman (1941) initiated the principal components analysis of categorical data, and he further demonstrated the enormous flexibility of the approach in Guttman (1946), where he introduced PCA of variables defined on *pairs of objects*, instead of the objects themselves as units, including various forms of restrictions on $\mathbf{x}$.

9.3.4 Ordinal variables: isotonic regression

An ordinal variable is a rule for identifying ordered classes of individuals or other objects of study. Although further subdivisions are possible, we shall take 'ordered' only to mean that all individuals within a class are equivalent (Kruskal's 1964 secondary approach to ties), and that the classes themselves are *strictly* ordered (i.e. any class is either above or below one particular other class). Because an ordinal variable carries more constraints than a nominal one, there is no objection against having a relatively large number of classes, and even $k_j = N$ (one individual per class) is a practical possibility, especially if m is large relative to N. In the case of a Likert item, it is the semantic ordering from 'strongly disagree' to 'strongly agree' that induces a classification of individuals into ordered classes.

Intra-class regression under the additional requirement that the classes should be ordered is called *isotonic regression* (Barlow *et al.* 1972). It minimizes the sum of squares in (9.17) with the specification that ϕ_j should keep the order in $\mathbf{h}_j$ unchanged (their elements are pulled in the same direction; hence the term isotonic). There are several ingenious algorithms for calculating the isotonic regression $\mathbf{q}_j^*$ of $\mathbf{x}^*$, which all lead to the same solution. What is shared with all regressions is the variance reducing property $s^2(\mathbf{q}_j^*) \leqslant s^2(\mathbf{x}^*)$, but what is specific for regression of the isotonic kind is the fact that for two consecutive individuals i and k one always has: if $h_{ij} > h_{kj}$ *and* $x_i < x_k$, *then* $q_{ij} = q_{kj}$. If individual i has an observed value higher than k, but the ith object score is lower than the kth object score, then the best isotonic transformation will give them equal values (create a *tie*). Here it was assumed for simplicity that each individual forms one category, but with a limited number of categories the principle remains the same, except that the tie is created in the mean object scores of all individuals in the two different categories.

9.4 MULTIPLICITY OF SOLUTIONS

When we are not satisfied with one single centre for the distribution of variables, because we anticipate different areas of concentration, or when we are interested in a kind of residual analysis of the major deviations from the first component, the question of multiplicity of solutions arises – unlike the relatively simple situation in linear PCA. Multiplicity means that there are several ways to proceed, which can only be briefly sketched here. For a more extensive discussion of some advanced proposals in this area, the reader is referred to Bekker and De Leeuw (1988).

9.4.1 Multiple quantification and HOMALS

For the case of simple rescaling, it was shown in (9.14) how one may determine multiple principal components in the present framework. A straightforward generalization is to replace the scalar quantities a_{js} by a multidimensional mapping $\phi_{js}(\mathbf{h}_j)$, resulting in the *multiple loss of homogeneity* function

$$\sigma^2(\Phi_1, ..., \Phi_m, \mathbf{X}) = (Nmp)^{-1}\Sigma_j\Sigma_s \| \phi_{js}(\mathbf{h}_j) - \mathbf{x}_s \|^2. \tag{9.18}$$

By far the best known special case of (9.18) occurs when, for all variables, $\phi_{js}(\mathbf{h}_j)$ is specified using *multiple nominal* quantification, where repeated intra-class regressions of the different components $\mathbf{x}_s$ are called for. Multiple nominal quantification was the approach taken in Guttman's seminal 1941 paper, and the ALS method for calculating the minimal multiple loss of homogeneity forms the core technique of the Gifi system (Gifi 1990), called HOMALS.

A convenient display of the results of a HOMALS analysis is to make a joint scatter plot (or *biplot*) of the object scores and the category quantifications for pairs of components $(\mathbf{x}_s, \mathbf{x}_t)$. Thus the joint plot contains two sets of points, one for the categories and one for the individuals. The optimal category points will be in the centre of gravity of the object points that share the same category. For each variable, the k_j categories partition the total configuration of points into subconfigurations, and when the k_j centres of gravity of these subconfigurations are far apart, the variable discriminates well. Good discrimination is again expressed in a discrimination measure, which now equals the weighted mean squared distance of the category points towards the origin. The discrimination measures can be interpreted as the relative contribution of each quantified variable to the total variation of the individuals as expressed in the components $\mathbf{x}_s$. In view of (9.15), they are equal, dimensionwise, to squared correlations of the object scores with the quantified variables.

It can be shown (Heiser 1981) that minimization of (9.18) with multiple nominal quantifications amounts to *multiple correspondence analysis*, i.e. correspondence analysis on a special type of binary table, called *indicator matrix*, which is partitioned by variable, and which codes for each of the m variables to what category each individual belongs. The results will be exactly equal if in both approaches the solution is normalized so that $s^2(\mathbf{x}_s) = 1$ and $c(\mathbf{x}_s, \mathbf{x}_t) = 0$, the standard normalization in HOMALS. A similar relationship exists with *dual scaling* of contingency tables, the Anglo–Saxon precursor of correspondence analysis (Nishisato 1980). There is also an inverse connection between correspondence analysis and HOMALS, in which the former is regarded as a special case of the latter, since the solution of a simple correspondence analysis can be reconstructed by appropriate renormalizations of a HOMALS with two variables (cf. De Leeuw 1973; for still other connections, see Israëls 1987). Unfortunately, these connections are often erroneously interpreted as equivalences, whereas − as recently emphasized by Greenacre (1991) − the methodological rationale of correspondence analysis is quite different from the rationale of homogeneity analysis as discussed here.

The indicator matrix has been the inspiration for proposals to use a so-called *fuzzy coding* of nominal variables (Van Rijckevorsel 1987), which incorporates the estimated probability of an object being connected with any one of the categories into the analysis. For variables that carry more information than a partitioning into classes, Gifi (1990, p. 169) − relying on an argument by Guttman (1959) − has explained that multiple quantification has to have a special form, called single transformation, to which we turn next.

9.4.2 Single transformation, but multiple weighting: PRINCALS

Suppose that we split off − as before, in the discussion of principal components − a scaling factor from the quantifications, writing $\phi_{js}(\mathbf{h}_j) = a_{js}\phi_j(\mathbf{h}_j)$, with $\| \mathbf{a}_j \|^2 = p$ for all $j = 1, ..., m$ serving as identification constraints, and with $s^2(\mathbf{x}_s) = 1$ for all $s = 1, ..., p$. So each variable is transformed once, by ϕ_j (hence the term single), but contributes by multiple amounts a_{js} to the components. Substituting this restriction into (18) for all variables, multiple loss of homogeneity reduces to

$$\sigma^2(\phi_1, \ldots, \phi_m, \mathbf{A}, \mathbf{X}) = (Nmp)^{-1}\Sigma_j\Sigma_s \| a_{js}\phi_j(\mathbf{h}_j) - \mathbf{x}_s \|^2. \qquad (9.19)$$

Single transformation with multiple weighting cleverly combines the idea of optimally re-expressing the variable as $\phi_j(\mathbf{h}_j)$, in the process of comparing it with the other variables, with the aim of identifying multiple areas of concentration in a multidimensional, non-uniform distribution – an aim that was presented in the present framework as the prime justification for PCA.

Computationally, all the ALS steps for the minimization of the ordinary PCA function (9.14) remain valid for the minimization of (9.19), as long as they manipulate some temporarily fixed, feasible transformation $\mathbf{q}_j^*$ in the place of $\mathbf{h}_j$. Then, finding better transformations given the current best guesses $\mathbf{A}^*$ and $\mathbf{X}^*$ of the weights and the object scores, respectively, can be based upon the decomposition

$$\sigma^2(\phi_1, \ldots, \phi_m, \mathbf{A}^*, \mathbf{X}^*) =$$
$$(Nmp)^{-1}\Sigma_j\Sigma_s \| a_{js}^*\mathbf{x}_j^* - \mathbf{x}_s^* \|^2 + (Nm)^{-1}\Sigma_j \| \mathbf{x}_j^* - \phi_j(\mathbf{h}_j) \|^2. \qquad (9.20)$$

Here the choice of the identification constraints $\| \mathbf{a}_j \|^2 = p$ gives the weights the geometrical interpretation of being equal, up to a factor p, to the *direction cosines* that serve to indicate, in the space of the principal components, the direction of the variable $\mathbf{x}_j^*$ defined as $\mathbf{x}_j^* = p^{-1}\Sigma_s a_{js}^*\mathbf{x}_s^*$, a mixture of the components $\mathbf{x}_s^*$. The decomposition in (9.20) is a special application of *Huygens' theorem*, which in general asserts that the weighted mean squared Euclidean distance of an arbitrary multidimensional point towards a number of given points equals the sum of the weighted mean squared Euclidean distance between the given points and their weighted *centre of gravity* and the squared Euclidean distance between the weighted centre of gravity and the arbitrary point in consideration, multiplied by the sum of the weights. To obtain (9.20), Huygens' theorem has to be applied m times, for each variable separately; the given points are $\mathbf{x}_s^*/a_{js}$, the weights are a_{js}^{*2} (so the sum of the weights is p), the weighted centre of gravity is $\mathbf{x}_j^*$, and the arbitrary point is $\phi_j(\mathbf{h}_j)$.

Huygens' theorem gives the decomposition of a weighted sum of squared distances into two useful terms. The second term on the right-hand side of (9.20) shows that, for variable j, the component mixture $\mathbf{x}_j^*$ is regressed on the space of transformations, instead of the single component $\mathbf{x}^*$ in (9.17). The first term on the right-hand side of (9.20) shows that $\mathbf{x}_s^*$ is one of the p descriptors of the distribution of the $\mathbf{x}_j^*$s. Another interpretation of the same two terms will be given in the next section.

PRINCALS (De Leeuw and Van Rijckevorsel 1980, Gifi 1985, SPSS 1990) is a program that calculates all the quantities mentioned above, according to user-specified classes of transformations, which may be multiple nominal for some variables and single nominal, ordinal, or numerical for others. For the case that all variables are single, it can be shown that PRINCALS searches for transformations that yield a correlation matrix with maximal *sum of the first p eigenvalues*. Of course, the second and further eigenvalues (up to p) may become relatively larger than when we would have maximized only the first one.

9.4.3 Single transformation with analysis of residuals: generalized PRIMALS

It is often a good idea to perform a multidimensional analysis even when the optimal transformations are defined as those that maximize only the largest eigenvalue of the correlation matrix. Suppose that the largest eigenvalue is maximized by a one-dimensional HOMALS, by a one-dimensional PRINCALS, or by using similar options in a similar program. Then it is possible to proceed with an *analysis of residuals* to study the actual distribution of the transformed variables. For there are many interesting ways in which the transformed variables still may deviate from their first principal component.

More formally, assume that, starting with optimizing (9.20), the best $\phi_j(\mathbf{h}_j)$ are found to be $\mathbf{q}_j^*$. The optimal value of (9.20) can be written just as (9.8) with (9.9) inserted, but now with $\mathbf{q}_j^*$ and $\mathbf{q}_l^*$ in the position of $\mathbf{h}_j$ and $\mathbf{h}_l$:

$$\sigma^2(*, *) = 1 - m^{-1}\lambda = 1 - m^{-2}\Sigma_j\Sigma_l\hat{a}_j\hat{a}_l r(\mathbf{q}_j^*, \mathbf{q}_l^*). \tag{9.21}$$

The largest eigenvalue λ of the correlation matrix with elements $r(\mathbf{q}_j^*, \mathbf{q}_l^*)$ is determined by the optimal weights $\hat{a}_j$ and $\hat{a}_l$, and the residual analysis simply consists of determining further weights and further components of the same collection of variables $\{\mathbf{q}_1^*, ..., \mathbf{q}_m^*\}$. The structure of the correlations *after* transformation $\{r(\mathbf{q}_j^*, \mathbf{q}_l^*)\}$ may then also be compared with the structure of the correlations *before* transformation $\{r(\mathbf{h}_j, \mathbf{h}_l)\}$. When all transformations are nominal, this approach has been called PRIMALS (because it focuses on the PRImary eigenvalue, computed by ALS; cf. Van de Geer and Meulman 1985). The only thing that matters, however, to support the rationale of this method as an analysis of residuals, is that the class of transformations must be general enough to be written as $a_j\phi_j(\mathbf{h}_j)$ with $s^2(\phi_j(\mathbf{h}_j)) = 1$. Otherwise, the nature of $\phi_j(\mathbf{h}_j)$ does not matter; hence it is suggested here to use the term *generalized PRIMALS* whenever the family of transformations is more general than one-to-one mappings that preserve class membership.

Determining q principal components of the quantified variables $\{\mathbf{q}_1^*, ..., \mathbf{q}_m^*\}$ implies optimizing the PRINCALS loss function (9.19) with fixed numerical variables $\mathbf{q}_j^*$. To distinguish this case from PRINCALS, the PRIMALS components will be called $\mathbf{z}_s$, and the PRIMALS weights will be denoted by b_{js}.

The mixture variable x_j becomes $z_j = p^{-1}\Sigma_s b_{js} z_s$, and analogously to (9.20), the use of Huygens' theorem now yields the decomposition

$$\sigma^2(\mathbf{B}, \mathbf{Z}) = (Nmp)^{-1}\Sigma_j\Sigma_s \| b_{js} z_j - z_s \|^2 + (Nm)^{-1}\Sigma_j \| \mathbf{q}_j^* - p^{-1}\Sigma_s b_{js} z_s \|^2. \quad (9.22)$$

Here z_j is the linear combination of the components that is closest to $\mathbf{q}_j^*$. The first term on the right-hand side of this equation shows how much the unit length of z_j must be adjusted to get it closest to each component z_s separately, and this term never vanishes (in fact, it can be shown to be constant if the components are uncorrelated). The second term on the right-hand side of (9.22) shows the way in which $\mathbf{q}_j^*$ is predicted from the q components, and can be made exactly equal to zero by choosing q large enough. Because the first principal component that optimizes $\sigma^2(\mathbf{B}, \mathbf{Z})$ and the eigenvalue in (9.21) relate to the same variables $\{\mathbf{q}_1^*, ..., \mathbf{q}_m^*\}$, it must be true that the best choices are $b_{j1} = \hat{a}_j$, $b_{l1} = \hat{a}_l$, and $z_1 = \mathbf{x}^*$, so that a vanishing second term in (9.22) implies, for all j,

$$(\mathbf{q}_j^* - p^{-1}\hat{a}_j \mathbf{x}^*) = p^{-1}\Sigma_{s \neq 1} b_{js} z_s, \quad (9.23)$$

and this shows which residuals are accounted for in generalized PRIMALS. If the components $\{z_2, ..., z_q\}$ exhibit systematic trends, which can be studied by plotting the coefficients $\{b_{js}\}$ — either with, or without the first series $\{b_{j1}\}$ — then there is *a posteriori* evidence of heterogeneity, even though the quantifications were found by initially assuming homogeneity.

9.5   APPLICATION OF HOMOGENEITY ANALYSIS TO CONTROVERSIAL ISSUE VARIABLES

The concepts that have been discussed in this chapter will be illustrated using an example from a survey study on opinions towards a number of controversial issues. When real observations enter the stage, statistical questions of stability and generalizability often arise. Generally speaking, the statistical evaluation of the results of a homogeneity analysis is not straightforward, because the number of parameters and the different combinations of options is usually large, and there are two types of inference to be considered: from a sample of subjects to a population of individuals, and from a sample of variables to a domain of items or tasks. Asymptotic results based on simple multinomial sampling can be obtained (Lebart 1976, De Leeuw 1984), but not much is known about the quality of the approximations involved. Here, resampling techniques (Efron 1982) and permutation tests (Edgington 1987), which work with a minimum of assumptions, may offer a way out. We start the analysis of the example by assuming that the distribution of variables has one dominant component, and study the effect of optimizing over nonlinear

TABLE 9.1

Six statements about abortion and five statements about sexual freedom with response categories and marginal frequencies.

AB-1   A woman, 45 years of age, when menstruation fails to come, thinks menopause has started, and does not worry. Later she appears to be pregnant. She has a family with grown-up children. Until which month of pregnancy do you feel that abortion in this special case is still justified? Or in your judgement is abortion in this case not justified?

AB-2   A girl of 15 – unmarried – suspects she is pregnant. She is scared to talk about it with the family doctor or with her parents. As a result it takes much longer for her than necessary to enlist for medical aid. Until which month of pregnancy do you feel that abortion in this special case is still justified? Or in your judgement is abortion in this case not justified?

Responses: 1 = 'justifiable until 3 months', 2 = 'until 4 months', 3 = 'until 5 months', 4 = 'until 6 months', 5 = 'after six months', 6 = 'not justifiable'.

AB-3   It is the woman's right to have abortion when she wants it.
AB-4   Medical practitioners who perform abortion are no better than murderers.
AB-5   People who agree with abortion have little respect for life.
AB-6   Abortion is justifiable under no circumstances.

Response from 1 = 'agree completely' to 5 = 'disagree completely'.

SF-1   I don't object to children below the age of ten walking around on the beach naked.
SF-2   If sexual intercourse was separated from procreation it would soon become pure egoism.
SF-3   Parents should forbid children to have sexual play.
SF-4   Young people who have sexual intercourse before marriage do not have respect for each other.
SF-5   Parents should impress upon their children that it is better to have control over yourself and not to indulge in masturbation.

Response from 1 = 'agree completely', to 5 = 'disagree completely'.

Label	Marginal frequencies					
	1	2	3	4	5	6
AB-1	221	48	18	8	21	259
AB-2	213	62	31	12	33	224
AB-3	178	115	36	93	153	
AB-4	41	32	77	111	314	
AB-5	114	60	69	117	215	
AB-6	43	54	62	110	306	
SF-1	130	85	56	99	205	
SF-2	84	67	85	117	222	
SF-3	124	109	101	114	127	
SF-4	49	42	56	126	302	
SF-5	124	97	92	88	174	

transformations. Then the use of permutation tests will be demonstrated, particularly for deciding on the question of the right dimensionality.

9.5.1 Data description and initial analysis

The data were collected in 1974, in a Dutch survey among 575 subjects, on opinions towards a number of controversial issues (Veenhoven and Hentenaar 1975). In Gifi (1990, Chapter 13), several subsets of the survey questions have been analyzed; here we only used a subset of Likert items. Table 9.1 gives the description of the 11 questions that were used; the first six are statements about the *abortion* issue, and the next five are statements about the issue of *sexual freedom*. Table 9.1 also gives the marginal frequencies of the response categories. (The original data contain a few missing entries; because the standard option for missing data in HOMALS cannot be used easily for a study of the correlation matrix, the missing values have been replaced in the present analyses by the median response obtained from the remaining subjects, after recoding variables AB-1 and AB-2.)

There is a peculiarity in the response categories of variables AB-1 and AB-2, i.e. the first five categories indicate an increasing tolerance with respect to abortion, but category 6 denotes rejection of abortion under all circumstances. Also, categories 1 up to 5 of AB-3 and SF-1 indicate increasing intolerance, while for all other variables the order of the response categories implies decreasing intolerance. To compute a correlation matrix before transformation and its mean correlation (see Table 9.2), recoding of these variables is necessary. In a HOMALS analysis, the optimal quantification is supposed

TABLE 9.2
Correlation matrix original variables, with mean correlation 0.414, and eigenvalues.

	AB-1	AB-2	AB-3	AB-4	AB-5	AB-6	SF-1	SF-2	SF-3	SF-4	SF-5
AB-1	1.00	0.67	0.44	0.39	0.48	0.41	0.32	0.27	0.24	0.28	0.22
AB-2	0.67	1.00	0.51	0.42	0.49	0.46	0.27	0.18	0.22	0.29	0.23
AB-3	0.44	0.51	1.00	0.50	0.57	0.49	0.25	0.14	0.18	0.26	0.21
AB-4	0.39	0.42	0.50	1.00	0.71	0.68	0.27	0.27	0.28	0.36	0.30
AB-5	0.48	0.49	0.57	0.71	1.00	0.70	0.28	0.27	0.28	0.39	0.34
AB-6	0.41	0.46	0.49	0.68	0.70	1.00	0.23	0.26	0.25	0.37	0.38
SF-1	0.32	0.27	0.25	0.27	0.28	0.23	1.00	0.18	0.31	0.25	0.30
SF-2	0.27	0.18	0.14	0.27	0.27	0.26	0.18	1.00	0.36	0.39	0.38
SF-3	0.24	0.22	0.18	0.28	0.28	0.25	0.31	0.36	1.00	0.37	0.53
SF-4	0.28	0.29	0.26	0.36	0.39	0.37	0.25	0.39	0.37	1.00	0.50
SF-5	0.22	0.23	0.21	0.30	0.34	0.38	0.30	0.38	0.53	0.50	1.00
λ	4.657	1.512	0.959	0.797	0.642	0.600	0.503	0.445	0.321	0.295	0.270

W. J. Heiser and J. J. Meulman

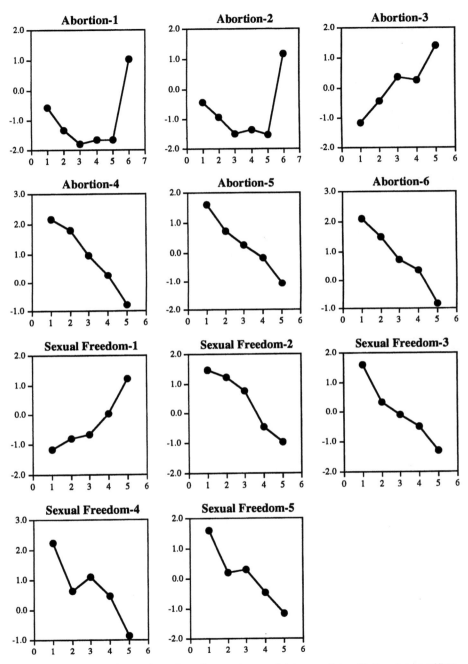

FIGURE 9.2 Transformation plots for controversial issue data. Optimal quantifications (vertical axes) versus original categories (horizontal axes).

to take care of these anomalies automatically. From the transformation plots in Figure 9.2 it can be seen that proper recoding is achieved, apparently due to the internal consistency with the other variables. The optimal standardized quantifications (vertical axes) are plotted against the original response categories (horizontal axes); AB-4, AB-5, AB-6, SF-2, and SF-3 show a decreasing function, so that a high value on the transformed variable indicates intolerance. AB-3 and SF-1 obtain increasing functions (high values remain high, indicating intolerance), AB-1 and AB-2 obtain decreasing functions, except for the sixth category, which obtains the highest value, indicating absolute intolerance. The transformations for SF-4 and SF-5 are basically decreasing, with a small increase for the third categories.

The correlation matrix after transformation, using optimal quantification in one dimension, is given in Table 9.3. The largest eigenvalue of this correlation matrix equals 5.00, which is m times the eigenvalue as given by the output of programs like HOMALS and PRINCALS. The latter quantities are indicated with $m^{-1}\lambda$ in this paper, and are displayed in Table 9.4. Thus $5.00 = 11 \times 0.455$; this eigenvalue is equivalent to $\alpha = 0.88$. Apart from the largest eigenvalue, the subsequent eigenvalues were also computed; these are important for the study of the residuals. If the largest eigenvalue is maximized to see whether a one-dimensional structure fits the data, the components associated with the second and following eigenvalues should *not* display a structural pattern. It is possible to perform a permutation test to see whether or not this is the case for our example.

TABLE 9.3
Correlation matrix variables after optimal quantification in one dimension, with mean correlation 0.443, and eigenvalues.

	AB-1	AB-2	AB-3	AB-4	AB-5	AB-6	SF-1	SF-2	SF-3	SF-4	SF-5
AB-1	1.00	0.64	0.52	0.49	0.57	0.51	0.32	0.31	0.28	0.35	0.29
AB-2	0.64	1.00	0.58	0.52	0.59	0.58	0.27	0.24	0.25	0.36	0.29
AB-3	0.52	0.58	1.00	0.54	0.58	0.51	0.26	0.15	0.19	0.28	0.24
AB-4	0.49	0.52	0.54	1.00	0.73	0.68	0.29	0.28	0.28	0.40	0.32
AB-5	0.57	0.59	0.58	0.73	1.00	0.71	0.30	0.29	0.30	0.41	0.36
AB-6	0.51	0.58	0.51	0.68	0.71	1.00	0.27	0.28	0.25	0.40	0.40
SF-1	0.32	0.27	0.26	0.29	0.30	0.27	1.00	0.15	0.30	0.28	0.29
SF-2	0.31	0.24	0.15	0.28	0.29	0.28	0.15	1.00	0.36	0.43	0.39
SF-3	0.28	0.25	0.19	0.28	0.30	0.25	0.30	0.36	1.00	0.38	0.54
SF-4	0.35	0.36	0.28	0.40	0.41	0.40	0.28	0.43	0.38	1.00	0.51
SF-5	0.29	0.29	0.24	0.32	0.36	0.40	0.29	0.39	0.54	0.51	1.00
λ	5.000	1.477	0.861	0.692	0.635	0.560	0.464	0.435	0.344	0.274	0.257

W. J. Heiser and J. J. Meulman

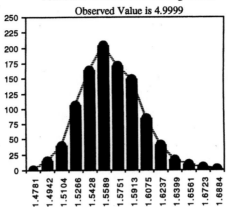

Permutation Distribution First Eigenvalue
Observed Value is 4.9999

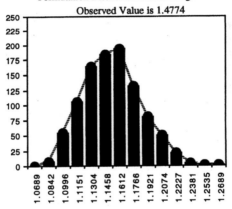

Permutation Distribution Second Eigenvalue
Observed Value is 1.4774

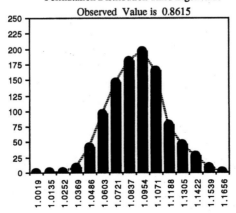

Permutation Distribution Third Eigenvalue
Observed Value is 0.8615

9.5.2 A permutation test for choosing the number of eigenvalues to be maximized

Under the null hypothesis that all variables are independent, the distribution of the various statistics that we compute can be approximated by generating a permutation distribution, which mimics independent sampling of subjects. The reader is referred to De Leeuw and Van der Burg (1986) and Ter Braak (1992) for an extensive discussion of the rationale of permutation methods.

A new feature of the present application is that we do not focus on the significance of the HOMALS eigenvalues that follow the first one, but on the residual eigenvalues of the quantification matrix (cf. (9.23)). The question is, what dimensionality (p) to choose for determining the optimal transformations, and this question is answered by estimating the probability to obtain eigenvalues for the correlation matrix of the transformed variables Q that are as large or larger than the observed values, under random assignment of the category responses. Thus the permutation distributions of the q eigenvalues of generalized PRIMALS are employed.

The category responses of the 11 variables were independently redistributed among the subjects, with the number of permutations set equal to 1 000; for each of the permuted data sets, one-dimensional homogeneity analyses were performed, with nominal treatment of all variables, and the optimal quantifications were used to obtain the permutation distributions of three eigenvalues $\lambda_{R(1,2,3)}$. The permutation distribution of the three eigenvalues is depicted in the form of a histogram in Figure 9.3. For the first eigenvalue we have $P(\lambda_{R(1)} \geqslant 4.9999) = 0.000$, for the second eigenvalue $P(\lambda_{R(2)} \geqslant 1.4774) = 0.000$, and for the third $P(\lambda_{R(3)} \geqslant 0.8615) = 1.000$, so there is a clear structure in the first two PRIMAL components and a random residual pattern in the third. These results suggest that the right dimensionality (i.e. the number of eigenvalues to be maximized) is 2. To obtain a maximized sum of the first *two* eigenvalues, a PRINCALS analysis was done, with a single nominal transformation. It can be seen from Table 9.4, that the second PRINCALS eigenvalue ($11 \times 0.146 = 1.608$) is considerably larger than the mean of the permutation distribution of the second PRIMALS eigenvalue ($11 \times 0.104 = 1.146$), while the third eigenvalue ($11 \times 0.076 = 0.840$) is considerably smaller than the mean of the distribution of the third PRIMALS eigenvalue ($11 \times 0.099 = 1.085$). This fact confirms our conclusion that for the controversial issue data the maximization should be carried out over two dimensions.

FIGURE 9.3   Permutation study of the eigenvalues of the correlation matrix after optimal quantification.

TABLE 9.4
$(m^{-1} \times)$ eigenvalues compared to mean correlations $(r_{..})$.

$r_{..}$	All variables 0.414			Abortion 0.607		Sexual freedom 0.486	
	Dim-1	Dim-2	Dim-3	Dim-1	Dim-2	Dim-1	Dim-2
1	0.423	0.137	0.087	0.609	0.150	0.492	0.169
2	0.455	0.134	0.078	0.656	0.109	0.512	0.162
3	0.448	0.146	0.076	0.644	0.126	0.481	0.201
4	0.142	0.104	0.099	0.229	0.173	0.262	0.203

1 Eigenvalues principal components analysis original variables.
2 Eigenvalues PRIMALS (first dimension optimal).
3 Eigenvalues PRINCALS (first and second dimension optimal).
4 Mean permutation distribution eigenvalues PRIMALS.

Two-dimensional component loadings for the variables are displayed as vectors in Figure 9.4, which contains from top to bottom: PCA of the original (recoded) variables, PCA with optimization over one dimension (PRIMALS), and PCA with optimization over two dimensions (PRINCALS). As could be expected from the study of the eigenvalues, there are no major differences, only subtle ones. Compared to the PCA at the top, PRIMALS minimizes the mean squared distance of the endpoints of the vectors towards the horizontal axis. In PRINCALS, on the other hand, the endpoints have a smaller mean squared distance towards the second dimension. The PRINCALS solution shows that the SF-1 variable really belongs to the SF-subset.

Having established the bimodal character of the distribution of the variables, the abortion and sexual freedom variables were also analyzed separately. The results are given in Table 9.4. First of all we note that the two mean correlations are higher (0.607 and 0.486, respectively) than the grand mean 0.414, and this remarkable result remains true when we compare the dominant PRIMALS eigenvalues (0.656 and 0.512) with the first two PRINCALS eigenvalues (0.448 and 0.146). Next, differential weighting has a slightly larger effect for the sexual freedom (0.492) than for the abortion variables (0.609). Third, the permutation results indicate that in *both* subsets of variables, the second eigenvalue is *not* larger than the one obtained by random assignment to the response categories (both P-values are < 0.0005). Finally, when the sum of the first two eigenvalues is maximized by PRINCALS, the second PRINCALS eigenvalue ($5 \times 0.201 = 1.004$) is not larger than the mean of the permutation distribution of the second eigenvalue

FIGURE 9.4   Three different two-dimensional representations of the variables in the analysis of the controversial issue data.

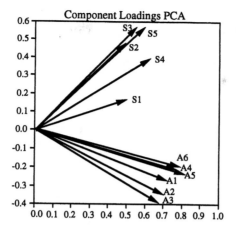

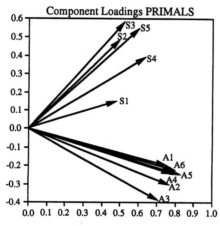

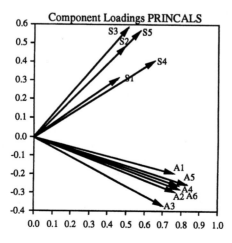

of the sexual freedom variables ($5 \times 0.203 = 1.016$), and considerably smaller ($6 \times 0.126 = 0.755$) than the mean ($6 \times 0.173 = 1.039$) for the abortion variables, so it is concluded that maximization over only one dimension is justified for the two subsets separately.

9.5.3 Substantive interpretation

What does this analysis tell us about the controversial issues themselves? First, leniency versus strictness in sexual freedom apparently involves different beliefs and values than the abortion controversy, since they constitute two separate components that are only weakly correlated. The implication is that persons in the sample who take a tolerant stand in the abortion issue may still be either strict or lenient on sexual freedom, while persons in the sample who think abortion is never justifiable also may be either strict or lenient on sexual freedom. Second, the statements that build up the abortion component are all about equally important, and different selections of them could be chosen as the same indicator of someone's position on this ethical dilemma. For the sexual freedom component, statement SF-1 is a bit odd, perhaps because it requires a double negation to express strictness here.

The analysis suggests using two new variables in further studies of a similar nature. To actually obtain these indicators of leniency toward sexual freedom and tolerance towards abortion, one would code the responses with the category quantifications (as given here on the vertical axis of Figure 9.2), and add the resulting coded variables within the two identified groups. Such indicators (equal to the object scores) will be more reliable and enable finer discrimination than each of the items alone. As an example, the object scores of the abortion component are plotted in Figure 9.5, in such a way that the five subsamples generated by the five categories of statement AB-5 are clearly visible as partially overlapping distributions. Note that the regression is linear after transformation (Figure 9.5 is a plot of x versus q_5, with the category centroids specially marked; the discrimination measure is equal to the squared correlation in this scatter plot). So the group of respondents on the right, labelled with 1, have all stated that they are fully convinced that 'people who agree with abortion have little respect for life', whereas the group of respondents on the left, labelled with 5, have all declared that they completely disagree with this statement. Although the partitioning of the object scores displays a clear shift of the means (heterogeneity), it is also true that there is considerable overlap (absence of perfect discrimination). Apparently, people who are extreme in terms of 'respect for life' (AB-5) need not be similarly extreme in terms of 'woman's right' (AB-3), for example, and vice versa. As long as we are looking at the level of an individual variable, we have to be careful for random disturbances (be it measurement error, response bias, temporary switch of opinion, or the like). Nevertheless, on average, the

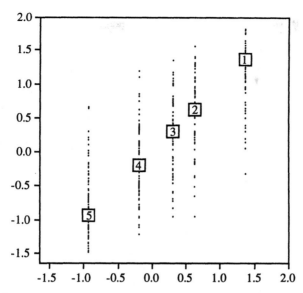

FIGURE 9.5 Object scores of the one-dimensional homogeneity analysis of the abortion statements plotted against the quantified variable AB-5.

abortion statements discriminate the respondents consistently from *tolerant* to *adverse*. The sexual freedom statements discriminate them differently, on average, yielding an order from *liberal* to *authoritarian*. It could be suggested that the former is a personal and moral distinction, while the latter is a social and normative one.

9.6 DISCUSSION

Homogeneity analysis was presented as a way to explore a distribution of variables and their nonlinear relationships. A uniform distribution occurs when the variables are equicorrelated, with uniformly zero correlation as a special case. More often, the variables have one or more areas of concentration, just as empirical univariate distributions often have one or more peaks. The centres of these areas may be determined by generalized principal components analysis, and are candidates for further practical use, since their reliability will be higher than any single variable, due to the effect of averaging out the measurement error. In fact, their reliability as measured by Cronbach's α will be maximal for new samples (of subjects *and* variables) that are similar to the one used to establish homogeneity.

What are the limits in the applicability of homogeneity analysis? Two kinds of limits should be mentioned. In the first place, certain types of nonuniform distribution are hard to describe by this form of nonlinear PCA (Heiser 1986), for instance when the data are translation families of functions with at least one inflection point. Such cases, which require a method with more complicated basis functions than the central vectors **x** used here, are not easy to detect, except by careful examination of bivariate scatter plots and some prior knowledge of the processes involved. In the second place, homogeneity analysis − as standardly defined − focuses on the univariate and bivariate marginal, or the means, variances and covariances. If one expects the presence of higher order interactions (in the categorical case) or non-zero higher order moments (in the numerical case), special action is required to take them into account. In the categorical case, it is just a matter of redefining the variables involved into an *interaction variable* (Gifi 1990, p. 376; Van der Lans 1992, Ch. 4), which is a variable whose categories correspond to the cells of the two- or higher-dimensional contingency table of the original variables. Thus this limit can be transgressed within the method itself.

It is also possible to present homogeneity analysis in an entirely different way, which starts from an identification of the individuals as points in m different spaces, of dimensionality $k_j - 1$, and which proceeds by finding some kind of consensus configuration in low-dimensional space that optimally represents the distances between individuals. Because this approach is extensively documented elsewhere, the present chapter has emphasized a different line of development. The reader is referred to Meulman (1986, 1992) for more information on the so-called *distance approach*. When an *a priori* partitioning of variables into K sets is available, we can profit from the nesting to study the homogeneity of the so-called *canonical variables*, which are themselves linear combinations of variables within sets. Then all possibilities of variable weighting and nonlinear transformation are open to us (Gifi 1990, Ch. 5), but an extensive discussion of these would lead too far. The reader is referred to Van der Burg *et al.* (1988) for a discussion of this situation, and the related technique called OVERALS.

Resampling techniques such as the bootstrap have been applied before in homogeneity analysis (Meulman 1982, Greenacre 1984, Saporta and Hatabian 1986, Van der Burg and De Leeuw 1988, Markus and Visser 1992), but applications of permutation tests await further development. In our example the results were encouraging: it could definitely be decided that two homogeneous clusters of variables are to be distinguished, with slightly correlated central components, and with a joint Cronbach's α considerably higher (0.895 and 0.762, corresponding to the mean squared correlations of 0.656 and 0.512 in Table 9.4) than the standard PRINCALS two-dimensional homogeneity analysis that is based on uncorrelated principal components (0.877 and 0.415, corresponding to the mean squared correlations of 0.448 and 0.146 in Table 9.4).

NOTE

The programs HOMALS, PRINCALS and OVERALS, mentioned in the text and/or used in the applications, were originally developed in the Department of Data Theory of the University of Leiden, and are available in the SPSS package Categories, documented in SPSS (1990).

ACKNOWLEDGEMENT

The authors are indebted to the editors, Shizuhiko Nishisato, Norman Verhelst, Niels Veldhuijzen, Ivo van der Lans, and Jacques Commandeur for many helpful comments on an earlier draft of this paper.

Visualizing Solutions in more than Two Dimensions

Jože Rovan

10.1 INTRODUCTION

Profiles, i.e. vectors of relative frequencies of rows and columns of a data matrix, can be geometrically represented as points in a multidimensional space, using their coordinates with respect to principal axes, or principal coordinates. Since the dimensionality of the profile space can be high, we are faced with the problem of its practical graphical representation in a plot. For this reason the profile points of the original profile space are usually projected onto an optimal subspace of a lower dimension. According to the objective of correspondence analysis, the first few dimensions usually represent most of the variability of the original data matrix. If this is the case, the position of the projections of profile-points in a subspace of lower dimension is a good approximation of the position of profiles in the original multidimensional space.

In the standard graphical approach, profiles are represented as points in a two-dimensional plot, using the first two principal coordinates. Alternative ways of graphical representation of the profiles are possible, for example the three-dimensional plot and the so-called Andrews' plot. These methods are described and illustrated using the results of a MCA. The common starting point of the MCA is an indicator matrix or a Burt matrix (see Chapter 7).

10.2 DIMENSIONALITY OF MULTIPLE CORRESPONDENCE ANALYSIS SOLUTION

Let us illustrate the use of the three types of plots mentioned above on an example. We use a socioeconomic sample survey relating to the 'Living conditions and aspirations of the French' (Lebart and Houzel van Effenterre 1980). The subsample of 1000 respondents (Lebart *et al.* 1984) is representative of the population over 18 years. We shall pay attention to seven variables ($Q = 7$), containing a total of 25 categories ($J = 25$) (*Sex*: SEXM − male, SEXF − female; *Level of education*: EDU1 − none, EDU2 − grade school, EDU3 − some high school, EDU4 − high school graduate, EDU5 − some college; *Home ownership*: LOD1 − mortgage, LOD2 − owner, LOD3 − tenant, LOD4 − rent free; *Ownership of stocks and shares*: STO1 − yes, STO2 − no; *Ownership of real estate*: HOU1 − yes, HOU2 − no; *Age*: AGE1 − under 25 years, AGE2 − 25–34 years, AGE3 − 35–49 years, AGE4 − 50–64 years, AGE5 − 65 years and over; *Size of town*: SIZ1 − less than 2000, SIZ2 − 2000–50 000, SIZ3 − 50 000–100 000, SIZ4 − 100 000–500 000, SIZ5 − more than 500 000). The data of our example are displayed in the form of the Burt matrix given in Table 7.2.

Correspondence analysis of the Burt matrix results in a diagonal matrix of principal inertias $\mathbf{D}_\lambda$ (i.e. eigenvalues) and the matrix of standard coordinates $\mathbf{Y}$ (i.e. eigenvectors) (see Chapter 7). The number of nontrivial principal

TABLE 10.1
The values of inertias λ_k, the percentages of inertia and the cumulative percentages of inertia.

	Principal inertia	Percentage of inertia	Cumulative percentage of inertia	2	4	6	8	10%
1	0.2500	9.72	9.72	************************				
2	0.2184	8.49	18.21	********************				
3	0.1847	7.18	25.40	******************				
4	0.1746	6.79	32.19	****************				
5	0.1610	6.26	38.45	***************				
6	0.1580	6.15	44.59	**************				
7	0.1542	6.00	50.59	**************				
8	0.1457	5.67	56.26	*************				
9	0.1409	5.48	61.74	*************				
10	0.1344	5.23	66.96	************				
11	0.1290	5.02	71.98	************				
12	0.1253	4.87	76.85	***********				
13	0.1223	4.76	81.61	***********				
14	0.1177	4.58	86.18	**********				
15	0.1060	4.12	90.31	*********				
16	0.0957	3.72	94.03	********				
17	0.0869	3.38	97.41	*******				
18	0.0666	2.59	100.00	*****				
	2.5714							

$1/Q = 1/7 = 0.1429$

inertias is $J - Q = 18$. The inertias, the percentages of inertia and the cumulative percentages of inertia are presented in Table 10.1.

Because MCA includes fitting of the diagonal submatrices of the Burt matrix, the total inertia is inflated and thus the proportions of the first few principal inertias as part of the total inertia are reduced (see Chapter 7). One of the ways to address this problem, proposed by Benzécri (1979), is to consider only those principal axes, whose inertias are higher than $1/Q$ (i.e. 0.1429, see Table 10.1). Let us calculate these inertias according to Benzécri's formula

$$\tilde{\lambda}_k = \left[\frac{Q}{Q-1} \left(\lambda_k - \frac{1}{Q} \right) \right]^2 \qquad k = 1, 2, ..., 8 \qquad (10.1)$$

The values of inertias $\tilde{\lambda}_k$, the respective percentages of inertia and the cumulative percentages of inertia are presented in Table 10.2.

The number of inertias has been reduced from 18 to only 8, with 91.7% of the total inertia explained by only the first three principal axes. Thus a three-dimensional solution can serve as a good basis for the display of the profiles. According to our experience, however this is just partially true. We believe, that for thorough analysis every single profile should be well represented and therefore some profiles should be represented by a higher number of principal coordinates. We shall return to this question later.

We have already calculated the matrix $\mathbf{Y}$ of standard coordinates. Next we need to transform the first eight columns of $\mathbf{Y}$, denoted by $\mathbf{Y}^*$, into the principal coordinates using the modified inertias given by formula (10.1)

$$\tilde{\mathbf{G}}^* = \mathbf{Y}^* \tilde{\mathbf{D}}_\lambda \qquad (10.2)$$

where $\tilde{\mathbf{D}}_\lambda$ is the diagonal matrix of the first eight modified inertias. The matrix $\tilde{\mathbf{G}}^*$ is given in Table 10.3.

TABLE 10.2

The values of modified inertias $\tilde{\lambda}_k$, the respective percentages of inertia and the cumulative percentages of inertia.

	Principal inertia	Percentage of inertia	Cumulative percentage of inertia	10	20	30	40	50%
1	0.015631	55.65	55.65	*************************				
2	0.007760	27.63	83.27	*************				
3	0.002378	8.47	91.74	****				
4	0.001372	4.88	96.62	**				
5	0.000450	1.60	98.22	*				
6	0.000313	1.12	99.34	*				
7	0.000174	0.62	99.96					
8	0.000011	0.04	100.00					
	0.028089							

TABLE 10.3

The matrix of principal coordinates $\bar{G}^*$.

		1.PC	2.PC	3.PC	4.PC	5.PC	6.PC	7.PC	8.PC
1	SEXM	0.0354	−0.0308	−0.0163	−0.0077	0.0196	0.0144	0.0031	0.0041
2	SEXF	−0.0313	0.0272	0.0144	0.0068	−0.0173	−0.0127	−0.0028	−0.0036
3	EDU1	0.0662	−0.0643	−0.1156	0.0348	0.0387	0.0046	−0.0043	−0.0072
4	EDU2	0.1084	−0.1130	0.0132	−0.0449	−0.0154	0.0058	−0.0040	0.0014
5	EDU3	−0.0994	−0.0002	0.0340	0.0523	−0.0154	−0.0581	−0.0069	0.0041
6	EDU4	−0.1365	0.0860	0.0912	0.0763	−0.0023	0.0274	0.0069	0.0008
7	EDU5	−0.0649	0.2350	−0.0157	−0.0883	0.0020	0.0128	0.0140	0.0012
8	LOD1	−0.0276	−0.1335	0.1796	−0.0326	0.0444	0.0272	−0.0043	−0.0009
9	LOD2	0.2891	0.0127	−0.0071	0.0124	−0.0098	−0.0049	0.0037	0.0019
10	LOD3	−0.1450	0.0212	−0.0283	−0.0148	−0.0058	−0.0109	−0.0068	−0.0014
11	LOD4	−0.0839	0.0189	−0.0753	0.1323	0.0098	0.0640	0.0491	0.0046
12	STO1	0.2452	0.3338	0.0545	−0.0019	0.0074	−0.0020	−0.0140	0.0021
13	STO2	−0.0338	−0.0460	−0.0075	0.0003	−0.0010	0.0003	0.0019	−0.0003
14	HOU1	0.3442	0.3153	0.1066	0.0503	−0.0077	0.0043	−0.0207	0.0009
15	HOU2	−0.0307	−0.0282	−0.0095	−0.0045	0.0007	−0.0004	0.0018	−0.0001
16	AGE1	−0.1802	0.0028	0.0238	0.2997	0.0437	−0.0406	0.0156	0.0014
17	AGE2	−0.2308	0.0497	−0.0178	−0.0051	−0.0180	0.0167	−0.0139	0.0047
18	AGE3	−0.0184	−0.0384	0.0710	−0.0347	0.0125	−0.0085	0.0191	−0.0002
19	AGE4	0.1624	−0.0646	−0.0165	0.0142	−0.0532	0.0036	−0.0171	−0.0058
20	AGE5	0.2379	0.0795	−0.1109	−0.0061	0.0487	−0.0008	−0.0047	−0.0003
21	SIZ1	0.3198	−0.1962	−0.0247	0.0395	−0.0798	0.0096	0.0324	0.0089
22	SIZ2	0.1220	−0.0638	0.1289	0.0307	0.0144	0.0325	0.0001	−0.0148
23	SIZ3	0.0664	−0.0548	0.0263	−0.0183	0.0430	−0.0547	0.0030	0.0033
24	SIZ4	−0.0649	−0.0273	−0.0075	0.0103	0.0072	0.0181	−0.0342	0.0029
25	SIZ5	−0.0841	0.1239	−0.0346	−0.0188	−0.0139	0.0000	0.0246	−0.0030

The eight dimensions of Table 10.3 are now considered as representing the (modified) full space. As in CA we calculate the quality of each profile by accumulating squared correlations for successive dimensions of this space, i.e. for dimension 1, then for dimensions 1 and 2, and so on (Table 10.4).

Our experience is that the quality of an individual profile, based on the appropriate number of principal coordinates, should exceed 90% for a good approximation of the position of the profile.

10.3 TWO-DIMENSIONAL MAPS

Profiles are usually displayed as points in a two-dimensional map. If the proportion of inertia of the first two dimensions as part of the total inertia is relatively high, then most of the profiles are well represented by their projections onto a plane. However, it frequently happens that this proportion is relatively low and therefore most of the profiles are poorly represented by their

TABLE 10.4
Cumulative proportions of inertias of profiles as part of the total inertias of profiles.

		1.PC	2.PC	3.PC	4.PC	5.PC	6.PC	7.PC	8.PC
1	SEXM	0.3985	0.7002	0.7848	0.8037	0.9262	0.9916	0.9948	1.0000
2	SEXF	0.3985	0.7002	0.7848	0.8037	0.9262	0.9916	0.9948	1.0000
3	EDU1	0.1777	0.3454	0.8865	0.9355	0.9963	0.9972	0.9979	1.0000
4	EDU2	0.4352	0.9081	0.9146	0.9893	0.9981	0.9993	0.9999	1.0000
5	EDU3	0.5661	0.5661	0.6325	0.7893	0.8029	0.9963	0.9990	1.0000
6	EDU4	0.4550	0.6354	0.8383	0.9804	0.9805	0.9988	1.0000	1.0000
7	EDU5	0.0621	0.8761	0.8797	0.9946	0.9947	0.9971	1.0000	1.0000
8	LOD1	0.0140	0.3402	0.9306	0.9500	0.9861	0.9997	1.0000	1.0000
9	LOD2	0.9940	0.9959	0.9965	0.9984	0.9995	0.9998	1.0000	1.0000
10	LOD3	0.9265	0.9463	0.9815	0.9912	0.9927	0.9979	0.9999	1.0000
11	LOD4	0.1893	0.1989	0.3512	0.8220	0.8246	0.9348	0.9994	1.0000
12	STO1	0.3439	0.9815	0.9985	0.9985	0.9988	0.9988	1.0000	1.0000
13	STO2	0.3439	0.9815	0.9985	0.9985	0.9988	0.9988	1.0000	1.0000
14	HOU1	0.5099	0.9379	0.9869	0.9978	0.9981	0.9982	1.0000	1.0000
15	HOU2	0.5099	0.9379	0.9869	0.9978	0.9981	0.9982	1.0000	1.0000
16	AGE1	0.2564	0.2565	0.2609	0.9700	0.9850	0.9981	1.0000	1.0000
17	AGE2	0.9362	0.9795	0.9851	0.9856	0.9913	0.9962	0.9996	1.0000
18	AGE3	0.0390	0.2093	0.7919	0.9311	0.9493	0.9577	1.0000	1.0000
19	AGE4	0.7717	0.8936	0.9016	0.9075	0.9901	0.9905	0.9990	1.0000
20	AGE5	0.7289	0.8103	0.9686	0.9691	0.9997	0.9997	1.0000	1.0000
21	SIZ1	0.6794	0.9352	0.9392	0.9496	0.9919	0.9925	0.9995	1.0000
22	SIZ2	0.3921	0.4991	0.9362	0.9610	0.9665	0.9943	0.9943	1.0000
23	SIZ3	0.3315	0.5574	0.6093	0.6344	0.7732	0.9985	0.9992	1.0000
24	SIZ4	0.6312	0.7426	0.7512	0.7670	0.7749	0.8240	0.9987	1.0000
25	SIZ5	0.2856	0.9050	0.9532	0.9675	0.9753	0.9753	0.9996	1.0000
Total		0.5565	0.8327	0.9174	0.9662	0.9822	0.9934	0.9996	1.0000

projections. This can be partly misleading when making conclusions about the explored phenomena. When appropriate, though, the principal advantages of the two-dimensional plot are:

- the projections of all profile points are conveniently presented on a single sheet of paper, even in the case of numerous profiles;
- the evaluation of the distances between points is simple and accurate;
- software and hardware required to produce this type of plot are widely available.

Let us represent the profiles in a two-dimensional map (Figure 10.1) with their first two principal coordinates. On the basis of the profile positions in this map, we might conclude that AGE1 and LOD1, for example, are well represented because of their outlying positions. In fact, the quality of AGE1

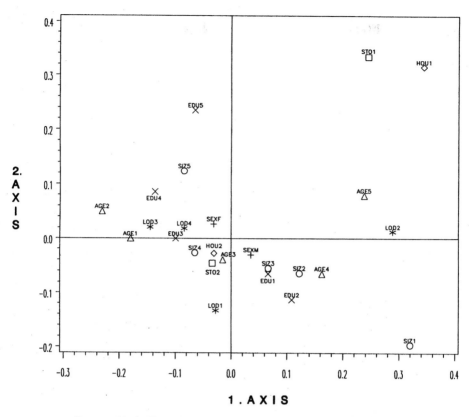

FIGURE 10.1   Two-dimensional plot (data from Table 10.3).

is only 25.7% and of LOD1 34.0% so that their true positions are far away from this plane. On the other hand, HOU2, being near the centre, might be poorly represented in the map, but its quality is actually 93.8%.

10.4   THREE-DIMENSIONAL MAPS

In order to take into account the third dimension, three specific planar projections, i.e. projections of the profile points on the planes spanned by the axes 1 and 2, 1 and 3 and 2 and 3, can be performed. These projections give a first view of the three-dimensional solutions.

A better view of the three-dimensional subspace can be achieved by a three-dimensional map, sometimes called a perspective plot. The principal features

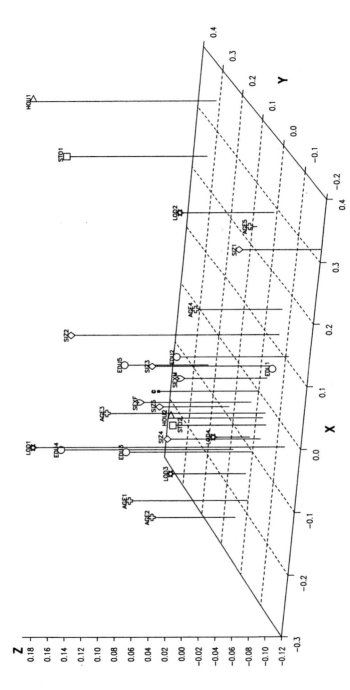

FIGURE 10.2 Three-dimensional plot of first three principal coordinates in Table 10.3.

of this plot are:

- the positions of the projections of the profile points are less conveniently presented but may be visualized better if we make a few plots from different angles (Figure 10.2) provided the profiles are not too numerous;
- the evaluation of distances between points is relatively complicated and only approximate, since the impression of the three-dimensional space is achieved by the use of perspective;
- software and hardware required to produce this type of plot are not always available in standard packages.

A preferable alternative to these static representations of the three-dimensional maps is provided by dynamic rotation of the display on the computer screen. A set of points which appear close together, can be safely treated as a cluster if their relative proximities are stable during the rotation of the display in all directions (Figure 10.3). Satisfactory evaluation of the interpoint distances is also guaranteed using this technique.

By including the third dimension the quality of some points which were poorly represented in the two-dimensional solution is highly improved, for example EDU1, AGE3 and SIZ2. Such points are explained to a large degree by the third dimension.

For some profiles, however, the proportion of inertia explained by the three-dimensional solution is still relatively low, for example LOD4 and AGE1 (see Table 10.4), again leading to possible misinterpretation. These points are explained by higher dimensions.

10.5 ANDREWS' CURVES

During the past 30 years a considerable number of graphical techniques have been developed to present multivariate data exactly in the plane: stars, glyphs, profiles, Andrews curves, Chernoff faces, etc. (e.g., du Toit *et al.* 1986). Of these methods, Wang (1978, pp. 123–141) finds that Andrews' curves usually fare best. As we shall demonstrate shortly, Andrews' curves are particularly suitable for factorial methods such as correspondence analysis. We have chosen them for the presentation of the position of profiles in cases when more than three dimensions should be taken into account.

Andrews (1972) proposed the use of trigonometric functions for graphical presentation of multivariate data points. He introduced the function

$$A(t) = z_1/\sqrt{2} + z_2 \sin t + z_3 \cos t + z_4 \sin 2t + z_5 \cos 2t + \dots \quad (10.3)$$

(a certain type of Fourier series) defined for the values of t on the range $-\pi < t < \pi$. In this way each observation vector $\mathbf{z} = (z_1, z_2, ..., z_k)$ is mapped into a function $A(t)$ of a single variable t by defining $A(t)$ as a linear

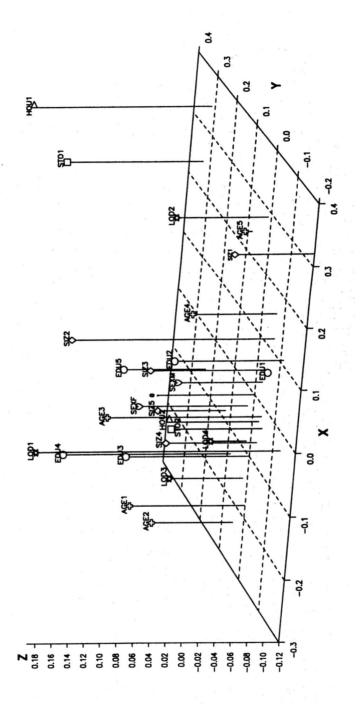

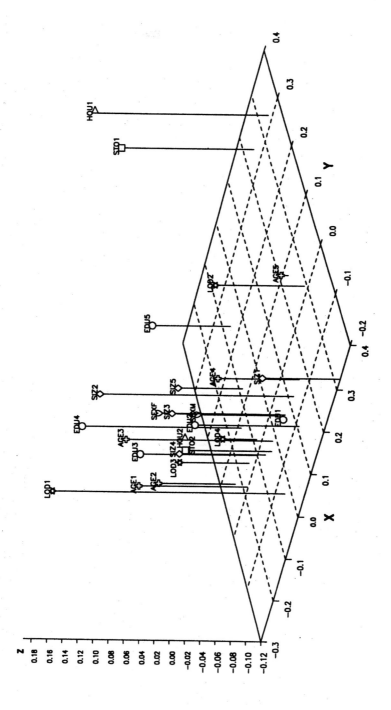

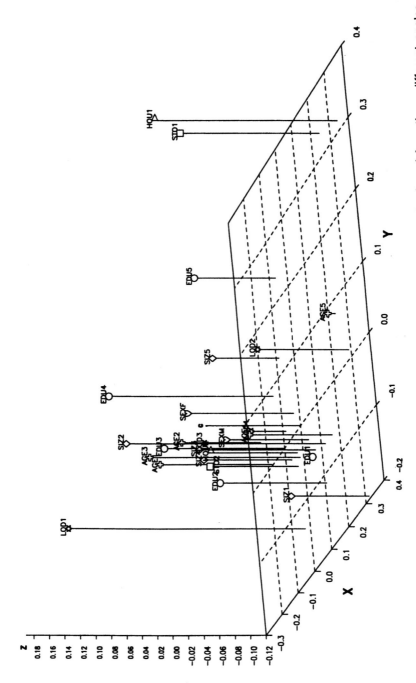

FIGURE 10.3   Three-dimensional plots of first three principal coordinates in Table 10.3, viewed from three different angles (20°, 45° and 70°).

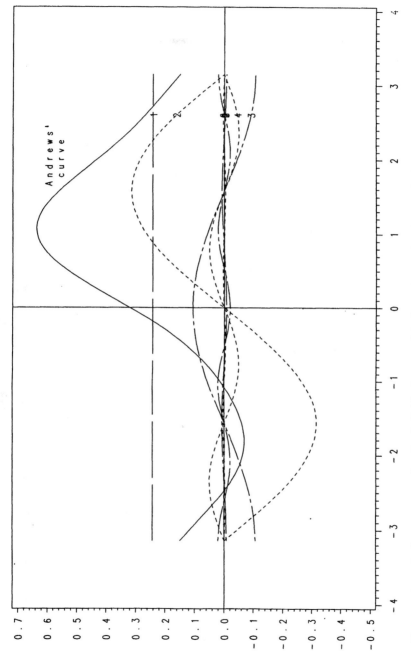

FIGURE 10.4 Andrews curves for the individual terms (labels 1–8) and for the whole function of the profile HOU1.

combination of orthonormal trigonometric functions $\sin t$, $\cos t$, $\sin 2t$, $\cos 2t$, etc. The coefficients of the function $A(t)$ are the observed values (the elements) of the vector z. The multivariate vector z is thus graphically represented by the curve in a two-dimensional plane.

For example, Andrews' function of the modified principal coordinates of category HOU1 (0.3442, 0.3153, 0.1066, 0.0503, -0.0077, 0.0043, -0.0207, 0.0009) (see Table 10.3) is

$$A_{\text{HOU1}}(t) = 0.3442/\sqrt{2} + 0.3153 \sin t + 0.1066 \cos t + 0.0503 \sin 2t$$
$$+ 0.0077 \cos 2t + 0.0043 \sin 3t - 0.0207 \cos 3t + 0.0009 \sin 4t$$

and is represented by the solid line in Figure 10.4. On the other hand, each of its eight terms is represented by the dashed lines labelled 1 to 8 . The solid line is the sum of all the dashed lines.

As mentioned before, we have selected a proportion of inertia of at least 90% as a criterion for a good approximation of a position of profile. The basis for this empirical criterion also arises from Andrews plot. If we plot a set of Andrews curves of the same profile, the first one representing only the first term of Andrews function, the second one representing the first two terms, the third one representing the first three terms, etc., we notice, that the shape of Andrews curve does not change substantially when adding new terms after achieving 90–95% proportion of inertia of the profile. For example, the shape of Andrews curves of the above mentioned type for the profile HOU1 presented in Figure 10.4 does not change substantially after including the first two or three terms, thus achieving 93.8% or 98.7% of the total inertia of the profile (Table 10.4 and Figure 10.5).

We now consider a set of Andrews curves $A_1(t)$, $A_2(t)$, ..., $A_n(t)$ for a set of K-variate observation vectors $z_1, z_2, ..., z_n$; these may be presented in the same plot and the observation vectors can be directly compared. When a similarity of a subset of curves in a plot can be noticed, this is an indication of the similarity of the appropriate observation vectors. If the number of cases is large (more than 10–15), however, it is difficult to trace the course of the individual curves and to compare their shapes. In this case the curves should be plotted separately in Andrews' plots and then compared with one another.

Andrews function $A(t)$ possesses many useful properties relevant for the study of multivariate data (Andrews 1972, p. 131). Two of them are most important:

- The function representation preserves means. If $\bar{z}$ is the mean vector (centroid) of a set of the observed vectors $z_1, z_2, ... z_n$, then the function of the mean vector $\bar{A}(t)$ is the pointwise mean of the functions $A_1(t), A_2(t), ..., A_n(t)$

$$\bar{A}(t) = \frac{1}{n} \sum_{i=1}^{n} A_i(t)$$

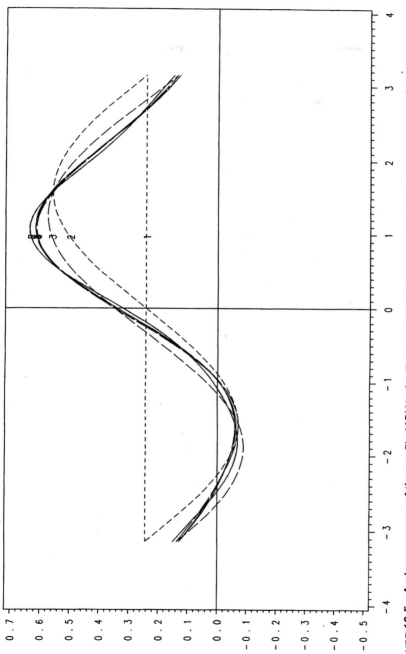

FIGURE 10.5   Andrews curves of the profile HOU1, the first one (label 1) representing the first term of the function, the second one (label 2) representing the first two terms, etc.

- The function representation preserves distances. It can be proved that the distance between two functions $A(t)$ and $B(t)$ (corresponding to observation vectors **a** and **b**), defined as

$$\| A(t) - B(t) \| = \sqrt{\int_{-\pi}^{\pi} [A(t) - B(t)]^2 \, \mathrm{d}t}$$

is proportional to the familiar Euclidean distance between **a** and **b**:

$$\| A(t) - B(t) \| = \sqrt{\pi \sum_{k=1}^{K} (a_k - b_k)^2}$$

The properties mentioned above are the basis for the conclusion that Andrews' curves are useful in clustering the observation vectors into homogeneous groups and for comparing observation vectors with the mean vector (centroid). According to the nature of Andrews' function, variables entering into the first terms are observed more easily than the last ones. For that reason Andrews recommends associating the most important variables with the first terms. In general, the importance of the variables is a subjective decision and it has a strong impact on the shape of the curves. Fortunately, in the case of visualizing correspondence analysis solutions, where the principal coordinates play the role of the variables, the sequence of variables is defined objectively. The first principal coordinate is chosen as the first variable, the second principal coordinate as the second variable and so forth.

The shape of Andrews' curves depends on:

- the number of coordinates of the observation vector;
- the absolute value of coordinates;
- the sign of coordinates.

For an accurate interpretation and comparison of the shapes of Andrews' curves it is essential to be aware of the role of every individual term in Andrews' function. The first term of the function (10.3) defines the position of the horizontal axis of the curve. The curve is mainly above the abscissa if the first coordinate is positive and below the abscissa if it is negative. The magnitude of the absolute value of the first coordinate defines the distance between the abscissa and the horizontal axis of the curve (see Figure 10.4, dashed line with label 1).

The even terms of Andrews' function of the form $z_{2j} \sin jt$ $(j = 1, 2, ...)$ define sinusoids, with the coordinate z_{2j} as the amplitude. The change of the sign of z_{2j} causes a reflection of the sinusoid across the abscissa. The sinusoid cuts the abscissa for every multiple of the number j.

The odd terms in Andrews' function of the form $z_{2j+1} \cos jt$ $(j = 1, 2, ...)$ also define sinusoids, but these sinusoids are shifted to the left by $\pi/(2j)$. This sinusoid cuts the abscissa at every odd multiple of $\pi/(2j)$.

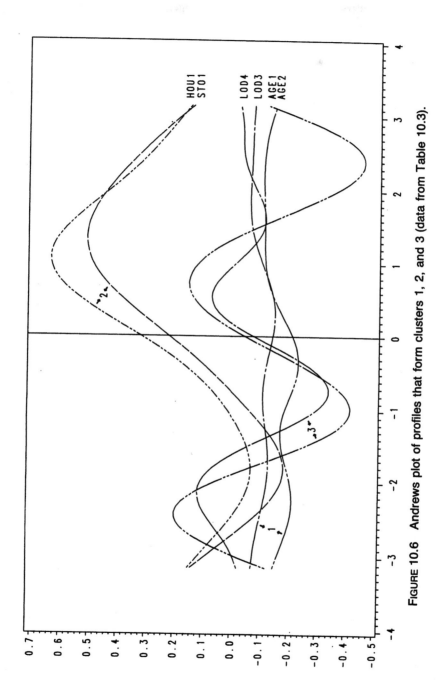

FIGURE 10.6 Andrews plot of profiles that form clusters 1, 2, and 3 (data from Table 10.3).

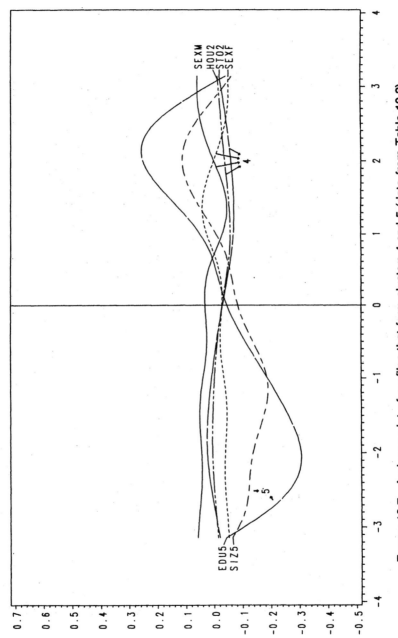

FIGURE 10.7 Andrews plot of profiles that form clusters 4 and 5 (data from Table 10.3).

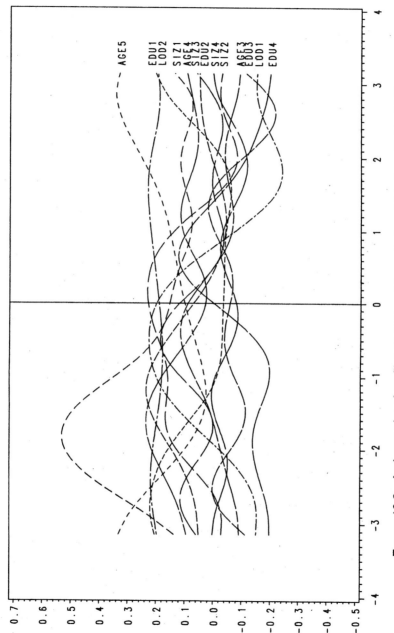

FIGURE 10.8 Andrews plot of profiles that do not form clusters (data from Table 10.3).

The shape of Andrews' curve is, of course, the result of the additive effect of all the terms. If Andrews' function consists of numerous terms the shape of the curve can become relatively complicated. If one of the vector coordinates strongly predominates, the shape of Andrews' curve is primarily influenced by the term with the dominating coordinate.

Let us return to our example, where we now use Andrews' curves to represent the eight-dimensioned coordinates in Table 10.3 exactly. If we represented all 25 Andrews' curves simultaneously, we would be unable to trace the shape of the individual curves and to compare their shapes. For that reason we needed to plot every single curve on a separate plot and then compare them to one another. Based on this exercise, we identified the following five clusters (see Figures 10.6 and 10.7):

(1) AGE2, LOD3
(2) STO1, HOU1
(3) AGE1, LOD4
(4) SEXM, SEXF, STO2, HOU2
(5) EDU5, SIZ5

The remaining curves which do not form clear clusters are shown in Figure 10.8.

Let us add some comments on the given set of clusters:

● The two curves of the variable 'sex' belong to the same cluster. Their shapes differ just slightly from the centroid line, the line that coincides with the abscissa. That means that there is little difference between the sexes with respect to the other six variables ('level of education', 'home ownership', etc.) under consideration. This is in accordance with the generally low inertias of the contingency tables of the variable 'sex' cross-tabulated with the other six variables.

● According to the two- and three-dimensional maps we would not conclude that the categories AGE1 and LOD4 form a cluster. Using Andrews' curves, however, we find these two categories together because both are explained well in the fourth dimension (see Table 10.4) and are on the same side (see Table 10.3).

10.6 CONCLUSION

Whenever one tries to visualize multivariate data, one is faced with the dilemma between simplicity and quality of representation. For visualizing correspondence analysis solutions, two- and three-dimensional maps and Andrews' curves have been proposed. On the one hand, scatterplots are preferred because of their simplicity, especially in the case when the approximation of the profiles in two or three dimensions is good. On the other hand, in many cases, we may prefer to use exact graphical methods such as Andrews' curves to improve the quality of a display.

Hence, the main advantage of Andrews' curves is their ability to give an exact representation of the points, based on their principal coordinates. The similarity of a subset of curves is an indication of the similarity of the appropriate principal coordinate vectors. The disadvantages are that the evaluation of the similarity of Andrews' curves is to some extent subjective, and that, for comparing a large number of curves, a high number of plots has to be drawn.

We use Andrews' curves simply as a graphical visualization of the principal coordinate vectors. Their shape does not have any substantive interpretation, they merely reflect the signs and the magnitudes of the principal coordinate vectors and cannot be directly interpreted in relation to the original data, i.e. specific profiles. As we have mentioned before, the similarity of the considered subset of curves clearly reveals the similarity of the corresponding profiles. On the other hand, when two or more curves have dissimilar shapes one cannot make any conclusions about the interrelations of the positions of their profile points in multidimensional space, except that they do not form a cluster in this space. In other words, on the basis of the dissimilar shapes of a subset of curves, one can conclude that the positions of the corresponding profile-points do differ substantially in a multidimensional space, but we cannot judge in what way they differ.

Despite the above mentioned deficiencies, we believe that in many cases, when more than three principal coordinates are needed for a good approximation of the position of the profile-points, an improved judgement about profile similarities can be made on the basis of Andrews' curves.

Part 3

Analysis of Longitudinal Data

Analyzing Event History Data by Cluster Analysis and Multiple Correspondence Analysis: An example using data about work and occupations of scientists and engineers

Bernd Martens

11.1 INTRODUCTION

11.1.1 *Methodological background*

In the last two decades social scientists have often come to discover the importance of adopting a dynamic perspective of social problems. Since they are interested in social processes, they increasingly collect data which report durations of episodes and changes in time, so-called 'event history data'. There are special procedures to analyze such dynamic information by survival analysis or regression models which are based on the concept of hazard rates (Tuma and Hannan 1984).

In this chapter, a different approach is outlined, using exploratory multi-variate methods. Here, the underlying concept is a special type of matrix, a *super indicator matrix.* These matrices can be used to analyze event history data which are regarded as *sequences of events* or as a panel survey with many waves. In the following, it will be briefly explained how such matrices are constructed, according to the work of van der Heijden (1987). Two possibilities of analyzing super indicator matrices seem to be natural: multiple correspondence analysis (MCA) or homogeneity analysis (Gifi 1990) and cluster analysis. Only a few examples for classifications of event histories can be found in the scientific literature (Martens 1991, p. 185 gives an overview), whereas the usage of multiple correspondence analysis seems to be favoured (van der Heijden 1987).

In comparison to traditional models coping with longitudinal data, van der Heijden (1987, p. 12) mentions two advantages of correspondence analysis: (1) a large number of variables and categories can be used, also if the number of objects is comparably small; (2) the interpretation of parameters is easier, while the estimation of parameters of alternative models is more cumbersome. Therefore, van der Heijden (1987, p. 13) writes that correspondence analysis (of longitudinal data) 'can supplement or even replace the existing methods'. In the case of event history data, the connection of super indicator coding and exploratory techniques seems to possess a third advantage: it is possible to detect some types of sequences of events which are based on the whole course of time. This is comparable to stochastic process models with many states that can be reached several times (multistate–multiepisode models). Such models need data sets that are based on events rather than on persons. In order to compute reliable parameters from these data based on events, several assumptions must be fulfilled: the present state depends only on the previous one; two successive episodes of one individual are independent; and the effects of the independent variables (covariates) do not vary with the number of the episodes (Blossfeld and Hamerle 1989, p. 432). Additionally, time ought to be measured on a continuous scale. Exploratory data analyses of super indicator matrices do not need such restrictions. Thus, they seem useful in a broad range of cases, especially if one is interested in some classification of sequences of events, and if the data are coded in discrete time and states domains.

In the following, attention will be restricted to cluster analysis and multiple correspondence analysis of super indicator matrices, using a data set of job histories as an example. A comparison of cluster analytic solutions with stochastic models can be found in Martens (1991, p. 171).

11.1.2   Theoretical and empirical basis

In modern societies a growing importance of research and development (R & D) can be seen for the economy at national and international level. For instance, this is the reason why the European Community has established large political programs in order to sponsor special technological fields and to strengthen European competitive power. At the personal level, in management science, and in sociology, a long continuing discussion exists about the *people* who actually *do* industrial R & D, about work and occupations of engineers and scientists in industry. This work as professionals within commercial organizations leads to some problems. The scientific literature about business management and organizational sociology

- reports some tension between technical experts and managers (Connolly 1983),
- even mentions a 'clash of cultures' between scientific orientated professionals and managers (Raelin 1986),
- sees the necessity for professionals who work in R & D eventually to switch to management, otherwise a hierarchical advancement would be very difficult or impossible (Badawy 1982, p. 50),
- recommends for this reason some special forms of separated organizational hierarchies ('dual ladders' for scientists and managers, Goldner and Ritti 1967).

Regarding the careers of professionals working in industry, relevant topics are the *department* where the people work and the *company's size* (Martens 1991, pp. 127–132). Generally R & D departments are important as the first working place of engineers and scientists in industry (Hack and Hack 1985 p. 478). However, there is some evidence that especially large companies are interested in a switch of R & D staff to other departments during the course of time. In Germany, like in other countries, the R & D activity also varies with the company's size in such a way that most R & D is undertaken by large companies (Martens 1991, p. 132). For these reasons the possibility of staying at R & D departments seems to be restricted, and it can be assumed that this perhaps causes some special career patterns of R & D staff between companies.

Till now only limited information about career patterns of engineers and scientists in industrial R & D exist (Gerpott *et al.* 1988), in contrast to the growing relevance of this work force. The scientific debate about engineers and scientists in industrial R & D and the lack of information about their actual careers constituted the underlying background of a social survey which was conducted in 1984 (Kossbiel *et al.* 1987; Martens 1991, pp. 54–60). A sample of 352 people, working in R & D units of German companies were asked about their occupational careers. According to assumed transitions during the careers, the resulting data set consists of event histories where the transitions between different departments, companies, branches of industries, and hierarchical positions are recorded yearly. Most of the occupational careers began in the 1960s and 1970s, and the mean of their durations is 15.5 years. There are in total 874 episodes (i.e. about 2.5 episodes per person), which last on average 6 years.

The research question is, whether there are some *typical patterns of careers*, that means some typical transitions between the variables in the course of time.

Unfortunately, the scope of the data set is confined. Because the survey was conducted via the R & D departments, it is biased. Data about management-orientated people are lacking and they should not be neglected. Due to its confinement, the survey will primarily be used as an illustration of the methodology that is applicable to event histories.

11.2 DATA

In order to answer the research question by means of cluster analysis and multiple correspondence analysis, it is necessary to transform the original event histories into a super indicator matrix. The data set is normally built of a variable number of events for each person (i) and some additional descriptive variables. In general, events are defined in empirical social research, as two vectors of variables (v_j, v_{j+1}): one for the beginning of an episode ($v_j^T = [t_j, s_{j1}, ..., s_{jm}]$) and one for the end ($v_{j+1}$). Here t_j denotes the time point, whereas $s_{j1}, ..., s_{jm}$ are variables describing the state of the object at time t_j. Additionally, it is necessary to determine how long each episode lasts ($d_j = t_{j+1} - t_j$). Figure 11.1 depicts the following transformation of event histories to super indicator matrices.

First, one has to choose a period of time for which the investigation is to be undertaken. In this example the period goes on for the first 10 years after the beginning of the career. According to some theoretical and empirical research about jobs and occupations of scientists and engineers in industry, this early stage of a career is a very important one (Martens 1991, p. 132; Rosenbaum 1984). Therefore, investigations have been restricted to this time frame.

Further, it is necessary to decide if the analyses are merely based on complete (or uncensored) data, by which one understands data that fully cover a certain time span. In this case, only persons with careers lasting for a minimum of 10 years remain in the data set. Otherwise, so called 'right-censored events' are also a subject of interest. If the data are right-censored, the period of investigation exceeds the actual duration of the career. Due to this, the data set contains a variable number of missing values. For reasons of comparison, uncensored data as well as right-censored ones were analyzed.

Censored and uncensored data must be altered in such a way that a variable number of episodes builds a *sequence* of events. Therefore, the period of investigation is divided into a number of points in time (T), with S different state variables at each time point (t), and each state variable possesses different categories, where the total number of categories is K. In doing so, the careers are represented by a multiway table having at least three dimensions. The resulting three-way table can be coded as slices (Z_t) of an indicator matrix, where $z_{ikt} = 1$, if person i is in category k at time t, but 0 otherwise. If there

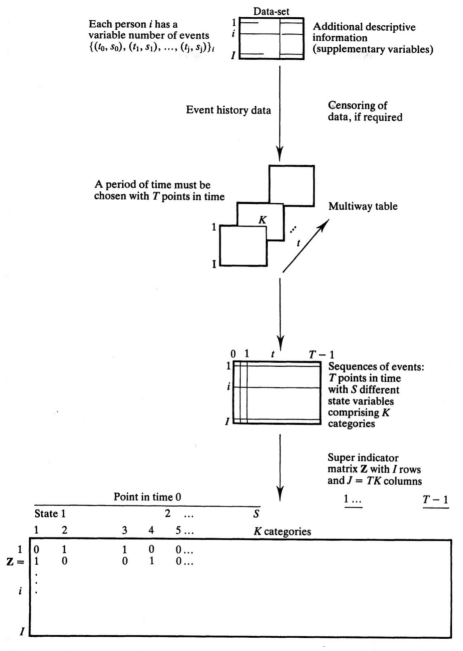

FIGURE 11.1 Scheme of the transformation of event history data to super indicator matrices.

is more than one state variable ($S > 1$) the resulting indicator matrix will be called a 'super indicator matrix', consisting of concatenated indicator matrices (van der Heijden and Meijerink 1989, p. 199).

As described by van der Heijden (1987, p. 168) the three-way matrix (z_{ikt}) can be transformed by concatenation into a two-way table in three different ways: (1) $z_{i(kt)}$: the matrix is concatenated over the rows. The points in time and the variables are coded interactively. (2) The second solution $z_{(ik)t}$ has T columns and K rows for each person (i). (3) The three-way table $z_{(it)k}$ is concatenated over the columns, while each person (i) is represented by T rows. The first one is called a *broad* matrix, whereas the third is named a *long* matrix (van der Heijden and de Leeuw 1989, p. 62, van der Heijden and Meijerink 1989, p. 197). The most interesting analysis is the first case of the matrix $z_{i(kt)}$. This one seems to be the only one which is useful in respect to dynamic analysis of (super) *indicator* matrices (van der Heijden 1987, p. 168) whereas the other concatenated matrices are valid, too, in the instance of multiway *frequency* tables. An example for the latter one is given in Table 11.4 and Figure 11.2. The coding of event histories by broad matrices is also equivalent to some graphically oriented efforts to analyze sequences of events (Schulz and Strohmeier 1985).

In the usual notation of multiple correspondence analysis (Greenacre 1984, p. 138), the super indicator matrix Z consists of Q separate indicator matrices with J_q categories. In the instance of sequences of events, for each state variable as well as for each point in time one indicator matrix is needed. Since the parameters are constants, this results in $Q = TS$. The matrix Z comprises $J = \Sigma J_q$ columns and I rows. Each row represents the career of a person i and it contains Q 1 s and ($J - Q$) 0 s. According to this, Z is equivalent to T concatenated indicator matrices, each one comprises K categories which leads to $J = TK$. Censored events must be coded by a special separate category for each state variable. By doing that, it is also possible to tailor super indicator matrices to problems where many state variables are involved. Nevertheless, if the number of columns is too large in comparison to I (that means $J \geqslant I$) the computations will become unstable (van der Heijden 1987, p. 237).

The analysis was restricted to the first 10 years of the careers. The interval between the points in time amounts to 1 year (or 2 years in the case of censored data), with the first time point denoted by $t_0 = 0$. The interval refers to the accuracy of the data. It is varied in order to assess its influence on the results. Due to special features of the data set (only few transitions), this influence is minor. There are two state variables: the department at which the scientists and engineers worked and the size of the company where he or she is employed. For each time point t, six categories exist (if censored data are analyzed, eight categories, respectively): R & D department, sales department, drawing office (censored); less than 100 employees, between 100 and 500, more than 500 employees (censored).

If one compares the distribution of the two state variables at different points in time $t = 0, ..., 10$ years (Table 11.1), the compilation will depict that, in general, only moderate changes occur between the categories. During the course of time, in the instance of censored data, the proportion between the categories alters slightly. The variations of the department variable are a bit more distinct. In the case of uncensored data the ratio of the categories without the censored one also remains roughly the same. However, the whole research endeavour aimed at the *whole* sequence of events, at its shape, and the type of transitions between departments and companies, at the level of each person i, rather than the marginal frequencies z_j. For this reason the matrix $\mathbf{Z}$ is used as the input to a cluster analysis in order to classify the sequences of events according to their *similarity*. In MCA, by contrast, each profile of a person i is compared to an average profile. Consequently, the MCA of $\mathbf{Z}$ shows how the individual profiles *depart from this average* (van der Heijden 1987, p. 168; van der Heijden and Meijerink 1989, p. 198).

TABLE 11.1

Percentages of the state variables at different time points. Censored categories occur at $t = 0$, because all cases are used.

Data set	Time	Size				Department			
		S	M	L	C	r	s	d	c
Un-	0	22	29	49		37	15	48	
cens.	2	19	30	50		43	14	43	
$n = 195$	4	24	26	50		43	14	43	
	6	26	30	44		46	13	41	
	8	24	31	45		47	13	40	
	10	22	34	44		47	12	41	
Cens.	0	25	26	48	1	47	15	34	4
$n = 352$	2	23	26	47	4	48	13	34	5
	4	24	23	43	10	44	13	31	12
	6	22	23	36	19	40	13	27	20
	8	19	22	33	26	36	12	25	25
	10	16	21	32	31	32	11	25	32

r = R & D department
s = business, sales department
d = the design department, drawing office
c = censored episodes (the sequence of events is briefer than 10 years)
S = the company's size is smaller than 100 employees
M = medium size, between 100 and 500 employees
L = the size is larger than 500 employees
C = censored data.

11.3 CLUSTER ANALYSIS AND MULTIPLE CORRESPONDENCE ANALYSIS

Cluster analysis or numerical taxonomy denotes a large variety of diverse methods which have different properties. In view of this diversity of different cluster analytic procedures, Gnanadesikan *et al.* (1977, p. 451) even recommend an algorithmic moratorium, because *sufficient* formal criteria for the goodness of cluster solutions are still lacking and ought to be developed. Seen from a *practical* point of view, the influence of the method chosen on the results can only be evaluated by using different algorithms and comparing the results with regard to criteria of their usefulness. In the instance of the data set described above, several methods were used (Martens 1991, pp. 133–138). Generally, the Ward algorithm, the flexible strategy, and the information gain statistics lead to comparable results (Wishart 1978), but the first and the second method show some disadvantages in comparison with the third one. Such drawbacks are equal-sized classes, general clusters of people who worked in companies of one size, independent of the department, or 'remaining' clusters which cannot be interpreted at all.

Thus, by considering the diverse cluster solutions and their evaluations in the light of their practical usefulness for classifying the data sets of sequences of events, attention was restricted to the classes produced by the information gain statistics. This method builds clusters which do not reveal the disadvantages mentioned above. The algorithm merges similar sequences into clusters. Additionally, the cluster solutions are relatively stable according to the criterion proposed by Rand (Hubert and Arabie 1985; Martens 1991, p. 135). Hence, the results presumably show some empirical structures of the data.

While cluster analysis algorithms compare and combine sequences of events according to their similarity, MCA *facilitates* the interpretation of such event history information in a much wider context of descriptive variables. Different approaches are possible:

- The super indicator matrix Z is analyzed and the objects (rows) are plotted. These dense clouds of points can be differentiated by some additional variables in order to picture clusters of points (Figure 11.3). (Madsen 1988 gives a large variety of interesting archaeological examples for this method.)
- The column categories of Z are shown in Figures 11.4 and 11.5. This graphical display is almost equivalent to that one of the Burt matrix $B = Z^T Z$, which can be regarded as a set of all transition tables. It represents the dynamic development of the categories between different points in time.

- Some additional information can be compiled with the sequences of events. Eight additional descriptive variables are used as supplementary variables with regard to **Z** in the following analyses.

11.4 RESULTS

11.4.1 Summary of the cluster analyses

The results can be summarized by two issues relating to the type of careers and their dynamic variations. The structure of the solution for uncensored data and that for censored data can be combined in five types of career patterns within the given time frame:

- scientific researchers, mainly working in large companies (R & D of the chemical industry), possessing university degrees (cluster 1 and 2 in Tables 11.2 and 11.3);
- people who are at the beginning of their occupational careers, indicated by the category of censored events (cluster 5 in Table 11.3);
- scientists and engineers who are engaged in R & D *and* in small companies, which they often managed as scientific or technical entrepreneurs (cluster 4 in Table 11.2 and class 7 in Table 11.3);
- employees, who are often the older ones, who work in production units or drawing offices, and do not possess academic degrees (clusters 3 and 6 in Table 11.2 and 3, 4, 6 in Table 11.3);
- people in non-technical units like sales departments (cluster 5 in Table 11.2 and cluster 8 in Table 11.3).

In principle, frequency tables of dynamic data (such as Table 11.1) can hide large fluctuations between categories, although the marginal frequencies of different years perhaps do not change at all. Cluster analysis avoids this

TABLE 11.2

Clusters of the super indicator matrix, two state variables, only uncensored data, $n = 195$.

Cluster	Percentage of cases	Description
1	24	R & D mostly in medium sized companies
2	14	R & D in large companies
3	24	Design in large and medium companies
4	8	R & D in small companies
5	10	Business units, sales departments in different companies
6	20	Primarily design in small but also large companies

TABLE 11.3

Clusters of the super indicator matrix Z, censored data, two state variables, n = 352.

Cluster	Percentage of cases	Description
1	21	R & D in large companies
2	13	R & D in medium companies
3	11	Design mostly in large companies
4	10	Design (and sales) in small companies
5	13	Beginners in R & D departments
6	17	Design in medium and large companies
7	6	R & D in small companies
8	9	Sales in large and medium sized companies

obstacle to a certain extent, because it merges sequences into clusters according to their similarity. Therefore, dynamic variations can be better estimated, also on the basis of frequency profiles, than in the case of ordinary contingency tables. Table 11.4 shows such aggregated output of a cluster analysis. The table contains frequencies of the two state variables differentiated with regard to clusters and points in time.

Generally, the dynamic variations are small. The scientists and engineers possess career patterns that are relatively stable within the first 10 years. There are no large fluctuations between companies of different types or between the diverse departments. In addition, further analyses of transition matrices also support that transitions chiefly occur between companies of the same size, changes between companies of different sizes are rare events (Martens 1991, pp. 171–172). This statement is also illustrated by simple correspondence analyses of Table 11.4. The input matrix consists of a set of six contingency tables: each cluster at the time points by companies' size and departments. According to the term used by van der Heijden (1987, p. 17), it is a *long* matrix. In Figure 11.2 there are only 4 out of 11 points in time (0–3–6–10) plotted because otherwise the plots would lose some clarity.

In Figure 11.2, the profiles of the six categories are concatenated in such a way, that their variations can be drawn in one picture. The plots of the profiles differentiated to clusters and time reveal that their spatial distributions do not disperse to a large extent. The dynamic alterations *within* and *between* the clusters are not very distinct (Martens 1991, p. 155).

The quality of the category points of the departments with respect to the first two principal axes is much better than the size categories (the concept of quality of display of points by means of principal axes is described by Greenacre 1984, p. 70). The sum of the squared correlations of the first group of categories with the first two axes is always greater than 0.9, whereas for the

TABLE 11.4
Occurrence of the two state variables differentiated with
regard to clusters and time, uncensored data ($n = 195$).

Cluster	Time	Size S	M	L	Department r	s	d
1	0	4	27	16	28	5	14
1	3	2	32	13	34	4	9
1	6	2	35	10	44	0	3
1	10	1	37	9	43	1	3
2	0	1	0	26	27	0	0
2	3	0	0	27	27	0	0
2	6	0	0	27	26	0	1
2	10	1	1	25	27	0	0
3	0	3	14	31	4	1	44
3	3	2	16	30	2	0	46
3	6	0	17	31	3	0	45
3	10	2	17	29	0	0	48
4	0	10	3	2	8	3	4
4	3	12	2	1	11	1	3
4	6	15	0	0	15	0	0
4	10	13	0	2	15	0	0
5	0	3	7	9	0	19	0
5	3	4	5	10	0	19	0
5	6	5	6	8	1	17	1
5	10	6	5	8	4	13	2
6	0	21	6	12	5	2	32
6	3	21	1	17	4	2	33
6	6	29	0	10	1	9	29
6	10	21	6	12	3	10	26

second group of variables the quality does not exceed 0.3. This also indicates a minor dynamic variation of the second type of profile.

11.4.2 Multiple correspondence analysis of sequences of events

Three multiple correspondence analyses were undertaken: (1) one of the *rows* of the super indicator matrix for *uncensored* data; (2) a second analysis of the same matrix but of its *columns*; (3) a third one which used the whole data set also including *censored* data. The computations of approximate percentages of inertia explained refer to the modification described in Chapter 7. According to the criterion of Benzécri (eigenvalues $> 1/Q$, Greenacre 1984, p. 145) eight principal axes, out of $J = 66$ or 48, respectively, are meaningful.

Case 1 – rows of the super indicator matrix, uncensored data. By means of plotting all rows of the super indicator matrix, within the coordinate system

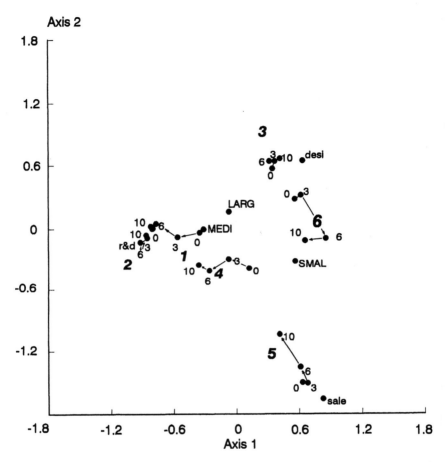

FIGURE 11.2 Correspondence analysis of profiles, the cluster analysis only takes uncensored data into account (n = 195). Eigenvalues: 0.368 (37%), 0.298 (30%). It is the symmetric plot, in which the abbreviations in capital letters denote the companies' size, whereas the small letters relate to the department. The rows of the input matrix are named by the point of time in years. The bold characters indicate the clusters. For ease of visualization only four points in time for each cluster are depicted (0, 3, 6, and 10 years after the beginning of the career).

of the principal axes, it is possible to examine the whole sample for some structures of the data. Figure 11.3 reveals for instance, the spatial distribution of the clusters, in the case of uncensored data (n = 195) and the first two principal axes. The distribution of the objects corresponds very well with the cluster division. People who possess similar careers, according to the cluster analyses, are also concentrated in separated areas of the graphical display of

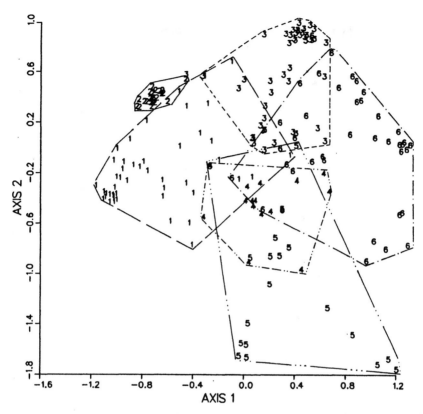

FIGURE 11.3   Plot of the rows of the super indicator matrix, MCA, uncensored data (*n* = 195), differentiated according the cluster solution. The numbers denote the clusters. Modified eigenvalue 1 = 0.184 (31.3%), modified eigenvalue 2 = 0.124 (21.2%).

the MCA. The R & D clusters 1 and 2 are located more on the left side of the first axis; the Design clusters 3 and 6 are more on the right side, whereas the Sales cluster 5 is situated at the negative part of the second axis, in contrast to all other classes. The variable company size can be used to distinguish between similar clusters (for instance between 1 and 2: R & D orientated careers in medium and large companies, respectively). In contrast to these clusters, the fourth one (R & D in small companies) does not show such a distinct pattern with regard to the first axis.

Furthermore, the plots show that the dispersion within the clusters varies. For example, cluster 2 – scientific researchers in large companies – has career patterns which are more similar to the patterns of other classes (for example cluster 1). In contrast to the assumption written above, these persons explicitly could remain in R & D.

Some persons of the sample have identical career patterns or coordinates. Therefore, the number of points that can be distinguished in a plot is smaller than the sizes of the clusters. This is especially true in the case of cluster 2 (24 identical points) and cluster 3 (27 identical points). As a remedy, a 'jittering' procedure is used (Chambers *et al.* 1983, p. 19): the points in Figure 11.3 are slightly perturbed in a random manner from their true positions, in order to avoid overlaying.

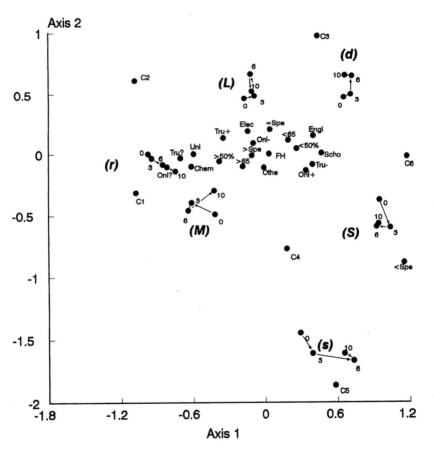

FIGURE 11.4 Plot of the columns of the super indicator matrix, MCA, only uncensored data (*n* = 195) and the supplementary variables. The active variables are denoted by a number counting the year after the beginning of the career. Only four time points are drawn. The bold letters indicate the categories as in Table 11.1.

The combined plots allow one to give a joint display of additional overall information about the structure and location of the clusters. However, for ease of describing patterns in the data, a combination of both techniques is recommended.

A meaningful spatial distribution also holds true for other additional descriptive variables, like differentiation according to the distribution of the departments where the engineers and scientists actually worked in 1984. In this case the categories R & D and Design are differentiated along the first axis (R & D at the left, Design at the right side), whereas the category Sales is distributed along the second axis. This distribution of points is in line with the depiction of the clusters in Figure 11.3. The correspondence between the sequences of events, which partly already began in the 1950s and 1960s, and the actual department at the moment of the survey (1984) can be seen as an additional hint that the dynamic variation of the state variables is confined.

Case 2 – columns of the super indicator matrix, uncensored data. By means of analyzing the columns of the matrix previously described, the picture can be enhanced by a set of eight descriptive variables which are used as *supplementary points* in Figure 11.4. (The abbreviations of the categories, which are used in Figure 11.4, are explained in Table 11.5.) These variables have a different status: a few of them have not shown any dynamic variations at all, within the first 10 years of the careers. Such variables are the educational

TABLE 11.5

List of supplementary variables used in the analyses and explanations of the abbreviations of Figures 11.4 and 11.5.

< Spe	= hierarchical position < specialist
= Spe	= hierarchical position = specialist
> Spe	= hierarchical position > specialist
Engi	= branch of the engineering industry
Chem	= chemical industry
Elec	= electric industry
Othe	= other branches of industry
Uni	= university degree
Scho	= school degree
FH	= college degree
Tru	= company is part of a trust (+ = yes, − = no, ? = missing value)
Onl	= company has only one plant (+ = yes, − = no, ? = missing value)
> 65	= job began after 1965
< 65	= job began before 1965
> 50%	= R & D activities amount to more than 50% of the daily working hours
< 50%	= R & D activities amount to less than 50% of the daily working hours
CC	= cluster of the censored data set (the number denotes the cluster)
C	= cluster of the data set which contains only uncensored information

degree (school, college, or university) and the year of the beginning of the career which is dichotomized (before and after 1965).

A second group of variables refers to the situation at the point of time when the survey took place. These characteristics comprise the branch of industry (engineering, chemical, electric, or other branch of industry), the organizational structure of the company where the respondent worked in 1984 (group of companies and subsidiaries), and the amount of R & D activities in relation to the daily working time. These variables are not available as event histories. The hierarchical position refers to the situation ten years after the beginning of the career. Additionally, the affiliation to the respective cluster is also indicated by a supplementary variable. The *active variables* of this analysis are the state variables Departments and Company Size differentiated to 11 time points. Consequently, the set of the stationary supplementary variables is referred to the principal axes of the event history data.

The dynamic variations between the categories are confined, as the drawings of the profiles by simple correspondence analysis has already shown. The category points are located in separated areas according to the first two principal axes of the MCA.

There are some similarities between the MCA plot of rows (as described above) and the one of columns, because the category points are − up to a scaling factor on each axis − also the centroids (centres of gravity) of the objects which possess the same categories (Gifi 1990, p. 119). Therefore, the axes in Figure 11.4 are differentiated by the categories R & D, Design, and Sales in such a way as already described. Additionally, the picture is mainly influenced by these categories: R & D, Design, Small, and Medium sized firms (r, d, S, M) are related to the first axis, whereas the categories Sales, Design, and Large firms (s, d, L) correlate with the second axis.

The clusters of sequences of events correspond to the state variables that both − active variables and cluster categories − are located at distinct positions in the plane.

Most of the other supplementary variables are distributed along the first axis, where the R & D clusters (1 and 2) and the Design clusters (3 and 6) are situated on opposite sides. This dichotomy can be used to describe some features of career in industrial R & D: on the one hand, there are researchers who are occupied in companies especially in the chemical industry (in Figure 11.4 the point is labelled by 'Chem'). These people mainly spend more than 50% of their daily working hours with R & D activities (>50%), they have academic degrees (Uni), and often began their careers after 1965 (>65). On the other hand, there are employees who often have no academic degrees (Scho), who are occupied in smaller companies of the engineering industry (Engi), which have also no ties to larger trusts (Tru-), and who are engaged in R & D to an extent less than 50% of their daily working hours (<50%). In particular, the difference between the amount of R & D activities describes the various careers. In contrast to these categories, some branch of industry and some educational degree (the points Elec, Othe, and FH) are only poorly represented by the plot.

It can be supposed that the divergences of careers correspond to some developments of the educational and industrial systems since the 1960s. There was a general tendency towards higher educational degrees, augmented efforts in industrial R & D, and increasing organizational structures in industry.

Case 3 – columns of the super indicator matrix, censored data. In comparison with the previous plot, Figure 11.5 of this analysis is only seemingly different. The main contrast occurs on the basis of censorship of data. Due to

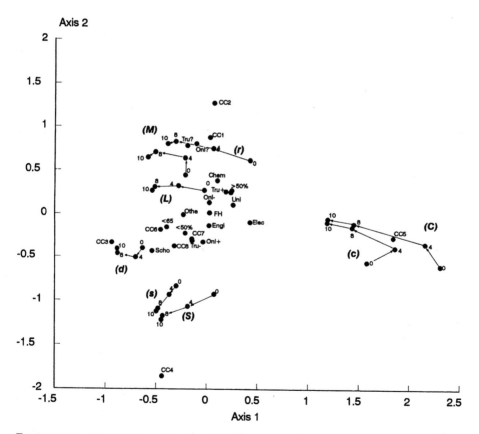

FIGURE 11.5 Plot of the columns of the super indicator matrix, MCA, censored data (*n* = 352) and the supplementary variables. Modified eigenvalue 1 = 0.257 (29.0%), modified eigenvalue 2 = 0.119 (13.4%). Further explanations are given in Figure 11.4.

this special coding, the points in Figure 11.5 are not situated as densely as in Figure 11.4, because of a greater amount of variability. The first axis of the plot is influenced by the censored categories (C, c) which correspond to cluster 5 – beginners in R & D departments. All other points are roughly distributed along the second axis. Here one can see a distribution of descriptive variables which has some similarities to that in Figure 11.4. There is also a differentiation between R & D and Design orientated careers which can be described in similar terms of personal and organizational features as in the case of uncensored data. If one takes additionally the third axis into consideration, the new axis will differentiate between Design and Sales as the second one does in Figure 11.4.

Due to geometric properties of MCA (Gifi 1990, p. 120), the increasing frequencies of the censored categories during the time frame of investigation cause a trend towards the centre of the coordinate system, whereas the points of the other categories (r, d, s, S, M, and L) move towards the periphery owing to diminishing frequencies. These characteristics restrict the interpretations of the censored data in terms of dynamic developments. However, in this case, censorship of data provides an additional perspective. Thus it is possible to depict the beginners who mainly use R & D departments as 'entrance' to companies. Additionally, the introduction of a perturbing category (the censored categories can be interpreted as a perturbation term) does not influence the results to a large extent. Analyses of censored and uncensored data lead to comparable consequences. Therefore the empirical patterns seem to be relatively stable.

In general, the main result (little changes during the careers) confirms findings of other studies, for example the investigation of Hutton and Lawrence (1981, p. 50) who stress a peculiar minor job mobility of German engineers.

11.5 CONCLUSIONS

The basis of this investigation is the concept of sequences of events, which can be regarded as super indicator matrices of the broad type. In general, super indicator matrices are a very flexible instrument to handle event history data, including censored events and many state variables. Cluster analysis algorithms allow one to combine these sequences in respect of their similarity. Correspondence analysis of the cluster profiles over time graphically reveals dynamic transitions between state variables in an easy way. MCA can be used to enhance and differentiate the resulting pictures of the sequences. This type of analysis of super indicator matrices can give an impression of structures in the data. Furthermore, it is the natural and obvious way to handle disaggregated longitudinal data (for example sequences of events), but, in the light of problems associated with the interpretation of MCA (Greenacre 1989; 1990a), it is necessary to evaluate the results in a multi-method approach in order to avoid artifacts.

In comparison to cluster analysis, MCA of sequences of events has the advantage that the results can be interpreted more quickly. It also allows a comprehensive outline of data more easily. In particular, if the sample is of medium or large size, determining the number of relevant clusters and their interpretations is a laborious process. In addition, the resulting clusters usually refer to aggregated data (tables), although the internal variations of the clusters are minimized owing to the method chosen. Since cluster analysis denotes a large variety of methods with different characteristics, careful evaluation of results is necessary, too. MCA depicts not only *disaggregated* information at the level of the objects, but also *aggregated* data, which seems to be of advantage. Furthermore MCA can also be used as a classification method, by differentiating objects according to their correlations with, and their positions on, principal axes.

These investigations aimed at the detection of career patterns of scientists and engineers in industrial R & D. The findings can be summarized as follows: on the one hand there are several types of careers; the R & D department as an 'entrance' to large companies, the contrast between R & D and design-orientated careers, the associations of these types with descriptive variables like the amount of R & D activities, the beginning of the job, or some organizational structures of companies. On the other hand, and in accordance with other studies, the scientists and engineers possess careers which are relatively stable within the given time frame. No large fluctuations between types of occupations appear in this workforce. On average, the number of transitions between different state variables is not large. In particular, changes between companies of different sizes or departments are rare events. Owing to this special feature of the data set, the career patterns are rather static in comparison to other event histories, which for example van der Heijden (1987) has used.

However, the depiction of event histories shows some important features of professionals' work and occupations in German industrial organizations. Exploratory methods provide a concise overview of various descriptive variables giving some insight into empirical patterns of careers.

The 'Significance' of Minor Changes in Panel Data: A Correspondence Analysis of the Division of Household Tasks

Victor Thiessen, Harald Rohlinger and Jörg Blasius

12.1 INTRODUCTION

This chapter discusses a simple yet detailed design for an analysis of panel data. Frequently, repeated measurements reveal only minor changes. Seldom are spectacular changes found, especially if only a short time period spans the given set of data. Our design might prove useful in such situations.

Social research methodology has adopted various forms of multivariate analysis appropriate for diachronic comparisons. Typically the central question in diachronic approaches is *how much* change has occurred, and with this emphasis, minor changes are usually ignored. In this chapter we focus on *how* a given, multifaceted domain changes over time. It is precisely the pattern of minor changes that is difficult to extract and hence is often neglected. Correspondence analysis is a promising technique to detect, summarize and visualize changes in trend or panel data, whether they be minor or major.

Our approach permits an analysis of changes over time, taking into account a number of indicators, with the findings integrated in one visualization. In doing this, no single attribute or characteristic is at the centre of interest; rather, a whole set of aspects of a phenomenon is viewed. Even if no single aspect has altered dramatically over time, the totality of minor changes, we argue, often provides an understanding of the nature of change in its individual manifestations. The proposed approach should be generally applicable to topics such as changes in opinions, orientations, attitudes, values and beliefs. For this reason, we present the general requirements of our approach in detail, in addition to a specific application. The topic we illustrate is change in the division of household tasks over the first few years of marriage; the tool we use is simple correspondence analysis (CA).

12.2 GENERAL DESIGN AND DATA STRUCTURE

Although CA makes no distinction between dependent and independent variables, the structure of our application becomes clearer if distinctions between three types of variables − row, column, and control[1] − are made. Figure 12.1 presents the formal structure of the data. Each section, diagrammed as a rectangular matrix, contains parallel information[2]. That is, the type of information contained in corresponding rows and columns for the different sections is identical. The rows of a given section are the attributes or aspects of a domain of interest. This domain of interest is analogous to a multiplicity of dependent variables. If the domain is voting behaviour, the rows might consist of a set of issues the researcher thought affected voting behaviour, on which respondents provided information. If the domain is the division of household labour, the rows might consist of a set of tasks, such as washing dishes, tidying, and minor household repairs.

As input for CA, the data of each section are concatenated, making sure to keep the column structure constant. In a study of voting behaviour, for example, responses on the importance of each of a number of issues might be cross-tabulated with occupation. Each column contains frequencies from one occupation; for example, the number of teachers who consider a given issue, such as foreign affairs, important. In CA the number of row variables as well as the number of time points is essentially unconstrained, while the number of column variables as well as the number of control variables is severely restricted by the sample size.

We use the first time point (section 1 in Figure 12.1) to find a primary CA solution. This solution, namely the geometric orientation, acts as a reference frame for additional time points. The second and subsequent time points are

---

[1] Control variables split the data into subsets. These subsets rest on important demographic or structural distinctions which influence the given row and column structure. If, for instance, attitude constellations differ by gender, where gender is used neither as column variable nor as row variable, it can be used as a control variable.
[2] Where a given row profile is unavailable, for any supplementary sections it can be omitted.

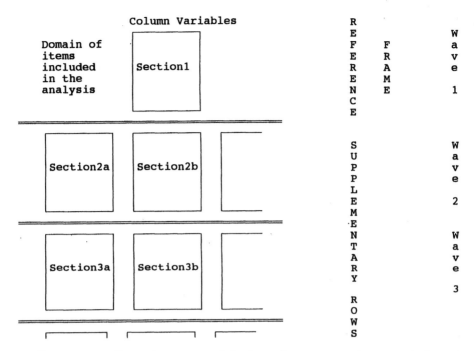

FIGURE 12.1 General structure of the input for correspondence analysis.

simply projected into this multidimensional space. They do not alter the defined space. This makes it possible to gauge the extent to which the configuration of points changes.

With the first time point as a reference frame, the possibility of examining the impact of control variables emerges at the second time point. As we apply the term, control variables are events that are experienced by some but not all cases between any two adjacent time points. For the political behaviour example, a control variable might be a televised debate which the researcher believes may alter voter orientation to the issues. However, not all voters would have watched the debate. Hence at the second time point, the sample would be split in two on the basis of whether the debate was seen. Likewise for housework, the birth of the first child is an event that is expected to alter the division of tasks. Nevertheless, it is an event which characterizes only some households within a given point in time. Hence at the second (and subsequent) time points the sample would be divided into those still childless, and those having children.

The defining features of our design can be seen clearly in an analogy with experimental design. In such a setting, the control variable is the treatment given. The data in the first time period precede the treatment and correspond to section 1 in Figure 12.1. At the second time period, subjects are assigned into control and experimental groups. The experimental group is given a stimulus whose effect is to be assessed at the second time point. In our CA approach, maturation effects would manifest themselves as changes in the configuration of points among the control subjects. The comparable changes among the experimental subjects comprise both maturation and treatment effects. In Figure 12.1, sections 2a and 2b might contain the responses of the control and experimental subjects, respectively.

12.3 THE USE OF SUPPLEMENTARY VARIABLES IN CORRESPONDENCE ANALYSIS

To show the significance in pattern of the sum of minor changes over time, we use a special feature in CA: the projection of supplementary information into the subspace of a prior solution. CA permits a distinction between variables which determine the geometric orientation of the vector subspace and variables used only for supplementary information. The latter variables have no effect on the geometric orientation of the axes; their profiles are treated as points with zero masses (see Greenacre 1984, p. 73).

In empirical applications supplementary variables are used in various ways. Some use them to provide additional information for the interpretation of the relationships being described (see van der Heijden 1987, van der Heijden and de Leeuw 1989, Blasius 1993). Van der Heijden and de Leeuw (1989), for example, used supplementary variables to add information about socio-economic status. This helped them to show that the conditions under which social security recipients notified officials of changes in their income depended on their living standard.

A few studies use supplementary variables to describe panel or trend data. Examples are given by Thiessen and Rohlinger (1988), and Scheuch and Blasius (1991) as well as in the application given in Chapter 13 in this book. Thiessen and Rohlinger (1988) analyzed the division of household labour in a three-wave panel. Their example forms the background for the empirical application in this chapter. Similarly Scheuch and Blasius (1991) used trend data to see whether the citizens of the countries of the European Community were growing more similar socio-psychologically.

12.4 THE DIVISION OF HOUSEHOLD LABOUR – AN EXAMPLE

For our application we use data collected by Eckart and Hahn[3] concerning married women's integration into the labour market. In the years 1977, 1978

---

[3] The authors wish to thank Roland Eckart and Alois Hahn for access to their data. The data set can be obtained from the Zentralarchiv für empirische Sozialforschung, study number 1562.

and 1980 young, dual-wage earning couples from five German cities were asked who was responsible for each of the following tasks: preparing breakfast (abbreviated by B), making dinner (D), preparing main meals (M), minor household repairs (R), shopping (S), laundry (L), taking care of insurance matters (I), automobile driving (A), washing dishes (W), taking care of financial matters (F), tidying the house (T), taking care of official matters (tax, etc.) (O), planning trips and vacations (V).

The study has two design features. First a dyadic unit of analysis was chosen to permit measurements of wife–husband agreement on a variety of topics, such as the division of household labour in the family. Second, a three-wave panel design was employed, permitting an evaluation of changes over time. Selection criteria required the couples to be newly-married, childless, and dual-wage earners at the first time point. Complete data for all three waves was obtained for 223 couples.

In the 1950s and 1960s only a minority of married women were in the paid labour market. This permitted functionalist theories of the division of labour, which assigned instrumental activities (external-to-the-household) to the husband, and expressive activities (internal-to-the-household) to the wife, to be taken seriously (see Parsons and Bales 1956). With the rapid rise in married women's employment in the paid labour market, a realignment of the division of labour should have occurred, but seems not to have materialized. The sampling design of the Eckart and Hahn study is particularly appropriate for assessing the new situation: young, urban, childless, dually-employed, and newly-married couples. If the division of household labour were to be renegotiated, these marriages should make an ideal 'laboratory' in which to observe and document the restructuring of housework. The three-wave panel provides a unique opportunity to observe changes in the division of labour over the first four years of marriages.

In our empirical application the 'big event' (the marriage) took place shortly before the first interview in the panel-study. From this event onwards one can expect ongoing change in the life of the young couples, especially with regard to the main focus of the primary investigators – the impact of this event on women's occupational careers (see Eckart et al. 1989). In principle, any time point could be used as an anchor, but there should be substantive reasons for the selection of the reference point. One could argue that other important events (for example, unemployment, new job, the death of a member of the family), might be a more judicious starting point.

Turning to our application, information on the 13 common household tasks was obtained separately and independently from both husbands and wives. To simplify the application we restricted our analysis to those instances where the spouses gave identical responses to the division of household tasks. Hence, four response categories remain indicating whether a given task was performed primarily by the husbands (HH), the wives (WW), jointly (JJ) or alternately

TABLE 12.1
Input data.

Reference variables	WW	JJ	AA	HH
B1	82	7	36	15
D1	77	13	11	7
M1	124	4	20	5
R1	0	2	3	160
S1	33	55	23	9
L1	156	4	14	2
I1	8	77	1	53
A1	10	3	51	75
W1	32	53	24	4
F1	13	66	13	21
T1	53	57	11	1
O1	12	15	46	23
V1	0	153	1	6

Supplementary variables	WW	JJ	AA	HH	Supplementary variables	WW	JJ	AA	HH
B2	60	1	15	15	BH2	24	0	7	3
D2	57	6	17	4	DH2	22	1	4	0
M2	80	3	11	8	MH2	31	0	5	0
R2	0	0	5	108	RH2	3	0	1	29
S2	27	27	21	9	SH2	16	10	2	2
L2	101	2	6	2	LH2	40	0	2	1
I2	5	39	0	36	IH2	4	13	1	17
A2	6	6	23	46	AH2	2	1	8	16
W2	19	32	16	5	WH2	17	7	5	0
F2	15	26	5	28	FH2	4	5	1	6
T2	40	30	5	2	TH2	26	3	5	0
O2	12	5	21	28	OH2	4	2	8	8
V2	0	109	2	2	VH2	0	38	3	0
B3	55	3	14	20	BH3	41	0	14	4
D3	55	8	16	3	DH3	43	5	10	0
M3	73	2	8	6	MH3	59	1	5	1
R3	0	0	3	100	RH3	1	0	3	54
S3	28	26	16	7	SH3	25	7	16	2
L3	105	3	6	2	LH3	78	0	2	0
I3	6	31	0	44	IH3	4	19	0	29
A3	2	12	25	31	AH3	3	4	23	25
W3	28	22	11	1	WH3	25	10	3	0
F3	16	19	8	31	FH3	13	10	3	16
T3	29	29	9	1	TH3	50	1	3	0
O3	10	5	26	30	OH3	8	1	16	23
V3	1	100	0	5	VH3	0	70	0	0

(AA) (double notation is used in the labels to emphasize the dyadic response). Restricting ourselves to these four combinations reduced the number of cases remaining in the analysis. The loss of information ranges from 21% (laundry) to 57% (taking care of official matters). This deletion does not affect the nature of 'movements' to be described in this chapter (see also Thiessen and Rohlinger 1988).

Table 12.1 gives the distribution for the division of household labour as reported by the couples for the set of 13 tasks. As mentioned previously, the 1977 data are used to determine the geometric orientation of the vector subspace (see section 1 in Figure 12.1). The supplementary variables are subdivided into four sections (see sections 2a, 2b, 3a, 3b in Figure 12.1) – on the basis of the two additional time-points (1978, 1980) and on the basis of the control variable – the wife's employment status (at home versus in the paid labour market). Therefore the full input information contains a concatenated matrix with four columns and 65 rows; the last 52 rows are treated as supplementary variables.

12.5 FINDINGS AND RESULTS

Table 12.2 shows the chi-square components. The sum of these components is the chi-square value for Table 12.1. The total inertia of 1.11 (the chi-square divided by the total frequency) indicates high variation in the data. The

TABLE 12.2
Chi-square-components.

CELLIN	WW	JJ	AA	HH
B1	23.77	28.06	11.95	7.94
D1	42.73	10.88	1.42	11.67
M1	96.75	37.01	0.23	24.17
R1	56.77	44.24	18.41	426.24
S1	1.66	11.39	1.75	11.31
L1	150.46	43.68	5.28	34.55
I1	33.16	32.72	18.29	16.87
A1	29.91	34.79	46.73	65.60
W1	1.22	12.15	3.46	17.33
F1	17.22	33.06	0.73	0.55
T1	2.90	12.85	2.58	24.69
O1	13.39	6.05	73.32	0.20
V1	55.05	241.99	21.35	23.98

following cells contribute most to the total variation: HH and 'minor household repairs', WW and 'laundry' and JJ and 'planning trips and vacations'.

Separate calculations of total inertia for each of the five sections (see Table 12.1) yield only slight differences. Where wives remain in the paid labour market, only a small decrease marks the scores for total inertia (from 1.11 through 1.03 to 1.02, corresponding to time points one to three, respectively). In households where wives no longer have paid employment, at the second time point the total inertia is 1.10. At time point three, total inertia increased to 1.31. In short, only between time points two and three and only for single-earner households does the total inertia increase somewhat. This implies that, considered over all household tasks, a specialization arises – perhaps in the direction of his or her jobs; perhaps in the direction of jointly, perhaps in the direction of alternately, or perhaps in some combination of these possibilities. For the other sections total inertia is essentially constant which leads to the conclusion that either there are no 'movements' or specializations to balance generalizations; for example, an increasing concentration to 'her' workloads offsets a decreasing concentration to 'his' tasks. In general, the magnitude of the total inertias in each section indicates that there is sufficient variation to warrant the type of analysis we performed (see Chapter 2 for more details on the interpretation of total inertia).

Figure 12.2 presents the 65 row variables and the four column variables. With only four columns, 100% of the variance is explained in three dimensions. The first axis explains 48.7% of the total variance and the second another 39.9%, while the third axis explains only 11.4%. Clearly, the obtained solution has two main axes.

The two-dimensional symmetric map is useful for a first inspection of the results. Starting with the four column-variable projections on the first axis, the contrast is dominated by two categories: agreement that the husband is responsible for certain tasks (HH), and agreement that the wife is responsible for others (WW). This suggests that the first axis can be described as a sex-role ascription of tasks. The second axis is associated with the column variable 'joint responsibility for a task' (JJ); 'alternating responsibility' (AA), which is close to the origin, can be used to label the third axis.

When using symmetric maps, distances between rows and columns should not be interpreted (see Greenacre 1984, 1989, 1993, Greenacre and Hastie 1987, Chapters 1 to 3 in this book). But it is permissible to compare the cosines of the angles between the variables and the principal axes (see Blasius and Rohlinger 1989, Chapter 2 in this book). Comparing the angles of the row variables and the column variables with the first axis reveals that 'minor household repairs' (R) is stereotypically the husband's task – both 'HH' and this task load on the same side of this axis. On the opposite side, 'laundry' (L) and 'preparing meals' (M) are 'her' chores. Turning to the second axis, 'planning trips and vacations' (V) seems to be an activity the spouses engage in jointly

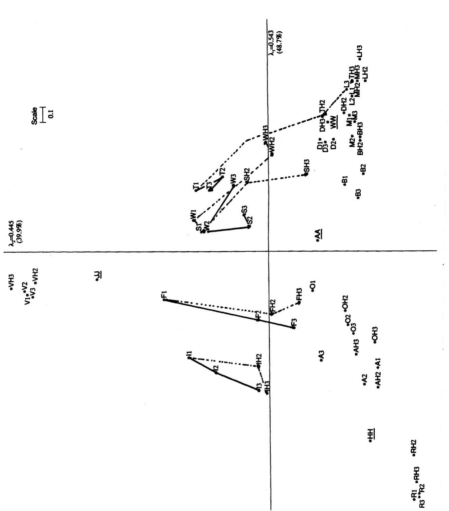

FIGURE 12.2 Graphical solution.

(the numbers in the labels indicate the panel waves and an 'H' signifies that the wife becomes a housewife).

The positions of the supplementary variables alter slightly with time and wife's employment status. To highlight the changes over time, arrows have been drawn into the graphical presentation of the vector subspace in Figure 12.2. The arrows point towards the final time point, i.e. 1980. The dotted lines show the changes in the division of labour for those couples that became single-wage earning families. Over time joint activities decrease (see the movements along the second axis away from 'JJ') while the sex-role ascription of the workload increases (see the outward movements along the first axis). This process is particularly noticeable when the wife's employment in the labour force is interrupted (the 'H'-labels are closer to the margins). The outside-the-home tasks such as taking care of financial, insurance and official matters, tend to become more and more 'his' tasks while the more time consuming daily routines such as shopping and tidying the house turn into 'her' chores over time.

The difference between single- and dual-wage earning households is more pronounced on 'her' side, with 'tidying the house', 'shopping' and 'washing dishes' moving from 'JJ' towards 'her' area. 'His' side also shows a lessening importance of a shared workload, but the wife's employment status has virtually no impact on 'his' tasks (all time points are close to one another). Only one small anticyclic tendency can be seen: automobile-driving. Note that movements are nothing more than differences in the profiles and therefore are to be understood as increased or decreased proportions in these profiles.

Table 12.3 provides the numerical parameters of CA. The squared correlations (see QCOR1, QCOR2) are equal to the (squared) cosines of the angles between the row (column) variables and the axes in the three-dimensional plane (see Chapter 3). With reference to the column variables, WW (QCOR1 = 0.802) and HH (QCOR1 = 0.772) have the highest squared correlations with the first axis. The sign of the LOC1 parameter shows the contrast between both with respect to the 13 tasks. The columns JJ and AA dominate the second and the third axis, respectively. Since the column variable AA loads only on the third axis (not shown in Table 12.3), its location is not properly visualized in a plot of the first two axes.

In the joint vector subspace the tasks 'making dinner' (QCOR1 = 0.777), 'laundry' (QCOR1 = 0.740) and 'preparing meals' (QCOR1 = 0.742) as well as the column variable WW (QCOR1 = 0.802) correlate highly with the positive part of the first axis. This implies that the spouses agree these are 'her' tasks. On the other hand only 'minor household repairs' (QCOR1 = 0.707) and the column variable HH (QCOR1 = 0.772) have high correlations with the negative part of the first axis.

'Planning trips and vacations' (QCOR2 = 0.962), 'taking care of financial matters' (QCOR2 = 0.837), 'shopping' (QCOR2 = 0.748) and to a lesser

TABLE 12.3
General statistics.

GENSTATS	MASS	SQCOR	INR	LOC1	QCOR1	INR1	LOC2	QCOR2	INR2
WW	0.344	0.954	0.270	0.838	0.802	0.445	− 0.365	0.152	0.103
JJ	0.292	0.998	0.282	− 0.149	0.021	0.012	1.027	0.977	0.691
AA	0.146	0.110	0.106	0.062	0.005	0.001	− 0.292	0.105	0.028
HH	0.218	0.980	0.342	− 1.161	0.772	0.542	− 0.602	0.208	0.178
B1	0.080	0.905	0.037	0.509	0.505	0.038	− 0.453	0.400	0.037
D1	0.062	0.930	0.034	0.693	0.777	0.055	− 0.308	0.154	0.013
M1	0.088	0.974	0.081	0.876	0.742	0.124	− 0.490	0.232	0.047
R1	0.095	0.933	0.281	−1.529	0.707	0.407	− 0.864	0.226	0.159
S1	0.069	0.811	0.013	0.118	0.064	0.002	0.403	0.748	0.025
L1	0.101	0.925	0.120	0.992	0.740	0.183	− 0.495	0.185	0.056
I1	0.080	0.885	0.052	− 0.647	0.576	0.061	0.474	0.309	0.040
A1	0.080	0.767	0.091	− 0.742	0.432	0.081	− 0.653	0.335	0.076
W1	0.065	0.764	0.018	0.189	0.118	0.004	0.442	0.646	0.028
F1	0.065	0.997	0.027	− 0.271	0.161	0.009	0.618	0.837	0.056
T1	0.070	0.975	0.022	0.394	0.440	0.020	0.434	0.535	0.030
O1	0.055	0.119	0.048	− 0.227	0.053	0.005	− 0.254	0.066	0.008
V1	0.092	0.992	0.176	− 0.252	0.030	0.011	1.435	0.962	0.425
B2	0.052	0.992	0.027	0.502	0.437	0.024	− 0.565	0.555	0.037
D2	0.048	0.993	0.028	0.699	0.755	0.043	− 0.393	0.239	0.017
M2	0.058	0.942	0.047	0.771	0.662	0.064	− 0.502	0.280	0.033
R2	0.065	0.950	0.186	− 1.502	0.706	0.269	− 0.882	0.243	0.113
S2	0.048	0.264	0.006	0.153	0.171	0.002	0.113	0.093	0.001
L2	0.064	0.892	0.082	1.007	0.709	0.119	− 0.510	0.182	0.037
I2	0.046	0.849	0.031	− 0.737	0.721	0.046	0.310	0.128	0.010
A2	0.046	0.901	0.044	− 0.802	0.604	0.055	− 0.563	0.298	0.033
W2	0.041	0.659	0.009	0.119	0.059	0.001	0.380	0.600	0.013
F2	0.042	0.827	0.009	− 0.431	0.812	0.015	0.059	0.015	0.000
T2	0.044	0.880	0.013	0.476	0.674	0.018	0.263	0.206	0.007
O2	0.038	0.714	0.020	− 0.447	0.335	0.014	− 0.474	0.378	0.019
V2	0.065	0.996	0.127	− 0.222	0.022	0.006	1.461	0.974	0.311
B3	0.053	0.980	0.020	0.343	0.282	0.011	− 0.540	0.697	0.035
D3	0.047	0.995	0.026	0.702	0.810	0.043	− 0.335	0.185	0.012
M3	0.051	0.922	0.047	0.829	0.665	0.065	− 0.515	0.256	0.030
R3	0.059	0.940	0.176	−1.527	0.702	0.254	− 0.889	0.238	0.105
S3	0.044	0.639	0.004	0.219	0.440	0.004	0.148	0.200	0.002
L3	0.067	0.888	0.084	1.001	0.714	0.123	− 0.494	0.174	0.036
I3	0.046	0.834	0.036	− 0.849	0.830	0.062	0.058	0.004	0.000
A3	0.040	0.620	0.032	− 0.670	0.512	0.033	− 0.308	0.108	0.009
W3	0.036	0.946	0.008	0.431	0.769	0.012	0.207	0.177	0.003
F3	0.042	0.942	0.009	− 0.457	0.852	0.016	− 0.149	0.090	0.002
T3	0.039	0.999	0.010	0.386	0.547	0.011	0.352	0.453	0.011
O3	0.041	0.614	0.030	− 0.489	0.294	0.018	− 0.510	0.320	0.024
V3	0.061	0.989	0.112	− 0.255	0.032	0.007	1.404	0.957	0.269

TABLE 12.3
(continued).

GENSTATS	MASS	SQCOR	INR	LOC1	QCOR1	INR1	LOC2	QCOR2	INR2
BH2	0.019	0.997	0.014	0.681	0.598	0.017	−0.556	0.399	0.014
DH2	0.015	0.989	0.015	0.931	0.799	0.025	−0.454	0.190	0.007
MH2	0.021	0.982	0.024	0.991	0.762	0.037	−0.532	0.220	0.013
RH2	0.019	0.923	0.044	−1.279	0.637	0.057	−0.856	0.286	0.031
SH2	0.017	0.814	0.004	0.439	0.747	0.006	0.132	0.067	0.001
LH2	0.025	0.883	0.034	1.025	0.685	0.048	−0.551	0.198	0.017
IH2	0.020	0.847	0.011	−0.708	0.841	0.019	0.058	0.006	0.000
AH2	0.015	0.903	0.017	−0.832	0.562	0.020	−0.648	0.341	0.015
WH2	0.017	0.994	0.006	0.632	0.992	0.012	−0.025	0.002	0.000
FH2	0.009	0.715	0.002	−0.365	0.712	0.002	−0.022	0.003	0.000
TH2	0.019	0.991	0.015	0.864	0.853	0.027	−0.347	0.138	0.005
OH2	0.013	0.510	0.007	−0.354	0.197	0.003	−0.447	0.313	0.006
VH2	0.024	0.999	0.042	−0.182	0.017	0.001	1.394	0.982	0.103
BH3	0.034	0.976	0.025	0.703	0.610	0.031	−0.545	0.367	0.023
DH3	0.033	1.000	0.025	0.840	0.853	0.043	−0.349	0.147	0.009
MH3	0.038	0.920	0.046	0.996	0.727	0.069	−0.513	0.193	0.022
RH3	0.033	0.952	0.089	−1.443	0.697	0.128	−0.872	0.255	0.057
SH3	0.029	0.613	0.013	0.504	0.504	0.013	−0.234	0.109	0.004
LH3	0.046	0.866	0.073	1.111	0.698	0.104	−0.545	0.168	0.031
IH3	0.030	0.834	0.024	−0.865	0.834	0.041	0.017	0.000	0.000
AH3	0.032	0.565	0.033	−0.634	0.342	0.023	−0.511	0.222	0.018
WH3	0.022	0.914	0.011	0.701	0.914	0.020	0.010	0.000	0.000
FH3	0.024	0.675	0.004	−0.291	0.491	0.004	−0.178	0.184	0.002
TH3	0.031	0.900	0.042	1.054	0.733	0.063	−0.503	0.167	0.018
OH3	0.028	0.781	0.022	−0.542	0.327	0.015	−0.637	0.453	0.025
VH3	0.040	0.993	0.087	−0.203	0.017	0.003	1.539	0.976	0.214

degree 'washing dishes' (QCOR2 = 0.646) and 'tidying the house' (QCOR2 = 0.535) are tasks at time point one for which the couples declare joint responsibility. These five tasks as well as the column variable 'JJ' correlate positively with the second axis. Only one task is connected with the third axis along with the column-variable 'AA': 'taking care of official matters (tax, etc.)' (QCOR3 = 0.881, not reported in Table 12.3). This is understandable if we take the single status of the spouses prior to their marriage into account, in this context 'alternating' means each partner looked after this outside-the-home task on their own.

In the next step the row and column contributions of the inertias of each axis as well as the total inertia of the model are interpreted. The column contributions of total inertia (INR) reveal the extent to which the geometric orientation of the total model is determined by each column variable. The geometric orientation of the axes is influenced highly by the column variable

'HH' (INR = 0.342) and by the row variable 'minor household repairs' (INR = 0.281); to a lesser degree by the row variables 'planning trips and vacations' (INR = 0.176) and 'laundry' (INR = 0.120). Among the columns 'AA' (INR = 0.106) is least important for determining the geometric orientation of the axes.

The parameters INR1 and INR2 give the contributions of the inertia to the first two axes, respectively. The inertia contributions of each variable on each axis is calculated by multiplying the squared distances of the vector endpoint projections to the origin (see LOC1 and LOC2) with their relative masses (see MASS), divided by the inertia of each axis (the eigenvalue). The geometric orientation of the first axis is determined by the column variables WW (INR1 = 0.445) and HH (INR1 = 0.542). Among the row variables, the tasks 'minor household repairs' (INR1 = 0.407), 'laundry' (INR1 = 0.183), and 'preparing meals' (INR1 = 0.124) have a high influence; these tasks clearly distinguish the gendered nature of housework at the first time point.

The complete set of coefficients is also available for supplementary variables. The algorithms for calculating the coefficients describing the supplementary variables are identical to those for the variables which determine the geometric orientation of the vector subspace. The squared correlations can be interpreted in the same manner as the reference variables; the inertia contributions are, however, purely hypothetical.

We can now focus on how to use the split in the control variable to interpret task changes. It seems that over time sex roles become a more important basis for the ascription of tasks to 'him' and 'her'. The number of jointly shared tasks generally decreases. The numerical output confirms that 'JJ' loads on the positive part of the second axis (LOC2 = 1.027, QCOR2 = 0.977). At the first time point couples jointly accomplished the tasks of 'shopping', 'washing dishes', 'taking care of financial matters', 'tidying the house', and 'planning trips and vacations' (all squared correlations are higher than 0.500)[4]. At the second time point where the wives remain employed, only 'washing dishes' and 'planning trips and vacations' were done jointly. At the third time point 'planning trips and vacations' remains the sole joint activity. Should she become a housewife only 'planning trips and vacations' remains a joint task. From the other prior joint tasks he takes responsibility for financial matters, especially if she remains in the paid labour market (comparing the QCOR1 coefficients from F2 and F3 versus FH2 and FH3, note the higher correlations with the negative part of the first axis).

Because the control variable has a greater impact on 'her' tasks, we should look at these changes in greater detail. If the wife remains employed there are only minor changes. In the second time point she gets the task 'tidying the

---

[4] As shown in Chapter 2 any value such as 0.500 is intrinsically arbitrary. The choice of the critical value is similar to that in principal components analysis.

house' (see the increased correlation coefficient). Two years later this task is still hers but not as clearly as for the second time point whereas 'washing dishes' is now also a part of 'her' workload. Should she become a housewife she receives the additional chores of 'shopping', 'washing dishes' and 'tidying the house'. If she remains employed at the second time point, the task 'shopping' changes first from 'jointly' to 'alternately' (not reported in Table 12.3, but see SQCOR = 0.264), whereas at the third time point it becomes more 'her' job (QCOR1 = 0.440).

Little change over time marks the task 'preparing breakfast', 'making dinner', 'preparing meals', 'laundry', 'minor household repairs', 'automobile driving', 'taking care of insurance matters', and 'planning trips and vacations'. The first four remain her tasks; the next three his; 'planning trips and vacations' remains a joint task.

12.6 CONCLUSION

Our design for inputting data in CA permits complex explorations of repeatedly measured domains. Once the data in the reference frame is interpreted, an inspection of the altered positions of the various tasks can show 'movements' over time points. Since each wave is projected into the reference frame, these movements can be interpreted substantively. Although movements of single variables are seldom dramatic, meaningful patterns sometimes emerge when they are viewed in the context of the other changes occurring in relation to the dimensional structure. Some items in the domain of interest simply remain constant over time, while others show fanshaped and/or radial movements.

For our empirical example, the nature of changes suggest that the family is a rather conservative institution in society. It preserves the gendered division of labour in a family household – in fact, it seems to intensify over time the traditional ascription of sex roles. While the description of the movements is unproblematic, its interpretation is more ambiguous. It remains an open question whether the sum of minor changes found is due to pressures placed on families by external societal norms, or simply to the lure of a more effective organization with the passage of time and the drudgery of repetitive chores. Our description of changes is consistent with both interpretations. Were we to attribute the changes to internal family organizational dynamics, it would nevertheless be remarkable that there is no general exchange of tasks over time between spouses: tasks that were traditionally hers remain hers, tasks that were traditionally his remain his. Efficiency implies not only that chores are less often done jointly; it would lead to a reassignment of the repetitive chores, such as washing dishes and tidying.

Another important aspect of this chapter lies in the general applicability of the design. We introduced a scheme for the organization of data that allows for simultaneous comparison of the structure of row and column profiles. With sufficient sample size, control variables can be introduced through subdivisions of the data within a given wave. Parallel sets of supplementary rows can be created for each value of the control event. Contrasting the movements over time for different values of the control variables has proved to be an important and powerful instrument in the toolbox of exploratory data analysis. CA can be used to visualize a sum of minor changes over time.

13

The Visualization of Structural Change by Means of Correspondence Analysis

Thomas Müller-Schneider

13.1 INTRODUCTION

Structural change is a very important subject of social science research. Social change can be defined as '... the significant alteration of social structures (that is, of social action and interaction), including consequences and manifestations of such structures embodied in norms (rules of conduct), values and cultural products and symbols' (Moore 1968, p. 366). This paper is concerned with the modelling and visualization of structural change by means of CA.

Normally CA is used to analyze a data matrix which contains absolute frequencies. However, there are applications which suggest the use of relative frequencies as input data. For example, this kind of application is convenient in order to compare different studies with different numbers of cases as well as different wording of questions. In this chapter relative frequencies will be used as input data, because they can be interpreted as group-specific indicators of social structures and their changes.

I will initially outline a novel approach towards the modelling and measurement of structural change (section 13.2). I refer to this approach as 'historic profiles' (see also Müller-Schneider 1994). A historic profile is a row of relative frequencies of a particular characteristic as it occurs within certain groups. These relative frequencies are then expressed as percentages of their row totals. The alteration of these percentages can be taken as a model of structural change. The measurement of structural change by historic profiles will be illustrated by survey data from 1953/54 up to 1987. The historic profiles show the altered structure of leisure activities in relation to several age and education groups.

Section 13.3 is concerned with the visualization of historic profiles by means of CA. Structural change will be expressed as movements of historic profiles within a graphical display provided by CA. There are two possibilities to visualize the structural change measured by historic profiles. One possibility is based on the display of the historic profiles derived at the first point of observation (1953/54). The profiles observed at later points of time can be displayed as supplementary profiles and the graphical shift of these profile points indicates the structural change of the data. The second possibility considered here shows the historic profiles observed prior to 1987 as supplementary points on the display based on the 1987 profiles. We shall see that different aspects of the data will be observed in the two alternative analyses.

13.2   THE CONCEPT OF HISTORIC PROFILES

The concept of historic profiles has particular relevance to the measurement of structural changes which occurred in the last decades. Historic profiles use the historically comparable data collected at different points of time. Furthermore, historic profiles constitute a model of structural change between (social) groups.

The starting point of the concept of historic profiles are the relative frequencies of comparable characteristics within fixed groups. Activities, interests or opinions which are derived from different studies can normally be compared. In the context of historic profiles only the occurrence of those characteristics will be of interest. In the following the basic idea of historic profiles will be illustrated by survey data[1] concerning the leisure activity 'attendance at theatres or concerts'.

Table 13.1 shows the relative frequencies of the characteristic 'attendance at theatres or concerts' with respect to three different points in time. For example, in 1953 14% of the respondents with a low education level visited a theatrical performance or a concert within a certain period of time. The analogous relative frequency within the medium education group was 37% and 45% within the high education group. For the subsequent time points the

---

[1] All findings are based on the following studies provided by the Zentralarchiv für empirische Sozialforschung: Bundesstudie 1953 (ZA-Nr.: 0145); Leseranalyse 1954 (ZA-Nr.: 1482); Leseranalyse 1963 (ZA-Nr.: 0259); Freizeit im Ruhrgebiet 1970 (ZA-Nr.: 0649); Media-Analyse 1977 (ZA-Nr.: 0854); Media-Analyse 1987 (ZA-Nr.: 1619). SimCA 1.2 (Greenacre, 1986) was used for data analysis.

TABLE 13.1

The relative frequencies of the item 'attendance at theatres or concerts' (upper categories) within three education levels observed at different points of time.

	Comparable levels of education		
Point of Time	Low level 'Hauptschule'	Medium level 'Realschule'	High level 'Abitur, Univ.'
1953	14%	37%	45%
1970	20%	48%	67%
1987	6%	17%	36%

rows of relative frequencies can be interpreted in the same way. Since the time period specified in each survey may be different, we are only interested in these relative frequencies.

The use of absolute frequencies is also not convenient when the group sizes can vary substantially within the time period. In 1953, for example, 86% of the respondents had a low education level; by 1987 this percentage had decreased to 60%, whereas the percentage of the groups with the medium and high education level correspondingly increased. Thus it would not be very meaningful to compare rows of absolute frequencies of the attendance at theatres or concerts, because they depend on the group sizes.

But even the rows of relative frequencies in Table 13.1 cannot immediately be compared. For all groups the attendance at theatres seems to increase from 1953 to 1970 and to drop from 1970 to 1987; this interpretation can be wrong, because the wording of the characteristic 'attendance at theatres or concerts' differs in all studies. For example, in 1953 the respondents were asked whether they 'frequently' attend a theatre or a concert (yes or no); whereas in 1987 they had to answer the question 'How often do you attend a theatre or a concert?' and were provided with response options such as 'once a month' or 'three times a year'. If the item observed in 1987 is dichotomized, we obtain the 'occurrence' of the respective characteristic. In 1987 13% of the total population attended a theatre or a concert at least every 6 months. In 1953 17% of the respondents attended a theatre or concert 'frequently'. It makes no sense to compare these percentages in the different years, but they are comparable across the three education groups within a particular year.

The data in Table 13.1 contains both, the comparable relations and the incomparable effects of wording and data modification. To straighten out these undesired effects the relative frequencies of the respective occurrences are expressed as percentages of their row totals. For example, the row total of the relative frequencies in 1970 is 135 $(20 + 48 + 67)$. The attendance at theatres or concerts of 20 (low education group) is 15% of this total. If this transformation is carried out for all frequencies in Table 13.1 we obtain the rows of percentages given in Table 13.2. Each of these sets of percentages is called a historic profile. The term 'historic profile' suggests that the percentages in Table 13.2 are historically comparable indicators of group characteristics; they are normalized row vectors, the percentages sum up to 100. In terms of correspondence analysis the masses of all historic profiles are equal.

As mentioned above historic profiles constitute a model of structural change. Structural change is modelled as the alteration of group specific indicators over time. Each percentage within a historic profile is a group-specific indicator. Thus, the altered distribution of row percentages presented in Table 13.2 indicates a structural change. Comparing 1970 to 1987 this structural change can be described as follows. The high education group has become more inclined to attend theatres or concerts, whereas for other education groups the attendance has decreased. There is hardly any difference between the profiles of 1953 and 1970. If we consider a situation in which all groups attend theatres at the same rate (33% within each education group), it can be seen that the historic profile of 1987 is more 'distant' from this situation of equality than the profiles of 1953 and 1970. In other words, the determination of the attendance at theatres or concerts by the formal education has increased from 1953 up to 1987.

With the help of the concept of historic profiles a great number of historically comparable data sets can be constructed. In the following I will present historic profiles which belong to the theoretical context of 'adventure society' (Schulze 1992) and its unfolding within the Federal Republic of Germany during the last four decades. Within this period the alteration of social structure can be described by two processes: the levelling of social strata and the aesthetic segmentation of social milieus identified by age and education (Müller-Schneider 1994). In this context it was important to develop a method to measure and represent the structural change of social groups combined by age and education with respect to aesthetic preferences.

Table 13.3 presents historic profiles of several aesthetic preferences (leisure activities) with respect to different groups. The rows present the relative

TABLE 13.2
Historic profiles of the item 'attendance at theatres or concerts'.

| Point of time | Levels of education | | | |
	Low	Medium	High	Sum
1953	15%	38%	47%	100%
1970	15%	36%	49%	100%
1987	10%	29%	61%	100%

TABLE 13.3
Historic profiles of leisure activities.

| Year | Profiles | Groups | | | | | | | | | |
		G1	G2	G3	G4	G5	G6	G7	G8	G9	Sum
1953/54	a1	10.3	12.1	10.3	8.6	15.5	15.5	6.9	10.3	10.3	100.0
	b1	15.6	17.1	15.6	8.7	13.4	8.9	4.0	8.7	8.2	100.0
	c1	3.5	4.7	24.4	2.3	8.1	22.1	2.3	8.1	24.4	100.0
	d1	5.8	13.9	17.7	5.4	14.6	14.3	3.7	9.9	14.6	100.0
	e1	13.7	16.1	12.1	10.5	12.9	9.7	7.3	8.9	8.9	100.0
1963	b2	15.3	19.1	20.5	6.6	9.4	10.4	4.5	6.6	7.6	100.0
1970	b3	14.3	22.9	30.0	5.0	8.6	11.4	1.4	2.9	3.6	100.0
	d3	6.0	12.9	17.9	5.3	11.2	16.7	4.1	11.2	14.6	100.0
	e3	27.1	16.1	4.2	16.9	11.0	4.2	8.5	6.8	5.1	100.0
1977	a4	14.3	11.4	2.9	15.2	12.4	5.7	16.2	15.2	6.7	100.0
	e4	24.1	14.9	8.0	17.2	10.3	4.6	8.0	8.0	4.6	100.0
1987	a5	14.3	10.2	4.1	15.3	11.2	5.1	18.4	16.3	5.1	100.0
	b5	15.3	21.3	29.0	4.9	7.1	14.2	1.1	2.7	4.4	100.0
	c5	0.0	4.8	14.3	4.8	4.8	14.3	4.8	9.5	42.9	100.0
	d5	2.7	7.7	19.2	3.8	9.3	19.2	3.3	12.1	22.5	100.0
	e5	25.4	15.5	7.0	18.3	11.3	2.8	9.9	7.0	2.8	100.0

Note. The numerals denote the relative frequencies of the items (upper categories), expressed as percentages of their row totals, with respect to nine groups of age and education. These profiles were observed at five different points of time (1 = 1953/54; 2 = 1963; 3 = 1970; 4 = 1977; 5 = 1987). The items and groups are listed below.

Groups of age and education				Items
		Age (years)		
Education level	18–29	30–44	45/higher	a = Readership of *Das Neue Blatt*
				b = Attendance at cinema
Low	G1	G4	G7	c = Readership of *Rheinischer Merkur*
Medium	G2	G5	G8	d = Attendance at theatres or concerts
High	G3	G6	G9	e = Readership of *Wochenend*

frequencies within nine groups combined by age and education. The relative frequencies are expressed as percentages of their row totals. The leisure activities were observed at five points in time, but not all information was available for all points of time. The historic profiles include the attendance at theatres or concerts, the attendance at cinema and the readership of certain magazines: *Wochenend*, *Rheinischer Merkur* and *Neue Post*. The *Rheinischer Merkur* is a quality newspaper which takes a conservative line. The articles include contributions to politics, economics and culture. The *Wochenend* and

Das Neue Blatt are weeklies of the so-called yellow press. The *Wochenend* can be characterized by the plurality of erotic pictures and articles, while the content of the periodical *Das Neue Blatt* is determined by gossip about the affairs of the international 'high society'.

The nine groups of age and education (see caption of Table 13.3) were determined by all combinations of three age groups and three education levels. The historic profiles in Table 13.3 and the inherent change of structure can be interpreted in the same way as those in Table 13.2. The only difference is that structural change is now related to the variable which combines age and education. It is possible to visualize the complex structural change of the data in Table 13.3 by means of correspondence analysis.

13.3   THE VISUALIZATION OF HISTORIC PROFILES

This section will focus on the visualization of historic profiles and their 'movements' over time in a graphical display. In order to carry out this application the concept of supplementary profiles in CA has to be considered. Supplementary profiles are projections of additional row or column profiles on an existing display (see Chapters 1 and 12). Supplementary profiles neither influence the total inertia nor do they determine the principal axes. Treated as supplementary row profiles the historic profiles discussed in the chapters above can be visualized on an existing display. The geometrical movements of these profiles can be interpreted as a measure of structural change over time.

Thiessen and Rohlinger (1988) as well as Thiessen *et al.* (in this book) have used CA to visualize the structure of marital distribution of house-work and its change over a 4-year period. However, this kind of application of CA is burdened with a certain problem. The profile configurations observed at different points of time have to be related to one another in a reliable way (Thiessen and Rohlinger 1988, p. 651). The dimensionality of the solution may change from one point of time to the next one. The dimensionality of the solution denotes the number of dimensions which have to be retained in order to explain the profile variation sufficiently.

In order to show the effects of different dimensionalities of the solution two CAs of the data in Table 13.3 will be discussed. The first CA is based on the five historic profiles of 1953/54, whereas the second is based on the five historic profiles observed in 1987. The historic profiles which are not analyzed in the respective analyses are displayed as supplementary profiles. The two displays can be taken as different views of the same structural change as indicated by the respective profile movements.

13.3.1   The structural change based on the 1953/54 profiles

Figure 13.1 presents the optimal two-dimensional display of the historic

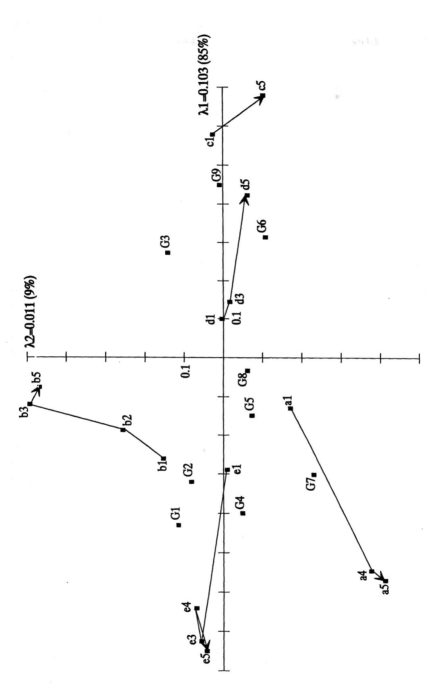

FIGURE 13.1 Two-dimensional display, by correspondence analysis, of the data of Table 13.3. The historic profiles of 1953/54 are active. The remaining historic profiles are displayed as supplementary points.

profiles observed in 1953/54 and the supplementary row profiles observed at later points of time. The supplementary row profiles will be neglected for the moment. The total inertia of the model (for interpretation of this value see Chapters 1 and 2) is 0.120, therefore the variation in the reference data is relatively low. In this solution the first axis explains about 85%, the second an additional 9% of this variation. It can be followed that only one dimension has substantive meaning.

The graphical locations of the rows (leisure activities) and columns (age and education groups) have to be interpreted relative to the principal axes. The row profiles c1 (readership of *Rheinischer Merkur*) and d1 (attendance at theatres or concerts) are located on the right, whereas the profiles a1 (readership of *Das Neue Blatt*), b1 (attendance at cinema) and e1 (readership of *Wochenend*) are located on the left. These locations along the first principal (horizontal) axis can be interpreted as expressions of a cultural dimension with the poles 'high-culture' on the one hand and 'mass-culture' on the other hand. The graphical locations of the column profiles express a hierarchical order of the education levels. The groups with high education levels (G3, G6, G9) are located on the right, the groups with low education levels (G1, G4, G7) are located on the left. The first principal axis can be interpreted as the manifestation of a hierarchical sociocultural segmentation of society in the 1950s.

The second axis is not quite as clear to interpret as the first axis. All 'youngest' groups (G1, G2, G3) are located in the upper half and two of the 'oldest' groups (G7, G8) are located in the lower half of the display whereas the oldest with a high formal educational level are located at the centroid of the second axis. We can label this axis as an age-related dimension, but this dimension is not expressed well.

The row and column contributions of this CA are presented in Table 13.4. Because of the definition of historic profiles all row profiles have equal masses. Referring to the first axis, one can see the high squared correlations of the low and high educational groups which refer to the opposite parts of the first axis. There is also a clear interpretation of the rows on the first axis: 'attendance at cinema' (b1) and 'readership of *Wochenend*' (e1) are related to the low educated groups (see the negative signs in the column $k = 1$ and the high correlations) whereas the 'readership of *Rheinischer Merkur*' (c1) is related to the high educated groups.

In contrast to the clear interpretation of the first principal axis the second axis cannot be interpreted substantively by the column variables: there are no patterns in the squared correlations as well as in the contributions. Of the leisure activities only the historic profile 'readership of *Das neue Blatt*' (a1) can be used to characterize the negative side, opposing the historic profile 'attendance at cinema' (b1) on the positive side.

By maintaining the age-by-education column structure, the same leisure activities observed at later points of time are projected onto the map based on

TABLE 13.4
Numerical output of the correspondence analysis of the 1953/54 profiles.

Row contributions

Profiles	QLT	MAS	INR	$k = 1$	COR	CTR	$k = 2$	COR	CTR
a1	966	200	80	−129	345	32	−172	621	550
b1	982	200	155	−261	730	133	153	252	436
c1	989	200	564	579	987	653	25	2	12
d1	338	200	50	101	338	20	3	0	0
e1	920	200	151	−289	919	163	−10	1	2

Supplementary row contributions

Profiles	QLT	MAS	INR	$k = 1$	COR	CTR	$k = 2$	COR	CTR
b2	794	200	210	−187	277	68	256	517	1211
b3	666	200	643	−120	37	28	492	628	4502
d3	603	200	59	145	595	41	−17	8	5
e3	760	200	1157	−725	755	1023	56	5	58
a4	720	200	1017	−545	487	580	−378	233	2645
e4	759	200	910	−641	750	800	68	8	85
a5	655	200	1256	−569	429	631	−413	226	3164
b5	625	200	599	−76	16	11	468	609	4067
c5	545	200	1442	679	532	900	−104	13	202
d5	910	200	333	423	890	347	−63	20	74
e5	787	200	1192	−750	785	1096	42	2	33

Column contributions

Profiles	QLT	MAS	INR	$k = 1$	COR	CTR	$k = 2$	COR	CTR
G1	910	98	177	−430	849	177	115	61	120
G2	924	128	125	−320	869	127	81	55	77
G3	998	160	126	273	790	117	141	209	294
G4	964	71	99	−399	950	110	−49	14	16
G5	690	129	43	−148	552	28	−74	138	66
G6	977	141	134	315	872	136	−109	105	156
G7	899	48	65	−301	567	43	−231	332	239
G8	622	92	6	−33	135	1	−62	486	33
G9	990	133	225	449	990	261	8	0	1

the 1953/54 profiles. These supplementary points do not influence the geometric orientation of the model.

The profile movements over time are expressed by arrows which connect the specific profile points. Two of the profile movements can be interpreted very well. The arrows of the profiles 'd' (attendance at theatres or concerts) and 'e' (readership of *Wochenend*) are moving mainly along the first principal axis. 'Mass-culture' and 'high-culture' became more separated over time. This structural change can be interpreted as an increasing segmentation of the education groups in relation to the profiles 'd' and 'e'. The profiles 'a' (readership of *Das Neue Blatt*) and 'c' (readership of *Rheinischer Merkur*) show various movements: The movements along the first principal axis indicate the structural change mentioned above; the movements along the second principal axis cannot be defined since this axis is not interpretable in a meaningful way. The movement of the profiles 'b' cannot be defined very well, because they move mainly along the second principal axis.

If we look at the supplementary row contributions in Table 13.4 we see that the total profile variation has increased over time. For example, the value of the hypothetical inertia contribution of the profile a4 (see column INR) is 1017, this means the variation of 'readership of *Das Neue Blatt* in 1977' is nearly the same as the sum of all 1953/54 profiles. Furthermore, the high hypothetical contributions of inertia of many of the supplementary profiles indicate a strong expansion of both axes over time. The relatively high correlations of the supplementary profiles 'b' (attendance at cinema) with the second principal axis indicate their increasing importance over time when interpreting this axis.

Because of the high hypothetical inertias in the model one can assume that the total inertia increases over time. Therefore it seems to be convenient to use the 1987 profiles as reference frame for the historic analysis. The interpretation of the second axis is expected to be more evident than in the model used above.

13.3.2 Comparison with the analysis based on the 1987 profiles

The second correspondence analysis based on the 1987 profiles is given in Figure 13.2. First of all the total inertia increased from 0.120 to 0.464, therefore the variation in the data is about fourfold. Most obvious is the pronounced strengthening of the second dimension of the solution, the inertia increasing from 0.011 to 0.134 and the percentage of inertia from 9% to 29%. This higher inertia should lead to a substantive interpretation of the second axis. The percentage of inertia on the first principal axis has decreased from 85% (1953/54) to 64% (1987). Both the strengthening of the second dimension and increased overall inertias improve the interpretation of structural change in the last decades.

In the two-dimensional map in Figure 13.2 the nine points of the age and education groups turn out to be strongly patterned: with low to high education levels from left to right on the first axis and with low to high age groups from top to bottom. According to Figure 13.1 the first principal axis is identified again as a hierarchical sociocultural dimension. Compared to the first analysis, the meaning of the second (vertical) axis becomes more evident: it is a clear age-related dimension. The positive pole of this dimension refers to 'action'-culture, indicated by 'attendance at cinema'. The importance of this dimension has increased in connection with the 'youth movement' in Germany in the last four decades, the negative pole of this axis is partly determined

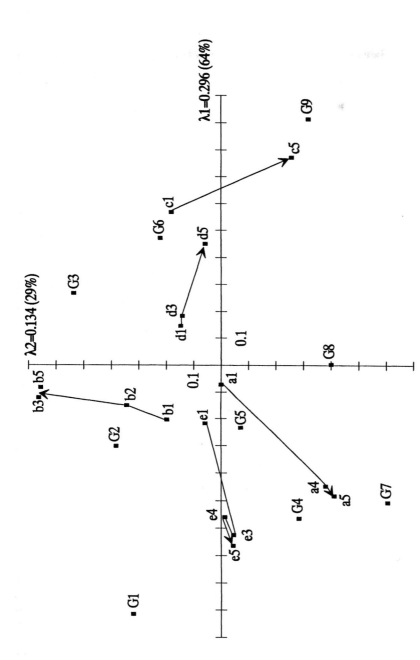

FIGURE 13.2 Two-dimensional display, by correspondence analysis, of the data of Table 13.3. The historic profiles of 1987 are active. The remaining historic profiles are displayed as supplementary points.

by the 'readership of *Das Neue Blatt*' which is a part of 'trivial'-culture (Müller-Schneider 1994).

Considering the numerical output (Table 13.5) it can be seen that most of the 'young' (G1 to G3) and 'old' groups (G7 to G9) are more highly correlated with the second principal axis than in 1953/54. This confirms that this axis can be interpreted as an age dimension.

Now the historic profiles observed prior to 1987 are projected on this existing map. Compared to Figure 13.1 there is a noticeable difference in the

TABLE 13.5
Numerical output of the correspondence analysis of the 1987 profiles.

Row contributions Profiles	QLT	MAS	INR	$k = 1$	COR	CTR	$k = 2$	COR	CTR
a5	915	200	189	− 481	528	157	− 412	387	253
b5	990	200	190	− 82	15	4	656	975	641
c5	955	200	301	773	860	405	− 257	95	99
d5	823	200	108	451	810	137	57	13	5
e5	899	200	211	− 662	896	297	− 44	4	3

Supplementary row contributions Profiles	QLT	MAS	INR	$k = 1$	COR	CTR	$k = 2$	COR	CTR
a1	48	200	45	− 71	48	3	−1	0	0
b1	675	200	51	− 203	348	28	197	327	58
c1	857	200	179	569	779	218	181	79	49
d1	349	200	53	147	175	15	147	174	32
e1	671	200	32	− 217	628	32	57	43	5
b2	930	200	64	− 149	149	15	342	781	174
b3	954	200	205	− 119	30	10	662	924	653
d3	504	200	46	184	317	23	141	187	30
e3	802	200	210	− 624	798	263	− 46	4	3
a4	943	200	157	− 446	545	134	− 380	398	215
e4	862	200	156	− 559	862	210	−14	1	0

Column contributions Profiles	QLT	MAS	INR	$k = 1$	COR	CTR	$k = 2$	COR	CTR
G1	931	115	159	− 756	896	223	151	36	20
G2	967	119	62	− 301	371	36	381	596	128
G3	980	147	116	268	197	36	536	784	315
G4	948	94	86	− 564	756	101	− 284	192	57
G5	728	87	15	− 233	665	16	− 72	63	3
G6	880	111	74	471	720	83	222	159	41
G7	920	75	110	− 505	377	65	− 606	543	205
G8	690	95	48	4	0	0	− 401	690	114
G9	952	155	330	915	849	439	− 319	103	118

profile movements and their interpretation. Since the second axis can be interpreted as an age dimension, all vertical movements can be interpreted substantively. These movements express the increasing segmentation within the age-related sociocultural dimension. This segmentation is mainly presented by the profile movements 'attendance at cinema' (b) and − to a lesser degree − 'readership of *Das Neue Blatt*' (a) and 'readership of *Rheinischer Merkur*' (c). All these changes in structure cannot be interpreted substantively if the display of 1953/54 is used as reference frame because the second dimension of the solution became prominent from 1953/54 up to 1987.

13.4 DISCUSSION

The starting point of the analyses were several studies concerning leisure activities observed between 1953 and 1987 in Germany. Because the measurements of equal leisure activities differ in the wording there was a need to normalize the data over time. In the given case historic profiles were constructed. These are rows of relative frequencies that each sum up to 100%. These historic profiles were used to show structural changes over time depending on age and education. These were visualized by using the data of one time point as reference frame and the data of the additional time points as supplementary variables.

In general, from a theoretical point of view there are at least two alternative choices of reference frames, either using the first time point for showing structural changes forward in time or using the last time point for showing the backward changes. Both analyses were performed and interpreted. The main differences refer to an approximately fourfold increase in the total inertia by using the last time point as reference data as well as the possibility to interpret the second axis substantively. It has been shown that CA is a useful tool for the visual description of structural changes which are modelled as altered distributions within historic profiles. The CA display of historic profiles can be interpreted easily, because it allows a direct impression of the structural metamorphosis.

Part 4

Further Applications of Correspondence Analysis in Social Science Research

Correspondence Analysis of Textual Data from Personal Advertisements

Helmut Giegler and Harald Klein

14.1 INTRODUCTION

This study concerns textual data from personal advertisements in a selection of German newspapers. Using CA we explore the relationship between newspapers, the gender of the advertisers on their different images on the one hand, and defined content analysis categories on the other.

The determination of social structures in modern societies is a central research field in sociology. The classical approaches of social dissimilarity were oriented to Marxistic class models (essential variable: ownership or non-ownership of means of production) or to social strata models (essential variables: education, income, status of job). In spite of the difference of both models they have in common that they only need a few variables: these are regarded as those crucial determinators that mark social groups in all living circumstances (biography, education, job, leisure time, family etc.) with a rather small chance for changes.

Recent approaches of social structure analysis cast doubt on this strong power of determination by these very few variables. They use terms such as 'individualization of life styles', 'destandardization of biographies', 'pluralization of life styles' and 'expressive dissimiliarity'. They all mean that space for action of people living in highly developed capitalistic industrial societies has grown compared with that in former times. The impact is at least a partial freedom from social surroundings, on the other hand loss of orientation and therefore uncertainty (see for example Beck 1983, Hradil 1987).

If the once influential social variables lose their importance, although social inequalities exist, the question arises in which forms do they occur. That is why the distinctive self-presentation becomes important: one presents oneself in public in order to gain acceptance from the social groups one thinks one belongs to and one separates oneself from other groups one thinks one does not belong to (Lüdtke 1989, pp. 73–80). An important medium of self-presentation is the socially identifiable and therefore sanctionable stylization of one's own life and its circumstances: these more or less consciously lived life styles can help to find 'social identity' in a manner that must not be underestimated in times when traditional social patterns of orientation lose their importance.

The following study is based on the implication that this 'self stylization' can be recognized clearly in personal advertisements where, with a minimum of words (every word costs money!) a maximum of oneself (self-image), of the imagination of one's partner (partner image) and of what one could do together (relationship image) is achieved to cause reaction in people who belong to the social group one belongs to or wants to belong to. This implies an articulation of group-specific orientation patterns that are regarded as valid. In this regard personal advertisements are social indicators, albeit highly stereotyped, that can be used in social groups to find out what their patterns of competence, motivation, attitude and action are without having to interview these persons.

In the long run, questions on the research into the direction and strength of the development of group-specific changes of values can be answered. The same applies to development of gender-specific changes of values that are currently being discussed.

This leads us to consider a content analysis of personal advertisements. This highly stereotyped kind of text favours a computer-aided content analysis that is strongly orientated by single words. This approach can be used for an efficient and reliable analysis of huge amounts of texts, probably read with scanners and processed with OCR (optical character recognition) programs. Because of this, the computer-aided content analysis of personal advertisements could become an excellent instrument for permanent observation of society.

14.2 DATA ON PERSONAL ADVERTISEMENTS

If personal advertisements are regarded as indicators for social heterogeneity, the question arises as to which content analysis categories and which types of newspapers are to be selected for this purpose. The content analysis categories are chosen on both theoretical and empirical grounds.

An important theoretical reference is Bourdieu's (1982) approach to grand-bourgeois life styles, and taking this as a yardstick for life styles of the 'new' petit-bourgeois, especially the three types of capital that form the center of this approach: 'cultural capital' (categories: 'cultural interests', 'intellectual mobility', 'academic professions'), 'economic capital' (categories: 'economic status', 'professional image') and 'social capital' (categories: 'social activities'). Furthermore, he supplies hints for the context of the content analysis categories to be considered, e.g. 'fitness of body' and 'body attributes', but also 'pleasure orientation'.

The choice of categories was based on theoretical considerations and from an extensive study of personal advertisements. Experiences from everyday life provided additional information for this choice of categories. Thus categories such as 'sex', 'social behaviour', 'social behaviour–erotic connotation', 'holiday, travel', 'nationality', and 'smoking, drinking negative' were established. Ambiguities were eliminated as much as possible.

The selection of the media to be analysed was guided by the criteria of 'theoretical sampling' (Strauss 1987) of qualitative social research. This means that those media were selected that are rather different from one another and also cover a wide range of the German population among their readership, including daily and weekly newspapers as well as cultural magazines. The media that were chosen are listed in Table 14.1. From each of these newspapers a random sample of 333 personal advertisements was chosen during the period 15 January to 15 May, 1988. The advertisements contained a total of 62 533 words.

TABLE 14.1
Identifiers, i.e. 'external variables'.

Identifier	Analyzed media
1	*Die Zeit* (Z) – intellectual, grand-bourgeois or new petit-bourgeois, social–liberal, nationwide distribution, famous for its pluralism of political opinions and quality of articles
2	*Westfälische Nachrichten* (WN) – catholic, Christian-democractic, Münsterland area, regional distribution, quality paper
3	*Westdeutsche Allgemeine Zeitung* (WAZ) – Social-democratic, Ruhr area, highest distribution in that area
4	*TIP* (TIP) – alternative cultural publication, Berlin area, mostly young readers, politically devoted to the alternative/green party
5	*Express* (EXP) – boulevard press, Rhine area, especially Düsseldorf, Cologne and Bonn, not devoted to a political party or group
6	*Heim und Welt* (H&W) – 'yellow press', national paper especially for women and elderly people, not devoted to a political party or group

14.3 COMPUTER-AIDED CONTENT ANALYSIS

Correspondence analysis needs data as frequencies, so a content analysis is required which transforms textual data into frequencies according to a system of categories. A category is made up of a set of search entries that belong together (for example the search entries Bach, Picasso, Rilke and opera form the category 'cultural interests'). Although most search entries are single words, there are many other types of search entries like sequences of words or any part of the word regardless whether it is at the beginning, at the end or somewhere within a word. The category system for the present study consists of 1367 search entries and 38 categories, of which 24 were used for CA (see Table 14.2).

TABLE 14.2
Categories of the content analysis.

1.	CI	cultural interests	Bach, Picasso, Rilke, opera
2.	IM	intellectual mobility	brain, quick-witted
3.	AP	academic professions	doctor, lawyer, dentist
4.	HEC		high economic status, villa, yacht, car
5.	FB	fitness of body	sport, strong, robust
6.	CLC	compassion, life crises	poverty, lonely
7.	SEX	sex	tolerant, hot, leather, lover, horny, randy, lingerie, games
8.	BA	body attributes	figure, attractive, slim, pretty, well developed
9.	HIP	high image profession	business man, head-clerk
10.	IV	inner values	honest, sincere, frank, heart
11.	SBE	social behav. erotic	flirt, candle light, romantic, necking
12.	SB	social behaviour	friend, life long, emancipated, eager to help, empathical
13.	FO	family orientation	mother, father, children, marriage, family
14.	HT	holiday, travel	abroad, mountains, nature, south, journeys
15.	NAT	nationality	African, Turk, foreigner
16.	30Y	17–30 years	27 y
17.	45Y	31–45 years	42 y
18.	60Y	46–60 years	56 y
19.	OLD	over 60 years	70 y
20.	PO	pleasure orientation	hedonistic, gourmet, wine, good meal
21.	SDN	smoking, drinking negative	abstinent, non drinker, cigarette free
22.	SOC	social activities	parties, friends, pub, cinema
23.	SIN	single	single, without partner
24.	DW	divorced, widowed	widower

The categories are shown with their SimCA-labels and examples. All examples are translated from German into English.

The most common problems involved in content analysis are (see Früh 1984):

- ambiguity: a search entry has more than one meaning;
- negation: a search entry is negated, the meaning may be the opposite intended;
- multiple coding: one part of the text is coded by more than one search entry.

Some ways to handle these problems are discussed by Klein (1992, pp. 483–487).

To start the content analysis we need to define text units. In our case a text file is used as a basis and each search entry is looked for in this file. If a search entry occurs in the text, its code is assigned and written to an output file. This file contains counters for every category and its records are the text units.

For the analysis of personal advertisements every advertisement was split into parts that formed a text unit according to self image, image of the desired partner, and image of the desired relationship, as well as the sex of the advertising person. The raw data that are processed by the content analysis can be divided therefore into two types: the data needed for identification (*identifiers*) and the text that is to be coded.

The identifiers are used for later identification of parts of the text and for the statistical analysis. Both identifiers and the text form a text unit which becomes a case in statistical analysis. Some (translated) examples are given in Table 14.3.

TABLE 14.3
Example text with reference data for a computer-aided content analysis.

$1(4001–1–1) Young man, 35 y, 176 cm, slim, with car, good income, looks for a $3(2) lovely, high-bosomed and well developed partner $3(3) for a common future.

$1(4002–1–1) Widower, 50 y, 174 cm, 70 kg, sportsman, cigarette free, children, house, garden looks for $3(3) a lovely female attendance. $3(2) If you are catholic $3(3) you can move to me $3(2) if you look for tenderness and secureness (marriage) $3(4) write to me (photo?).

$1(4003–2–1) Polish woman 55 y, good looking, well developed, good housewife $3(2) looks for an honest and true husband.

$1(4004–2–1) She, 45 y, 175 cm, would like $3(2) to make the acquaintance of an intelligent, self-assured, lovely and honest partner.

$1(4005–2–1) Simple lady, 62 y, seeks $3(2) a beloved partner $3(3) for friendship $3(4) in the Seeberg area.

The string starting with a $ is the control sequence which first separates the images of the personal advertisements and then sets the values of the external variables. The first control sequence in the text consists of three values in brackets, separated by two dashes. The first value contains more than one variable; the first digit is the *medium* (in this case the 'TIP'), the other three digits are the *running number* of the advertisement. The second variable is the *sex* of the advertising person, *1* for *man* and *2* for *woman*. The third variable represents the content of the text unit, the *image* (*1 = self image, 2 = image of partner, 3 = image of the relationship*). The $3(2) means that the value of the third variable (image) is set to the value 2 for the following text unit, so it is not necessary to repeat the values of all variables if only one is to be changed. Abbreviations such as 50 y mean 50 years old, 176 cm means that the person is 176 centimetres tall.

The result of a content analysis is a data matrix such as the one shown in Table 14.4. The first three columns are the three identifiers (medium and running number, sex of the advertising person, image); as one can see the last one can vary within an advertisement. The following columns are counters of the categories of the content analysis and show how often a code occurs within a text unit. For example the first line in Table 14.4 means that category 1 occurs 2 times, category 2 and 3 once each, categories 4 to 6 not at all, category 7 once, category 8 eleven times and so on.

Using Table 14.4 an aggregation of the frequencies for each content analysis category (24 columns) for the identifier combinations *media × gender × image* (6 × 2 × 3 = 36 rows) was performed leading to the rectangular matrix in Table 14.5. CA was applied to this table of frequencies.

14.4   RESULTS OF THE CORRESPONDENCE ANALYSIS

The first four axes of the CA explain nearly 80% of the total inertia of the frequency table (first axis 41.4%, second axis 15.7%, third axis 12.3%, fourth

TABLE 14.4
The coding results of a computer-aided content analysis.

1007	1 1	2 1 1 0 0 0 111 1 2 1 0 0 3 0 0 0 0 0 0 1 1 0 0
1007	1 2	2 0 0 0 1 1 0 0 1 1 0 1 0 0 0 0 1 0 0 0 0 0 2 0
1007	1 3	0 0 2 1 0 0 2 0 1 1 1 0 0 0 2 0 0 0 1 1 1 0 1 0
.		
.		
.		
6420	2 1	1 1 0 0 0 0 0 0 0 0 2 2 2 0 1 0 0 1 0 0 0 0 0 0
6420	2 2	0 0 0 0 0 0 4 0 0 0 0 0 0 2 2 2 0 0 0 0 0 1 0 0

axis 9.7%). The choice of a solution with four axes appears reasonable after inspecting the sequence of principal inertias, which shows a significant drop after the fourth axis. Maps of the first three axes can be found in Figures 14.1 and 14.2 while in Tables 14.6 and 14.7 we give the numerical results for all four dimensions. For reasons of clearness squared correlations of only those identifiers greater than 0.25 per axis are displayed in the figures; these are underlined in Tables 14.6 and 14.7.

Figure 14.1 shows the contrast between the categories of 'relationship images' on the right side of the first (horizontal) axis against the 'self images' on the left side. This distinction holds true for the media *Westdeutsche Allgemeine Zeitung, Express*, and *Westfälische Nachrichten*, whether the announcement is made by a man or a woman. For the other three media, the 'relationship images' still oppose the 'self-images' on the first axis, but the 'self-images' are better explained by the second axis. The 'relationship images' are closely related to the content category 'social behaviour'. Furthermore the 'relationship images' can be described by the categories 'social behaviour–erotic connotation' and 'family orientation'. All three content categories are indicators for social interactions. The 'self images' correspond with sociodemographic categories such as 'age between 31 and 45 years' and 'single' or visualizable attributes such as 'body attributes'.

The second axis is characterized by the main contrast of the readers of the *Zeit* against the readers of *Heim & Welt* and *Express*. Whereas the *Zeit* reader can be characterized by categories such as 'cultural interest', 'academic profession', 'intellectual mobility', and 'pleasure orientation', the *Heim & Welt* and the *Express* readers mentioned their 'compassion', their 'high socio-economic status', their age (46 years and older), and their parental status (divorced/widowed or single).

The third axis is mainly determined by the category 'inner values'. This corresponds with the 'partner images' for all media and for women as well as men. On the opposite side we find attributes such as 'academic profession', 'fitness of body', and 'holiday, travel'.

Considering the numerical output of the CA solution, we see that the rows of the matrix of input data (Table 14.5) have satisfactorily high (greater than 500) qualities (Table 14.6). This is not always the case for the columns of input data. The qualities of the categories 'nationality', 'pleasure orientation', and 'social activities' (marked by stars) are quite low; they are not very well explained by the four-dimensional solution (Table 14.7). One eye-catcher in Table 14.7 is that the category 'social behaviour' has the biggest share (23.3%) of the total inertia (column 'INR'). The most frequent content analysis categories are 'body attributes' and 'inner values' (column 'MASS').

Referring to the squared correlations and to the inertia contributions to the first axis in Table 14.6, the twelve rows of the 'relationship images' have on average higher values than the twelve rows of the 'self images'. This means

TABLE 14.5

Raw data matrix for SimCA.

	CI	IM	AP	HEC	FB	CLC	SEX	BA	HIP	IV	SBE	SB	FO	HT	NAT	30Y	45Y	60Y	OLD	PO	SDN	SOC	SIN	DW
ZFS	1	0	1	85	67	97	19	67	19	18	207	10	138	43	32	68	59	7	18	76	29	2	20	24
ZFP	1	0	2	33	80	9	3	11	4	9	37	7	149	59	69	38	12	3	1	14	16	5	13	8
ZFR	1	0	3	2	1	0	1	3	5	4		0	18	49	76	20	3	1	0	0	0	0	3	0
ZMS	1	1	1	72	44	99	14	50	10	11	101	23	87	30	22	21	44	18	6	54	20	6	9	19
ZMP	1	1	2	34	41	12	10	9	2	11	85	3	100	38	35	28	10	8	11	21	5	1	7	3
ZMR	1	1	3	8	6	0	2	12	3	5	3	0	23	41	69	33	4	0	0	0	0	0	4	0
TIPFS	2	0	1	6	9	7	0	4	4	3	42	0	23	12	8	11	0	0	21	28	3	0	1	3
TIPFP	2	0	2	5	11	2	0	6	1	3	19	0	25	24	25	12	3	0	8	8	2	0	1	0
TIPFR	2	0	3	2	1	0	0	1	0		0	0	2	23	21	3	0	0	0	0	0	0	0	0
TIPMS	2	1	1	15	8	14	4	17	7	18	107	5	39	21	20	14	10	5	46	68	4	0	3	10
TIPMP	2	1	2	6	19	1	0	4	6	38	59	0	41	31	28	17	5	5	31	20	1	1	4	2
TIPMR	2	1	3	5	2	0	0	5	0	23	3	0	14	67	38	20	2	0	2	0	0	0	4	0
WAZFS	3	0	1	7	14	9	16	9	3	3	102	2	29	9	6	29	11	1	16	48	33	7	7	4
WAZFP	3	0	2	9	14	3	4	0	2	0	22	1	76	14	13	10	4	3	3	12	15	5	1	3
WAZFR	3	0	3	1	0	0	0	1	2	0	0	0	2	7	13	10	5	0	0	0	0	0	0	0
WAZMS	3	1	1	5	13	31	24	36	8	2	103	21	54	26	8	38	18	0	47	95	38	13	9	16
WAZMP	3	1	2	5	17	0	8	9	7	10	76	3	122	31	36	27	4	6	18	32	13	7	6	3
WAZMR	3	1	3	0	0	0	0	2	3	1	0	0	8	24	56	21	16	1	1	0	0	0	0	0

Label																									
EXPFS	4	0	1	1	4	0	5	8	5	2	60	0	0	19	5	1	11	8	4	8	10	37	16	2	7
EXPFP	4	0	2	1	4	0	7	2	1	0	7	0	0	50	9	16	7	3	0	0	5	23	6	0	1
EXPFR	4	0	3	1	0	0	0	1	0	0	0	0	0	1	5	18	3	3	0	0	0	0	0	1	0
EXPMS	4	1	1	11	23	19	36	25	14	2	136	13	13	65	28	16	25	29	6	39	96	61	18	4	13
EXPMP	4	1	2	0	11	0	6	5	5	9	110	1	1	134	21	37	34	3	21	21	35	13	5	2	1
EXPMR	4	1	3	0	2	0	0	8	5	1	0	0	0	13	30	94	36	6	1	0	0	0	0	0	0
WNFS	5	0	1	3	12	12	23	20	9	0	82	1	0	49	4	5	39	14	1	24	45	44	7	5	3
WNFP	5	0	2	10	22	2	5	7	3	0	13	1	1	103	14	34	19	4	1	2	8	18	9	0	3
WNFR	5	0	3	1	0	0	0	2	2	0	2	0	0	3	7	46	4	4	0	0	0	0	0	0	0
WNMS	5	1	1	3	11	21	36	47	18	0	99	15	15	56	19	10	30	22	3	52	82	40	12	2	9
WNMP	5	1	2	11	10	0	6	9	3	5	72	3	3	129	25	26	29	2	4	19	21	11	0	3	1
WNMR	5	1	3	1	0	0	1	0	1	0	0	0	0	4	21	76	22	8	0	0	0	0	1	2	0
H&WFS	6	0	1	7	14	4	29	14	27	5	159	3	3	89	17	17	74	31	12	8	31	67	41	7	16
H&WFP	6	0	2	7	9	1	30	7	12	5	19	1	1	126	20	39	41	16	1	1	7	20	48	1	17
H&WFR	6	0	3	0	1	0	2	0	5	2	2	0	0	11	15	50	19	5	1	0	0	0	4	2	0
H&WMS	6	1	1	7	17	7	54	21	30	4	130	9	9	44	29	24	53	45	8	18	52	51	45	2	37
H&WMP	6	1	2	5	2	1	10	2	11	26	63	2	2	75	30	45	49	14	8	17	37	22	6	5	10
H&WMR	6	1	3	1	1	0	2	2	5	3	1	2	2	8	19	49	47	5	1	0	0	0	4	1	1

In this table the first three letters of the 5-letter row labels mean the medium (Z = *Die Zeit*, TIP = *Tip*, WAZ = *Westfälische Allgemeine Zeitung*, EXP = *Express*, WN = *Westfälische Nachrichten*, H&W = *Heim & Welt*), the 4th letter the sex of the advertising person (F = female, M = male), and the 5th letter is the image variable (S = self image, P = image of desired partner, R = image of desired relationship). See Table 14.2 for the explanation of the column labels.

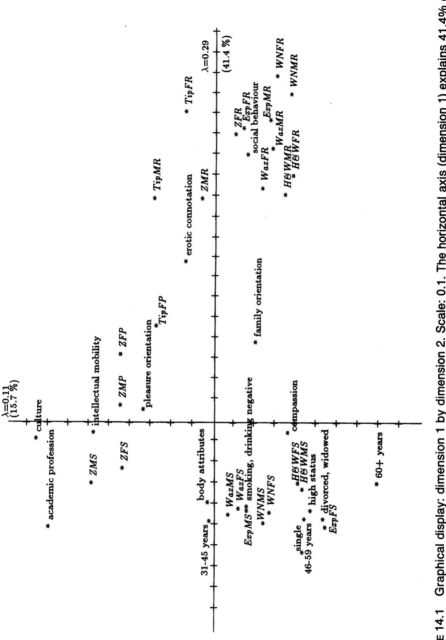

FIGURE 14.1 Graphical display: dimension 1 by dimension 2. Scale: 0.1. The horizontal axis (dimension 1) explains 41.4% of the variability; the vertical axis (dimension 2) explains 15.7% of the variability.

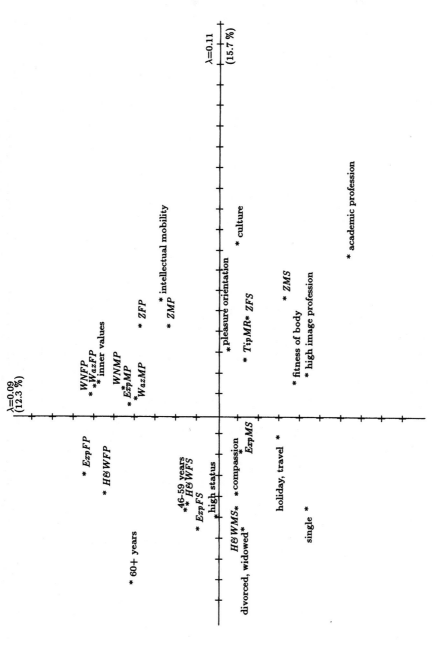

FIGURE 14.2 Graphical display: dimension 2 by dimension 3. Scale: 0.1. The horizontal axis (dimension 2) explains 15.7% of the variability; the vertical axis (dimension 3) explains 12.3% of the variability.

TABLE 14.6
Correspondence analysis: values of the rows of the four-dimensional space.

I	Name	QLT	MAS	INR	k=1	COR	CTR	k=2	COR	CTR	k=3	COR	CTR	k=4	COR	CTR
1	ZFS	896	88	45	-235	152	17	424	495	143	-201	111	41	-224	138	65
2	ZFP	768	45	39	323	171	16	432	307	76	362	215	68	-213	74	30
3	ZFR	937	14	44	1400	909	97	-116	6	2	-207	20	7	75	3	1
4	ZMS	922	60	65	-299	117	18	573	432	178	-338	150	79	-412	223	150
5	ZMP	807	36	15	70	16	1	432	631	62	218	160	20	-3	0	0
6	ZMR	924	17	32	1095	900	69	36	1	0	-173	22	6	-38	1	0
7	TIPFS	613	14	11	-207	83	2	248	120	8	-96	18	2	448	391	43
8	TIPFP	746	12	7	456	500	9	261	164	7	22	1	0	183	80	6
9	TIPFR	673	4	23	1515	629	35	121	4	1	-357	35	6	122	4	1
10	TIPMS	827	33	26	-273	138	9	230	98	16	-222	92	19	518	498	132
11	TIPMP	706	24	33	167	30	2	316	106	22	84	7	2	729	563	192
12	TIPMR	631	15	52	1092	478	60	273	30	10	-150	9	4	532	114	61
13	WAZFS	611	29	15	-423	491	18	-143	57	5	-74	15	2	133	49	8
14	WAZFP	837	17	11	-12	0	0	137	41	3	583	740	65	-160	56	6
15	WAZFR	792	3	9	1144	669	14	-249	32	2	-366	69	5	-207	22	2
16	WAZMS	734	50	30	-451	476	35	-90	19	4	-299	209	52	115	31	10
17	WAZMP	860	34	11	53	13	0	82	30	2	375	635	55	201	183	20
18	WAZMR	907	10	33	1333	799	63	-299	40	8	-369	61	16	-121	7	2

19	EXPFS	667	19	23	-512	302	17	-555	355	52	79	7	1	-49	3	1
20	EXPFP	643	11	14	95	10	0	-291	98	8	630	456	51	-262	79	11
21	EXPFR	797	3	12	1430	689	20	-157	8	1	-508	87	8	-194	13	2
22	EXPMS	833	55	22	-425	640	35	-179	113	16	-135	64	12	67	16	4
23	EXPMP	567	36	24	-21	1	0	46	5	1	411	367	72	299	194	48
24	EXPMR	948	15	53	1468	870	111	-289	34	11	-318	41	18	-95	4	2
25	WNFS	711	35	20	-441	479	23	-292	210	27	-89	19	3	30	2	0
26	WNFP	831	22	18	214	78	3	100	17	2	598	607	90	-276	129	24
27	WNFR	810	5	31	1693	717	54	-322	26	5	-481	58	15	-185	9	3
28	WNMS	745	51	40	-491	439	42	-263	127	32	-299	163	53	95	17	7
29	WNMP	677	30	15	10	0	0	130	47	5	440	539	68	180	91	14
30	WNMR	936	11	47	1602	825	94	-385	48	14	-414	55	21	-154	8	4
31	H&WFS	714	56	35	-301	207	18	-432	426	96	135	41	12	-133	40	15
32	H&WFP	693	33	40	87	9	1	-384	174	44	528	330	108	-390	180	75
33	H&WFR	929	9	23	1202	816	46	-401	91	14	-147	12	2	-131	10	2
34	H&WMS	721	57	40	-331	225	21	-453	422	106	-113	26	8	-151	47	19
35	H&WMP	506	35	11	111	55	1	-115	59	4	111	55	5	275	337	39
36	H&WMR	766	12	31	1109	665	49	-365	72	14	-203	22	6	-112	7	2

Rows = media (6) * gender (2) * image (3).
See Table 14.5 for the explanation of the abbreviations. The underlined values were taken for interpretation.

TABLE 14.7

Correspondence analysis: values of the columns of the four-dimensional space.

J	Name	QLT	MAS	INR	k = 1	COR	CTR	k = 2	COR	CTR	k = 3	COR	CTR	k = 4	COR	CTR
1	CI	936	28	38	-87	8	1	841	733	179	-112	13	4	-420	183	72
2	IM	661	37	35	-63	6	1	560	478	105	263	106	30	-218	72	26
3	AP	938	26	63	-514	158	24	785	367	148	-652	254	130	-518	160	105
4	HES	766	27	23	-428	303	17	-491	398	59	-8	0	0	-198	65	16
5	FB	704	32	17	-246	164	7	159	68	7	-393	417	57	-143	55	10
6	CLC	529	18	8	-60	11	0	-380	454	24	-109	37	3	-91	26	2
7	SEX	541	17	47	327	54	6	462	108	33	75	3	1	861	376	184
8	BA	758	145	51	-391	624	76	8	0	0	-6	0	0	181	133	70
9	HIP	527	9	13	-483	238	8	200	41	3	-451	208	22	-196	39	5
10	IV	912	145	77	20	1	0	156	65	32	559	840	527	-49	6	5
11	SBE	770	65	80	776	703	135	121	17	9	-81	8	5	188	41	34
12	SB	963	89	233	1304	924	519	-182	18	27	-188	19	36	-58	2	4

13	FO	595	72	34	377	436	36	−224	154	33	−14	1	0	−37	4	1
14	HT	632	33	16	−31	3	0	−106	32	3	−313	284	37	−328	313	52
15	NAT	137*	10	12	−166	32	1	166	31	2	235	63	6	98	11	1
16	30Y	845	33	39	−422	216	20	−20	0	0	−174	37	12	699	593	238
17	45Y	821	68	41	−478	539	54	2	0	0	−210	104	35	274	177	76
18	60Y	750	44	40	−473	354	34	−461	336	85	137	30	10	−136	29	12
19	OLD	742	20	43	−296	59	6	−810	441	120	396	106	37	−449	136	60
20	PO	290*	10	6	53	7	0	325	269	9	−74	14	1	−4	0	0
21	SDN	484	16	14	−415	290	10	−181	55	5	−84	12	1	−274	127	18
22	SOA	273*	11	11	280	112	3	269	103	7	−201	57	5	27	1	0
23	SIN	660	14	25	−630	324	20	−454	168	27	−454	168	34	4	0	0
24	DW	722	30	33	−467	282	23	−552	393	83	−135	24	6	−135	24	8

Columns = content analysis categories.
See Table 14.2 for the explanation of abbreviations. The underlined values were taken for interpretation.

'relationship images' have more distinctive profiles with respect to the describing variables, and this is true for both sexes.

One would probably expect that 'self images' and 'images of the desired partner' are opposites, but that is not usually the case: 'images of relationship' and 'self images' are much stronger negatively correlated. This result is based on a methodological effect and a substantive explanation. Due to the content analysis where the row-categories of 'relationship' frequently go together with the column-category 'social behaviour' the high correlation is expected. On the other hand, the above mentioned polarization shows the small amount of social, erotic and family orientation as characteristics for the presentation of one's self image. These – often quite vague – orientations are projected on a strong idealized togetherness, whereas the own person is characterized by unromantic, directly observed variables such as 'age', 'body attributes', and 'high economic status'.

The second axis (15.7% explained variability) reflects the contrast between the media *Zeit* (self and partner images) on the positive side and *Heim & Welt* (self images only) on the negative side (Table 14.6). This distinction holds true for men and women. 'Self images' of the 'female *Express* readers' loads on the same side of the second axis as the *Heim & Welt* readers.

The content analysis categories (Table 14.7) 'cultural interests' and 'intellectual mobility', and – slightly weaker – 'academic profession' as well as 'pleasure orientation' load on the positive side of the second axis. On the negative side of the axis we find the categories 'compassion', 'high economic status', '46–60 years', '60 years and older', and 'divorced/widowed'.

This axis does support the thesis by Bourdieu (1982, p. 405–619) about the contrasts between the upper classes ('sense for distinction') and the middle classes ('assiduity of education') on the one side, and the lower classes ('decision for the necessary') with social deficit living circumstances (compassion, divorced/widowed, older in age) on the other side (for more details on Bourdieu's theory see Chapter 16). From empirical readership studies it is known that the readers of *Zeit* belong to a high degree to Bourdieu's governing classes and middle classes. In contrast, *Heim & Welt* as well as *Express* are especially read by the members of the lower classes (Media Daten 1991).

The members of the lower classes compensate living circumstances such as 'age above 60 years', 'divorced/widowed' and 'compassion' by strong hints on good material stuff such as a 'high economic status'. This supports the results of Bourdieu's study where he shows that members of the lower classes relatively often mention their material goods. The members of the higher classes do not need to talk about these issues, it is a matter of course which does not need to be mentioned (Bourdieu 1982).

The third axis (12.3% explained variability) does not reflect a polarity between the categories of the identifiers (Table 14.6). The positive side of this axis is determined by the 'partner images' of *Westdeutsche Allgemeine Zeitung* and *Westfälische Nachrichten* as well as — but with lower values — of *Express* and *Heim & Welt*. As reported for the first two axes there is no clear distinction between the 'partner images' in respect to sex, with the possible exception of *Heim & Welt* for which the category '*Heim & Welt*, men, partner images' loads to a low degree (squared correlation of 0.055) on the positive side of the third axis.

On the level of the content analysis categories 'inner values' determine the geometric orientation of the third axis by more than 50% (Table 14.7). Negatively correlated with 'inner values' are the categories 'fitness of body', 'holiday/travel', 'academic profession' and 'high image profession'. This leads to the assumption that partner images are more to be regarded as symbols for petit-bourgeois idealizations and socially desired stereotypes than as realistic imaginations of the partner.

As we have shown for the first axis, the third axis visualizes the picture of a middle class basic family in an ideal manner. The family can be seen as a retreat providing security against a chaotic and hostile world outside. The family guarantees not only a 'regular outliving' of one's partner, erotic and intimate needs (symbolized by the first axis), but also security of values (symbolized by the third axis). The high divorce rates in western societies show that the relationships between partners are strained between demand and reality, in other words, they do not keep what they seem to promise. This is partially based on overdrawn expectations of happiness, that relatively often cannot be redeemed in principle under the given social circumstances. Many persons do not realize that their partnership conflicts are caused by the society; on the contrary they are strengthened by advertisements and by mass media products (soap operas, trivial novels) in an unrealistic expectation of happiness. In this context Beck (1986) speaks of a 'paradox socialization'. According to Beck the logic of the market (also the 'partnership market' is based on offer and demand!) implicates the successful lonely fighter in a permanent struggle and therefore the 'negation of social relationships'; in its last state it destroys also the 'implications of permanent togetherness', because 'the base figure of the enforced modernity is ... the single' (Beck 1986, p. 200). This polarity is expressed by the first and third axes; the strong idealized 'relationship' (positive side, first axis) and the 'partner image' (positive side, third axis) stand in contrast to the descriptions used for the self image (negative side, first axis).

The fourth axis (9.7% explained variability) for the identifiers (Table 14.6) is characterized by the magazine *Tip*, where — referring to the squared correlations — 'men' have higher values than 'women' (which is uncorrelated with 'relationship images'). For men this holds true for 'self and partner images' and for women for 'self images' only. In addition, the category *Heim & Welt*, men, partner images' loads on the positive side of this axis. The content analysis categories (Table 14.7) '17–30 years' and 'sex' are located on the positive side of this axis whereas the category 'holiday/travel' can be found on the negative side.

To evaluate these results one has to be aware that *Tip* and *Heim & Welt* are the only considered papers that publish personal advertisements with sexual contents. It is also interesting that in *Tip* − an alternative magazine mostly read by young people (Media Daten 1991) − only women write relatively often explicitly about sexuality in their self images. This can be evaluated as an indicator for an increasing self-conscious female sexuality (Kaufmann 1988, pp. 401 f.). For the *Heim & Welt* readers, who are frequently members of the lower classes, only the males stress the sexuality issues in the partner images.

14.5 CONCLUSION

Although the *substantive results* of the study show that personal advertisements are valid indicators of social structures in developed capitalistic societies, the main goal of this study is a *methodological* one of showing that CA is an excellent method for analyzing the results of a content analysis. There are three reasons why correspondence analysis is successful in this type of application:

(1) it is primarily applicable to frequency data − the data type that is the output of a content analysis;
(2) the processing of the rows and columns *simultaneously* is suited to the structure of a frequency matrix, where the rows are the identifiers and the columns the content analysis categories;
(3) it provides a compact, very informative and − compared with other competing multivariate methods (e.g. log-linear models) − a very clear presentation of the results. From the content analysis point of view it is useful that the identifiers can be interpreted relative to the content analysis categories they characterize.

Content analysis can be regarded as the quantifying missing link to the qualitative data acquisition methods. The deficits of standardized data acquisition methods can be avoided (see Giegler 1992) while still allowing a quantitative analysis. The combination of qualitative data acquisition methods, computer-aided content analysis and correspondence analysis of these data is a very promising research strategy for many hypotheses in the social sciences.

APPENDIX. SOFTWARE USED IN THIS STUDY

The computer-aided content analysis was performed with Intext/PC, a software package that performs textual analyses, e.g. word lists, concordances, indices, readability and content analyses. Interactive coding of ambiguous search entries, detection of negation and the very powerful feature of defining search entries are specific to Intext/PC (see Klein 1991). Computer content analysis packages like Intext/PC write the counters for each category together with the identifiers to a file that can be processed with most statistical software. In the present study the SPSS procedure AGGREGATE was used to prepare data required by the correspondence analysis program SimCA (Greenacre 1990b).

Explorations in Social Spaces: Gender, Age, Class Fractions and Photographical Choices of Objects

Ulf Wuggenig and Peter Mnich

15.1 SOCIOLOGY OF OBJECTS

Lately, neither psychology nor sociology has been sufficiently concerned with man in his role as *'homo faber'*, i.e. as creator and user of physical objects. Compared to the concern with intrapsychic processes as well as with inter-individual patterns of relations, the analysis of person–object relations has remained a sphere of research that is neglected. Exceptions in sociology are authors committed to pragmatism (cf. Csikszentmihalyi and Rochberg-Halton 1981) and sociosemiotics (cf. Baudrillard 1972, Bourdieu 1979). They have pointed out that objects, apart from their use value and exchange value, have 'sign value'. They are the embodiment of objectives and emotions, they mirror specific aspects of personality, are indicators of a particular life style, or symbolically represent convictions, achievements or social relations. Human personality also reflects in manifold ways those items with which it entertains interactive relations.

The study 'Aesthetic Attitudes and Lifestyle'[1] concentrates on objects within the sphere that people live in. It is objects within this sphere that − much more than those belonging to the external world − reflect the personality of their owners. Bourdieu (1979, p. 195) interprets 'interieurs' as a system of signs, in which social conditions of existence are symbolically expressed. 'Interieurs' serve to identify the life styles of social 'class fractions', which are based on dispositions he calls 'habitus' and 'taste'. Csikszentmihalyi and Rochberg-Halton (1981), moreover, point to the importance of objects as symbols of social integration while maintaining that empirical knowledge concerning the meaning of objects has so far remained scanty indeed. The same applies to the related topic of research on taste: 'The fact is that we virtually know nothing about people and taste ... Surveys of cultural activities and preferences produce data about general tendencies, which are made more general yet by currently fashionable forms of statistical analysis' (Gans 1985, p. 53). The exception is Bourdieu's empirical work (1979), which has stimulated a broad theoretical discourse, but so far only a small amount of empirical research (cf. e.g. Blasius and Winkler 1989, Blasius 1993).

In order to extend the narrow range of empirical knowledge within the 'sociology of objects' (Bourdieu), a newly developed surveying technique − photo questioning − was applied (cf. Wuggenig 1990). Combined with statistical tools that preserve much of the concrete richness and detail of the data (such as CA), photo questioning seems particularly suited to monitor 'person−object' relations and matters of taste. Since the data presented have been collected by means of this method, it is essential to present shortly the mode in which it was applied in order to understand the results.

15.2 PHOTO QUESTIONING AS A SURVEY METHOD

The method of photo questioning is based on the idea that research subjects are accorded an active role in the research process as amateurs taking photographs. Photographic tasks were embedded in a semi-structured oral interview that was essentially concerned with the apartments that the research subjects lived in, their life styles and their aesthetic competences and attitudes. All subjects involved in the study were assigned tasks which were formulated as follows: 'We ask you to take 8 photographs. First, take photographs of those two things or parts of the room which you like best in your living room, and then take pictures of those two objects or parts of the room that you dislike. When you are finished, do the same for the rest of the apartment − two photographs of the things or parts of the rooms that you like and two of those that

[1] The project 'Aesthetic Attitudes and Lifestyle', directed by Ulf Wuggenig, is supported by grants from the University of Lüneburg (FRG).

you dislike. It does not matter at all which room you choose.' If there was no living room, people were asked to choose the room which was most likely to be used in case of visitors.

As Castel (1981, p. 239) has pointed out, a photograph is the result of a choice and a classification – even if subject to ideological, ethic, or aesthetic norms: 'One does not ban everything and anything on film ... Whatever is perceived needs to be overvalued if it is to be accepted into the sacrosanct photographic order ... Behind every photograph there is a verdict on relevance'.

The restriction to a fairly small number of eight photos enhanced the selective character of photographic options. Whereas the total number of referential objects was maintained small in order to guarantee the selection of signs and objects that were strongly invested with emotions, the task was formulated in sufficient generality for the subjects to determine their own selection criteria.

Subjects were given a camera by the interviewers and instructed in its use. Once the task was explained, they were left to themselves to take the pictures, without being disturbed and, as far as possible, without being observed. Once they had finished their photographic activities, subjects were asked to indicate the objects chosen by means of photographs and to give reasons for their choice. The identification of the objects is necessary since the pictures may contain more than one particular bit of information.

The interviews that followed this exercise involved simple, open-ended questions concerning the reasons for the selection of each of the eight (groups of) objects of reference. Responses were expected to be as spontaneous as possible and not to be the result of lengthy reflection. The interviews were carried out in 1990 with 80 persons living in a middle-sized town in northern Germany.

15.3 BOURDIEU'S SOCIAL TOPOLOGY AS A GENERAL THEORETICAL FRAME

Correspondence analysis was chosen as a multivariate instrument for analysing the data, not only on technical but also on substantive grounds. One of the fruitful ways to conceive of sociology is to define it as 'social topology'. This general theoretical approach, which guides our empirical analyses, implies a relational conception of social reality. In view of this theoretical foundation, CA seems to be especially adequate, being a method that, as Bourdieu (1991, p. 277) put it, 'is essentially a relational procedure, whose philosophy corresponds completely to what in my opinion constitutes social reality. It is a procedure, that "thinks" in relations'.

Bourdieu (1979, p. 139) proposes consideration of three 'spaces' for sociological analyses. A 'structural space' of social conditions and positions, which is mainly organized by distribution of the volume (sum) and structure (composition) of the various kinds of capital, the most important being economic and cultural assets. The second is a space of life styles, also often called the 'symbolic space' and the third the 'space of habitus' (disposition, taste). Habitus is a system of schemes of perception and appreciation of practices, cognitive and evaluative structures which are acquired through the lasting experience of social conditions and positions. This system of propensities and capacities to appropriate (materially or symbolically) a given class of objects or practices, is the generative formula of the classes of practices and properties constituting a life style.

A life style is defined as a unitary set of distinctive preferences, which express themselves in the specific logic of symbolic subspaces, such as preferences for several kinds of furniture, cultural consumption, objectified cultural capital, clothing, language or body hexis (cf. Bourdieu 1979, p. 195ff.). People's choices in clothing, forms of entertainment, furniture, art, books, cars, etc., are related, because both their choices and the objects they choose reflect and express a certain taste or habitus (cf. also Gans 1985, p. 41). Life styles are symbolic expressions of social conditions of existence, with habitus or taste being the practical operator of the transmutations of things into distinct and distinctive signs. Through these properties and their distributions, the social world achieves the status of a symbolic system, which is organized according to the logic of difference. Differences function not only as distinctive signs but also hierarchically as signs of distinction.

The joint representation of structural and symbolic space, the 'social space' is theoretically constructed in such a way that the closer the units which are situated within this space, the more common properties they have; and the more distant, the fewer (cf. Bourdieu 1979, 1989a). Spatial distances coincide with social distances. People who are close in social space are supposed to have higher probabilities of association (e.g. to be friends, to marry, to be in the same organizations, etc.) than people that are far apart. Such social spaces are very well representable by means of CA.

The projective object–choices by photographic means constitute the symbolic space in our topological analyses. Occupations, gender and age are the social positions of primary interest. The joint space of structural positions and symbolic representations will be constructed and visualized with the help of CA. After introducing the categories of the symbolic space – the objects represented by photography – the general theoretical frame will be supplemented by some more specific assumptions.

15.4 THE OBJECTS OF THE SYMBOLIC SPACE

The first step of the empirical analysis was to identify and to categorize 730 objects, which had been photographed in the total of 640 (= 80 × 8) pictures and explicitly mentioned during the interviews. These 730 objects were reduced to a total of 37 categories, listed in Figure 15.1 (cf. Wuggenig and Kockot 1993). We distinguished between single objects in individual rooms (e.g. tables, media, plants, etc.) and parts of rooms (room corners, walls) or whole rooms (e.g. kitchen, bathroom). Objects or rooms of which photographs were taken only once, or rarely, were classified as residual categories 'other (parts) of rooms' (e.g. balcony, studio, sleeping room) or 'other objects' (e.g. ashtray, bowl for the cat).

With respect to content, about two-thirds of the primary objects may be subsumed under the following general categories:

(1) *Furniture*. This category includes furniture which, in a functional sense, is primarily required to make an apartment a habitable space. Approximately 30% of the objects belong to this category. These objects are: shelves; cupboards; closets; cabinets; drawers; chests; built-in-cupboards; chairs, armchairs and easy chairs; couch, lounge, suites, seating landscapes and seating corners; carpets and curtains; dining-tables; desks; other tables; and beds.

(2) *Objectified cultural capital*. Bourdieu distinguishes among three forms of cultural capital. It may appear (a) in embodied or internalized state, in the form of long-lasting dispositions of the mind and the body (like knowledge, taste or hexis), (b) in an institutionalized state, in the form of graduation or academic titles being proof of cultural competency, and (c) in an objectified state, i.e. 'in the form of cultural goods (pictures, books, dictionaries, instruments, machines etc.), which are the trace or realization of theories or critiques of these theories, problematics, etc.' (Bourdieu 1985b, p. 243). This category includes about 20% of the objects. These objects are: sculptures, designer's and craft objects; pictures (graphics, paintings, engravings, etc.); photographs and posters; books; visual media (tv, video); acoustic media (radio, hifi); musical instruments (mainly piano).

(3) *Rooms and parts of rooms*. In 17% of all cases it was not single objects that were portrayed but whole rooms or parts of them. They were primarily taken outside of the living room. The categories are: kitchens; entrance halls and corridors; bathrooms; parts of living rooms; children's rooms or corners; toilets; other (parts of) rooms (dining rooms, studios, libraries, bedrooms, balconies, basements, storerooms, verandas); and (parts of) walls.

(4) *Other objects*. The remaining third of objects have little in common: plants; lamps; office equipment; storage facilities; installations (radiator, pipes, ladder, meter); doors; windows; views from a window; household appliances; and chimney corners.

15.5 CLASSES AND CLASS FRACTIONS IN BOURDIEU'S THEORY

Since occupations are the ways people keep themselves alive, they are of fundamental importance. The hierachy of prestige strata and of economic classes have their roots in the occupational structure (cf. Blau and Duncan 1967, p. 7). In Bourdieu's theory occupations are also of central importance. Classes are often designated by the name of an occupation. In characterizing the concept of 'social space' Bourdieu (1989c, p. 379) classifies occupational groups in two dimensions: the 'volume' and the 'structure' of their resources or 'capital'. We will reconstruct this classification in order to show how the occupational groups in our sample are situated in Bourdieu's structural space. In the first of the two latent dimensions of this space occupational groups are classified according to the overall volume of their capital. In the second dimension, they are classified according to the composition of the two most important resources, that is the relative weight of economic and cultural assets. The spatial distribution runs from those occupations that are best provided with economic, cultural, social and symbolic capital (the bourgeoisie of ancient lineage), to those with low qualifications, a weak social network and having low prestige (unskilled workers and smaller farmers).

Bourdieu distinguishes three broad social classes on the basis of their overall capital. The material conditions of the upper classes (e.g. producers of culture, liberal professions, industrialists) engender a taste of freedom, a preference for cultural objects and practices that are removed from mundane material functions. This taste of freedom is contrasted with the taste of necessity of the lower classes (industrial and unskilled workers, small farmers), which only serves as a foil in the game of distinction. Being constantly concerned with the practicalities of material existence, these necessities become ingrained as taste, leading to a choice for things that are functional, unformalized and natural. The middle classes (e.g. cultural intermediaries, junior executives, shopkeepers) are characterized by cultural goodwill filled with reverence for culture and a taste of pretension. They aspire to bourgeois values and distinction, but have neither the resources nor the habitus to achieve them. Hence they seek to adopt a superficial life style not of their own, often borrowing the outward signs of legitimate culture of the dominant classes.

The three broad classes are horizontally differentiated according to the composition of their economic and cultural assets. These parts of the larger classes, which are called 'class fractions' have their own variants of habitus or taste. 'The basic opposition between the tastes of freedom and the tastes of necessity is specified in as many oppositions as there are different ways of asserting one's distinction *vis-à-vis* the working class and its primary needs' (cf. Bourdieu 1979, p. 204).

15.6   THE SPACE OF CLASS FRACTIONS AND PHOTOGRAPHED OBJECTS

In view of the exploratory character of the study, only a small sample in a northern German town of about 100 000 inhabitants was drawn. The quota target sample, which is not representative of the whole community, initially consisted of 85 persons in five class fractions in the sense of Bourdieu, two of them with higher (20 persons in each group) and three of them with lower occupational status (15 persons in each group). Five persons refused the photographic part of the interview, which gives a total of 80 persons for the analysis.

About one-half of the higher status group consists of people active in academic or intellectual professions (later abbreviated as 'ac'). The other half with higher status belong to the new industrial and post-industrial middle classes. These new middle class populations ('nm') of the sample include occupations involving presentation and representation and occupations in institutions providing symbolic goods and services. They all have a comparatively good education, but not academic educations in the traditional sense.

The category of the lower status occupations consists of three groups, which all are part of Bourdieu's lower classes: (a) old middle class, employees and civil servants, craftsmen, with low economic capital; (b) skilled and unskilled workers; and (c) smaller farmers or peasants. For the analyses presented old middle class and workers were classified in one group – lower middle class ('lm') – and separated from peasants and farmers ('fa'), because the main interest was to see whether peasants and farmers differ substantially from the other categories of the lower classes. It has to be added that homemakers without employment generally were classified according to their (marriage) partner's occupational position. The occupational groups of our sample do not exhaust the whole social space of Bourdieu. But there is sufficient vertical and horizontal social differentiation in the sample, to confront some of Bourdieu's class culture assumptions with the data.

A first general hypothesis is that culture and concern with cultural objects are symbolizing class differences. If this assumption is valid, there should be marked differences between the academic and the new middle class fractions on the one hand, and the lower class groups on the other hand regarding their choices of some of the objects of objectified cultural capital. Our projective data are well suited not only to see whether cultural class symbolization does exist, but also to investigate the validity of a specification proposed by Gartman (1991, pp. 429): 'The evidence for class differences is systematically stronger in the field of nonmaterial culture like visual art, music and literature than in fields of material culture like food, clothing, and furniture ... There do appear to be distinct differences in the consumption of nonmaterial culture ... But in the realm of material commodities, there exists a qualitatively indistinct mass culture, which distinguishes individuals solely by the quantity of their income'. And according to still more extreme mass culture theories (e.g. the early Francfort school) there should not be relevant differences in either of the spheres.

Figure 15.1 shows the two-dimensional display of the 'social space' based on the class-fractions introduced and the photographed objects. All objects were considered, irrespective of evaluation, since the general subjective

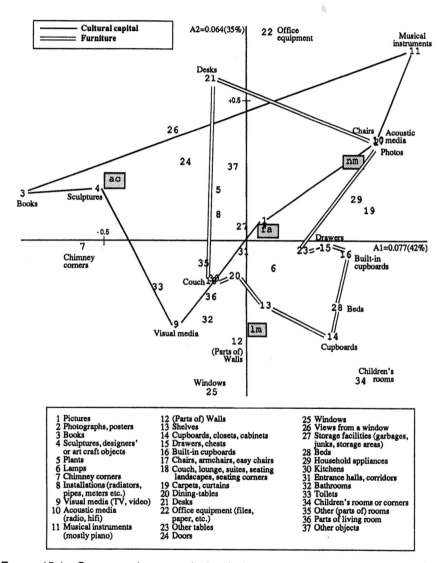

1 Pictures	12 (Parts of) Walls	25 Windows
2 Photographs, posters	13 Shelves	26 Views from a window
3 Books	14 Cupboards, closets, cabinets	27 Storage facilities (garbages,
4 Sculptures, designers'	15 Drawers, chests	junks, storage areas)
or art craft objects	16 Built-in cupboards	28 Beds
5 Plants	17 Chairs, armchairs, easy chairs	29 Household appliances
6 Lamps	18 Couch, lounge, suites, seating	30 Kitchens
7 Chimney corners	landscapes, seating corners	31 Entrance halls, corridors
8 Installations (radiators,	19 Carpets, curtains	32 Bathrooms
pipes, meters etc.)	20 Dining-tables	33 Toilets
9 Visual media (TV, video)	21 Desks	34 Children's rooms or corners
10 Acoustic media	22 Office equipment (files,	35 Other (parts of) rooms
(radio, hifi)	paper, etc.)	36 Parts of living room
11 Musical instruments	23 Other tables	37 Other objects
(mostly piano)	24 Doors	

FIGURE 15.1 Correspondence analysis of photographed objects with occupation (1st and 2nd axis). Academic professions (ac), new middle-class (nm), lower (middle) class (lm), farmers (fa).

TABLE 15.1

Correspondence analysis of photographed objects with occupation. Numerical output (for interpretation see Greenacre 1984, pp. 74ff. and this book, chapter 2). For meaning of row and column points ('NAME') see Figure 15.1.

Principal inertias (eigenvalues) and percentages of inertia

Axis	Eigenvalue	Percentage	Histogram
1	0.077133	42.10%	*************************************
2	0.063735	34.79%	******************************
3	0.042335	23.11%	********************

Row contributions to inertias

I	NAME	QLT	MAS	INR	k = 1	COR	CTR	k = 2	COR	CTR	k = 3	COR	CTR
1	1	1000	61	25	59	47	3	65	57	4	−257	897	96
2	2	1000	23	41	446	619	60	340	359	42	−84	22	4
3	3	1000	18	66	−782	895	141	160	37	7	−215	68	19
4	4	1000	46	81	−521	846	164	177	98	23	−135	57	20
5	5	1000	36	10	−102	194	5	170	541	16	119	265	12
6	6	1000	31	6	89	212	3	−96	245	5	−143	543	15
7	7	1000	18	46	−573	699	76	−22	1	0	375	300	59
8	8	1000	27	7	−103	218	4	85	150	3	175	632	20
9	9	1000	23	23	−248	344	19	−295	487	32	174	169	17
10	10	1000	18	50	443	376	45	341	223	32	456	400	87
11	11	1000	8	49	572	301	35	656	396	55	−573	302	64
12	12	1000	41	29	−46	16	1	−358	981	83	−19	3	0
13	13	1000	25	12	53	32	1	−228	593	20	−181	375	19
14	14	1000	60	65	285	410	63	−339	579	109	46	11	3
15	15	1000	34	14	259	891	30	−29	11	0	−86	98	6
16	16	1000	18	12	330	908	25	−50	21	1	92	71	6
17	17	1000	18	50	443	376	45	341	223	32	456	400	87
18	18	1000	66	16	−136	406	16	−142	441	21	84	153	11

J	NAME	QLT	MAS	INR	k=1	COR	CTR	k=2	COR	CTR	k=3	COR	CTR
19		1000	25	43	411	523	54	99	30	4	-379	446	84
20		1000	18	9	-54	33	1	-121	166	4	-266	800	30
21		1000	18	41	-141	47	5	558	738	87	-300	214	38
22		1000	14	44	56	5	1	724	889	112	250	106	20
23		1000	25	23	180	189	10	-37	8	1	372	803	80
24		1000	25	18	227	378	16	272	544	29	103	77	6
25		1000	15	28	-134	53	4	-530	832	66	-197	115	14
26		1000	22	26	-271	343	21	376	657	48	10	0	0
27		1000	12	1	-30	76	0	41	136	0	98	789	3
28		1000	19	17	302	567	23	-234	341	16	-122	92	7
29		1000	40	33	366	872	69	139	126	12	15	1	0
30		1000	31	14	-127	200	7	-130	210	8	218	590	35
31		1000	33	1	-24	90	0	-40	236	1	67	674	3
32		1000	22	28	-149	95	6	-280	336	27	364	568	68
33		1000	8	14	-323	323	11	-163	83	3	-438	595	37
34		1000	11	25	382	355	21	-490	583	41	-160	62	7
35		1000	36	11	-163	460	12	-78	105	3	-159	435	21
36		1000	18	6	-137	313	4	-195	637	11	-54	49	1
37		1000	40	16	-63	55	2	256	913	41	48	32	2

Column contributions to inertias

J	NAME	QLT	MAS	INR	k=1	COR	CTR	k=2	COR	CTR	k=3	COR	CTR
1	ac	1000	212	316	-472	816	612	211	162	147	-76	21	29
2	nm	1000	236	281	352	568	379	275	346	280	-137	86	105
3	lm	1000	367	213	18	3	2	-314	931	570	-84	66	61
4	fa	1000	184	190	55	16	7	33	6	3	430	978	805

relevance of objects was of primary interest for this analysis. The figure is the graphical result of a CA. It is a symmetric display (for the geometric interpretation see Chapter 1) of the clouds of the columns (class-fractions) and the rows (photographical choices of objects) of the frequency matrix.

The total inertia represented in the plot is 76.9%. Table 15.1 gives the numerical results of correspondence analysis in detail in the output format of Greenacre's SimCA. It contains the principal inertias (eigenvalues), the percentages of inertia, the quality of representation (QLT) of the points, their masses (MAS) and their inertia (INR) in the full space, the coordinates (k), and the relative contributions for the row and column points (CTR) for the three principal axes. The (squared) correlations (COR) of rows and columns with the principal axes, which indicate the part of the variance of a variable explained by a principal axis, can be used for the 'factor-analytical' interpretation of the results. All quantities concerning the row and column contributions are expressed in permills (cf. Greenacre 1984, pp. 69, 75). The numerical 'names' of the row and column points can be decoded with the help of the legend in Figure 15.1.

From Table 15.1 we can see that the horizontal axis (42.1% of total inertia) in Figure 15.1 is defined by the professional-academic elite (contribution to inertia CTR = 0.61) on the left side opposing the members of the new middle classes (CTR = 0.38) on the right. The vertical axis (34.8% of total inertia) shows the contrast between the lower middle classes (CTR = 0.57) and both the professional-academic elite (CTR = 0.15) and the new middle classes (CTR = 0.28), which slightly determine the positive side of this axis. The third axis (23.1% of inertia), which is not represented in Figure 15.1, separates all other fractions from the farmers, being strongly determined on the positive side by this class fraction (CTR = 0.81).

For purposes of better visualizing the results, objectified cultural capital and furniture were connected with lines. Figure 15.1 indicates, that the 'one culture' hypothesis of mass culture theory can be refuted. There is a sufficient amount of material as well as cultural objects, that are differentiated according to class. As correspondence analysis reveals, the four class fractions are situated apart from each other in 'social space'.

The two most specific objects for the academic-professional elite are books and sculptures. Numerically this can be seen when comparing the third row (books) and fourth row (sculptures) with the other rows of the 'row contributions' part of Table 15.1. Objects 3 and 4 have the highest squared correlations (0.90 and 0.85) of the objects which are located on the left side of the first axis.

Most specific for the members of the new middle class are musical instruments (mostly pianos) and photographs/posters, both loading on the first axis (0.30 and 0.62, respectively) and the second axis (0.40 and 0.36), which is also partly determined by the new middle classes. With squared correlations of 0.83

and 0.58, windows and children's rooms are the most specific choices for members of the lower middle classes on the negative side of the vertical axis. Very characteristic for farmers, who are represented on the positive side of the third axis (not shown), is the residual category of tables (not desks or dining tables) and the bathrooms (0.80 and 0.57), the latter perhaps being due to their kind of physical outdoor work.

The systematic differences between the two lower and the two higher class fractions is concerned with objectified cultural capital. Whereas books, sculptures, musical instruments and photographs/posters are of special relevance for the professional-academic elite and the members of the new middle classes, only the technical 'mass' media are modestly associated with the class fractions lacking economic and cultural capital: the visual media (tv, video) show associations with the lower middle classes (squared correlation of 0.49 on negative side of second axis), the acoustic media (radio, hifi) with the farmers (0.40 on positive side of third axis).

These results generally confirm Bourdieu's class culture hypothesis. The inspection of the data concerning other cultural objects yields further interesting results. There are marked contrasts between the professional-academic elite and the new middle classes. Books and sculptures have a strong academic, photographs and musical instruments a strong new middle class connotation. The professional-academic elite, being occupationally more concerned with words than with pictures, obviously reproduces this condition of existence also in the private context. On the other hand, the members of the new middle classes display some of the behaviours described by Lash (1988, p. 313 ff.), who contrasts a 'discursive' orientation, giving 'priority to words over images' with a 'figurative' (postmodernist) orientation, implying a 'visual rather than a literary sensibility'. The opposition shown by CA between the professional-academic elite and the partly postmodern new middle class fractions confirms this description. The new middle classes are much less concerned with words (books) than with posters and photographical pictures. The new middle class connotations of poster and photography correspond well to Bourdieu's (1981) characterization of photography as 'middle culture' and the description of the new middle culture by Gans (1985 p. 45), who sees some of its significant features in photographic realism and poster art.

Another result regarding cultural objects is worth mentioning, too. Two-dimensional pictures and visual art, when only crudely differentiated according to type (photographs, posters versus other pictures), but not to content, do not have a 'distinguished' connotation. Photographs and posters were already shown to be especially relevant for the new middle classes. Pictures do not load on the first and second axis, but are highly correlated with the third (COR = 0.90). This indicates that there are no differences between the two higher class fractions and the lower middle classes in this respect. It is the group of farmers which is much less concerned with pictures.

Three-dimensional sculpture, however, is highly distinctive for the professional-academic elite, even without any differentiations according to type of sculpture. Since outside the 'delimited field' of the artworld (cf. Bourdieu 1985c), sculpture is much less popular than painting, sculptures seem to be especially suitable as signs of distinction for the elite.

The specification of Gartman is not corroborated by the data in any convincing way. With regard to rooms, which clearly are 'material' objects, there are marked differences in the probability of photographing children's rooms (high in the lower middle classes, low in the elite) and moderate differences concerning bathrooms (more often the choice of farmers and members of the lower middle class than of the two higher fractions). Chimney corners, which belong to the material luxury object category, are most specific for the academic-professional elite (COR = 0.70, negative side of first axis), corroborating Bourdieu's assumption regarding the taste of freedom and luxury in the centres of the power field. A further falsification of Gartman's specification is the sharp contrast between desks (COR = 0.74) and office equipments (COR = 0.89) on the upper side of the second axis, and windows (COR = 0.83) and walls (COR = 0.98) on the lower side of the second axis. While desks and office equipments are specific for the professional-academic elite and for members of the new middle classes, reflecting their paperwork white collar occupations in the private sphere, these objects were only rarely photographed by members of the lower middle classes. And vice versa, walls and windows turned out to be objects, that are especially relevant for those in low positions, whereas they are quite irrelevant for the groups higher in the hierarchy of class.

Furthermore some furniture pieces apart from desks have clear class connotations. In Figure 15.1, desks and cupboards are furthest apart in 'social space', the latter having a lower middle class connotation. It was already shown, that 'other tables' are of specific interest only for farmers. Theoretically these results amount again to a refutation of Gartman's specification. From a methodological point of view they call in question the decision of Csikszentmihaly and Rochberg-Halton (1981) to lump together all sorts of furniture (with the exception of beds) into one category in their empirical study referring to the 'meaning of things' in the home. Based on a similar sample they report to have found no association between class and subjective concern with furniture (1981, p. 280), but this zero correlation might be a mere methodological artifact, because they did not pay attention to the 'smaller' distinctions between objects. A further reason might be the neglect of differences between class fractions, due to their crude dichotomous measurement of class.

The result, that home furnishings − a classical instance of Gartman's material culture − have class specific meanings, are supported by the theories of Bourdieu as well as Gans (1985, p. 52). Moreover, class and class fraction differences with respect to the furnishing of apartments, are also documented in several studies concerning the objective presence or absence of objects (e.g. Laumann and House 1970, Armaturo *et al.* 1987, Blasius 1993). The observation of Engel and Blackwell (1982, p. 134), that 'criteria used by families to furnish a home appear to be closely related to social class', is supported by a broad objective and subjective evidence. In view of all these data, Gartman's attempt to specify Bourdieu's theory does not seem to be viable or fruitful.

15.7 THE SPACE OF GENDER, AGE AND PHOTOGRAPHED OBJECTS

The most prominent concepts for characterizing male and female roles in sociology are based on Parsons and Bales's (1955) typology of instrumental and expressive functions. Csikszentmihaly and Rochberg-Halton (1981, pp. 106, 270) maintain that these distinctions also apply to the relationships to things, but they modify the typology of Parsons and Bales. Building on Hannah Arendt's famous differentiation between 'vita activa' and 'vita contemplativa', they propose two classes of objects: 'action objects' and 'contemplation objects'. This distinction is supposed to reflect on the level of symbolic household objects, the distinction between instrumental male and expressive female roles.

Action objects invite kinetic involvement and require some physical manipulation to release their meaning. Examples explicitly given for action objects that are also part of our list of photographed objects are: beds, musical instruments, television sets, stereos, radios, plants, appliances, fireplaces and bathrooms. Contemplative objects, on the other hand, 'do not require physical interaction'. Their 'use is mainly through reflection or contemplation'. The examples given refer to traditional objectified cultural capital in our sense: photographs, books, visual art (paintings) and sculpture.

According to the theory of Czikszentmihaly and Rochberg-Halton (1981, p. 105) men and women pay attention to different things in the same environment. Males will cherish 'objects of action' more frequently, whereas women will prefer 'objects of contemplation'. They assume that what objects are available, how one should react to them and why, are issues 'scripted' by the culture into which one is born, and that one of the most universal differences between culturally defined roles is the one ascribed to gender.

Age is considered as an additional variable in the second CA dealing with the issue of gender. Besides being interesting as an interaction variable, age — which has a 'double meaning' as temporal location in terms both of personal career and of history (Riley *et al.* 1988, p. 256) — should be important in its own right. The importance of age was also stressed by Csikszentmihaly and

Rochberg-Halton (1981, p. 96), who noticed a 'turn with age from action to reflection', with the young more often cherishing objects of action, and the old being more concerned with objects of contemplation.

Age was dichotomized, with 'younger' denoting all aged between 20 and 40 years, and 'older' those of 41 years and beyond. Interpreting as cohorts, 'younger' refers to the affluent post-war generations, while 'older' refers to those generations that experienced one or both of the world wars and the period(s) of crisis after the war(s). The cross-classification of age and gender thus leads to four groups: (a) men till 40 years, (b) women till 40 years, (c) men 41 years and older, and (d) women 41 years and older.

Figure 15.2 shows the two-dimensional display of the social space based on gender, age and the photographed objects. The age-gender groups form the column variable, the photographical choices of objects the row variables of the underlying matrix.

The total inertia represented in the plot is 79.9%. The first axis explains 51.6%, the second 28.3% and the third, which is not represented in Figure 15.2, 20.1% of the inertia.

The horizontal axis is on the left side defined conjointly by younger men (CTR = 0.24) and younger women (CTR = 0.21), and on the right side conjointly by older women (CTR = 0.41) and older men (CTR = 0.14). It can be interpreted as an age axis. The vertical axis is in the negative sphere determined by older women (CTR = 0.20) and by younger women (CTR = 0.18), in the positive sphere by older men (CTR = 0.50) and by younger men (CTR = 0.13). It can be interpreted as a gender axis. Axis 3 is determined in the positive sphere more by younger women (CTR = 0.28) than by older men (CTR = 0.17) and in the negative sphere more by younger men (CTR = 0.43) than by older women (CTR = 0.14). It can be interpreted as an 'interaction' axis.

Considering the first axis, the objects with the strongest connotations of 'young' are children's rooms (COR = 0.52) and beds (COR = 0.88) on the left side, while views from a window (COR = 0.67) have the strongest connotations of 'old' on the right side. On the second axis, most specific for male are chimney corners (COR = 0.98) and books (COR = 0.79) on the upper side, most specific for female are the choices of bathrooms (COR = 0.54), drawers (COR = 0.77) and children's rooms, which are loading only weakly on the lower side of this axis (COR = 0.21).

The points representing male and female till 40 years in Figure 15.2 are closer to one another than the points representing older men and women. Differences between men and women seem to have diminished in the younger generations, which reflects the 'gender revolution' in the advanced societies due to the joint effects of structural changes in the participation of women in the public sphere and the influence of women's liberation movements since the 1960s (cf. Giele 1989). Nevertheless, there remain some clear gender contrasts in the younger generations. Men are practically not concerned with children's

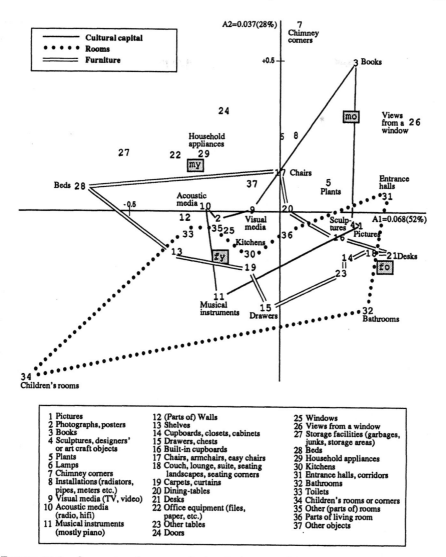

FIGURE 15.2 Correspondence analysis of photographed objects with gender and age (1st and 2nd axis). Male and 20–40 years (my), male and 41 years and older (mo), female and 20–40 years (fy), female and 41 years and older (fo).

rooms. Chimney corners, books and – as can be seen in Table 15.2, which gives the results for the third axis – storage facilities and musical instruments are more specific for younger men than for younger women. On the other hand, younger women – apart from children's rooms – are more concerned with drawers, photographs and posters.

TABLE 15.2

Correspondence analysis of photographed objects with gender and age. Numerical output (for interpretation see Greenacre 1984, p. 74 and chapter 2 of this book). For meaning of row and column points ('NAME') see Figure 15.2.

Principal inertias (eigenvalues) and percentages of inertia

Axis	Eigenvalue	Percentage	Histogram
1	0.067506	51.64%	**************************************
2	0.036938	28.26%	*********************
3	0.026282	20.10%	*******************

Row contributions to inertias

I	NAME	QLT	MAS	INR	k = 1	COR	CTR	k = 2	COR	CTR	k = 3	COR	CTR
1	1	1000	62	37	259	857	62	-43	23	3	97	120	22
2	2	1000	24	40	-199	176	14	-19	2	0	429	822	165
3	3	1000	18	40	244	207	16	478	793	111	-13	1	0
4	4	1000	32	23	234	568	26	-36	14	1	201	418	49
5	5	1000	36	12	156	572	13	89	188	8	101	240	14
6	6	1000	32	15	5	0	0	240	929	50	-66	70	5
7	7	1000	18	52	64	11	1	606	976	179	-69	13	3
8	8	1000	28	15	51	37	1	241	808	44	106	155	12
9	9	1000	24	2	-89	848	3	4	2	0	-38	150	1
10	10	1000	17	8	-250	974	15	14	3	0	39	23	1
11	11	1000	8	19	-207	142	5	-271	244	17	-430	614	59
12	12	1000	42	46	-315	687	61	-19	2	0	-212	311	71
13	13	1000	25	37	-343	602	43	-125	80	11	-249	318	59
14	14	1000	61	41	207	490	39	-141	229	33	-157	281	57
15	15	1000	35	33	-56	26	2	-308	770	89	158	204	33
16	16	1000	18	15	180	293	9	-79	57	3	-268	650	49
17	17	1000	18	2	-8	4	0	123	975	7	-18	21	0
18	18	1000	67	54	283	764	79	-132	165	31	-87	72	19

	QLT	MAS	INR	k = 1	COR	CTR	k = 2	COR	CTR	k = 3	COR	CTR
19	1000	25	9	-106	236	4	-174	637	20	-77	127	6
20	1000	18	4	14	8	0	7	2	0	162	990	18
21	1000	18	21	352	822	33	-134	119	9	94	58	6
22	1000	14	16	-346	787	25	177	207	12	30	6	0
23	1000	24	13	186	467	12	-193	502	24	48	31	2
24	1000	25	27	-189	250	13	322	719	70	66	31	4
25	1000	15	7	-174	475	7	-56	49	1	174	475	18
26	1000	22	46	426	668	60	283	294	48	102	38	9
27	1000	12	39	-511	646	48	189	88	12	-328	266	51
28	1000	19	72	-653	879	123	79	13	3	-228	108	39
29	1000	40	38	-254	517	38	184	271	37	163	212	41
30	1000	32	8	-107	373	5	-134	580	15	-38	48	2
31	1000	33	32	337	903	56	47	18	2	-100	80	13
32	1000	22	31	272	403	24	-314	535	59	-107	62	10
33	1000	14	12	-306	804	19	-69	41	2	-134	155	10
34	1000	11	112	-827	518	112	-532	214	85	595	268	149
35	1000	36	14	-213	917	24	-48	47	2	42	35	2
36	1000	15	1	12	24	0	-71	849	2	27	128	0
37	1000	40	5	-99	560	6	88	439	8	-5	1	0

Column contributions to inertias

J	NAME	QLT	MAS	INR	k = 1	COR	CTR	k = 2	COR	CTR	k = 3	COR	CTR
1	fy	1000	341	214	-205	514	213	-139	234	177	144	252	269
2	fo	1000	258	293	326	714	406	-169	193	200	-118	93	136
3	my	1000	204	246	-282	503	239	152	146	127	-235	351	430
4	mo	1000	197	247	220	297	142	305	568	496	149	135	166

The action-contemplation theory of Csikszentmihaly and Rochberg-Halton is only partially confirmed by the data. According to the theory the action objects should be located primarily in the younger-male upper left region of the 'social space' represented in Figure 15.2, the contemplation objects in the older-female lower right region.

Only two out of the eight action objects, appliances and beds, are located according to the theoretical expectations. Acoustic and visual media are placed very near the horizontal axis, chimney corners and plants in the upper right quadrant, bathrooms in the lower right and musical instruments load on the third axis (COR = 0.61) with negative sign, which means that they are associated with younger men and older women, which should definitely not be the case. Regarding the four contemplation objects only sculptures and pictures are situated in the expected older-female quadrant. Books are characteristic for men over 40 and photographs (COR = 0.82) for younger women as well as for older men.

When restricted to male–female contrasts on the second axis, the theory gives better predictions. Only one of the seven action objects (bathrooms) which are well represented in Figure 15.2, is located in the female sphere below the horizontal axis. Action object eight, musical instruments, is slightly more male–young than female–old. And also three of the four contemplation objects (sculptures, pictures and photographs) are (slightly) more female than male, which gives the restricted theory some plausibility. But in view of some strikingly disconforming cases (e.g. bathrooms should be (younger) male but are actually (older) female) the theory needs further elaboration and specification.

It was already demonstrated that the traditional sexual division of labour is reflected in the result of our projective test, that practically only younger women selected children's rooms or corners. But what about photographing the kitchen, which belongs to the most significant symbols of the traditionally defined female role? As expected, the kitchen turns out to be female (more young) and not male. But, perhaps surprisingly, the kitchen does not belong to the most specific female attributes. The part of the 'female habitus' (cf. Krais 1993) symbolized by the kitchen seems to be much less socially fixed than the affective or expressive role that refers to child rearing. Not only some furniture pieces (especially drawers) or photographs, but also bathrooms, the sphere where the preparations for 'the presentations of self' take place, are more strongly associated with female than is true for the kitchen.

The position of the polygon that circumscribes the rooms shows that rooms in general have a connotation that is slightly more female than male. This seems to reflect the greater responsibility of women for the home and the condition of the rooms. The same is true for furniture. Out of eleven objects only beds and chairs are (slightly) more male. Dining tables are highly positively correlated with the third axis (COR = 0.99), which means that they are photographed by younger women and older men, as well. The other eight furniture pieces have (most often older) female connotations, which corresponds to the results of marketing studies, which show that women tend to dominate the purchase decisions for furniture (cf. Engel and Blackwell 1982).

With only a few exceptions (chimney corners, books, children's rooms) age and generation, though only measured crudely, seem to be a more important factor for explaining the choice and meaning of objects in the sphere one lives in than gender (indicated also by the fact, that the age axis has much more weight than the gender axis). The age-related differences within the category of objectified cultural capital are not very marked, but in the categories of furniture (beds versus desks) and rooms (children's rooms versus bathrooms) there are distinct contrasts. But it was important to consider both variables, since gender turned out to be a relevant factor at least for a few choices and there could be demonstrated some weak, but systematic associations with this variable (the female connotations of rooms and furniture). With regard to some of the object choices, interaction between gender and age should not be ignored, as can be seen most clearly in the case of objects, which load on all three axes (children's rooms), on the first and the second (bathrooms) or highly on the third interaction axis (photographs and musical instruments).

15.8 SUMMARY

Correspondence analysis was applied in the neglected field of the sociology of objects and taste cultures. The data were collected by means of photo-questioning, a newly developed projective technique of data collection. A quota sample of $n = 80$ subjects was asked to photograph objects in the sphere they live in. The choices of objects were interpreted as elements of a symbolic space in the sense of Bourdieu, whose concept of 'social space' served as a general theoretical frame of reference for the empirical analyses.

The projective choices of objects were related first to class fractions, and then to age and gender. Correspondence analysis seemed especially well-suited as a statistical instrument, since relations and distances of age–gender groups and class fractions in 'social space' were of primary interest. According to Bourdieu's hypothesis that relevant distinctions between actors often are based on subtle differences in practices and choices, 37 categories of objects were distinguished and introduced as rows in CA. These objects were subsumed under theoretical concepts (e.g. objectified cultural capital, material versus nonmaterial culture, action versus contemplation objects) derived from some specific theories, which were introduced in supplementing the general theoretical frame.

Four occupational groups (professional-academic elite, new middle classes, lower (middle) classes, farmers) were distinguished in the sample according to volume and structure of economic and cultural capital and it was shown how these class fractions are situated in Bourdieu's (1989, 1979) macro-social structural space. Bourdieu's class symbolization theory was reconstructed and confronted with some alternative positions, which converge in the assumption that all classes participate more or less in the same mass culture and that class differences are not symbolized any more in a relevant way. Gartmann (1991) supposes that class culture symbolization is systematically stronger in the fields of 'nonmaterial' culture (the arts, objectified cultural capital), than in the fields of material culture (furniture, food, etc.).

Of all these assumptions the one of Bourdieu seems to be best corroborated by the data. The four class fractions are situated apart from each other in the representation of 'social space' (in three dimensions) obtained by means of CA. What systematically differs, for example, between the two lower and the two higher class fractions is concern with traditional objectified cultural capital (or nonmaterial culture). The professional-academic elite is contrasted with the new middle classes in the dimension of dicursive (books) versus figurative (posters, photographs) orientations. These contrasts reveal widely divergent object-choices, indicating relevant differences in class fraction cultures, as assumed by Bourdieu or Gans. They refute critical as well as neo-conservative and post-modern proclamations of a unified or 'imploded' culture. Also the specification of Gartmann is not confirmed in any convincing way with our subjective-projective data, since CA shows some sharp contrasts between objects of material culture, too (e.g. photographs of chimney corners and desks in the higher classes versus windows, walls and cupboards in the lower middle classes).

With regard to age and gender, Csikszentmihaly's and Rochberg-Halton's (1981) typology of action and contemplation objects was used. They assume that young and male are associated with action objects, and old and female with contemplation objects. Only when restricted to gender, some support could be obtained for this assumption. All in all age turned out to be more important than gender, but also some clear gender-related differences could be found, the most evident being that children's rooms or corners are highly specific object choices of (younger) women. This result supports theories that consider child–mother relations as being of special relevance for the formation and construction of female identity (cf. Chodorow 1989).

Apart from this evidence referring to theories, CA also brought forth some interesting heuristic results, regarding, for example, the different social-structural connotations of two-dimensional pictures (not specific for higher class fractions) and three-dimensional sculptures (specific for the professional-academic elite), or the different relevance of rooms (moderate, but systematic association with female).

On a more general level the results confirm that the social world does not present itself 'as a pure chaos, as totally devoid of necessity and liable to being constructed in any way one likes' (Bourdieu 1989a, p. 19). A considerable part of the photographical choices of objects are structured according to social positions and social conditions. The structuring of the projective object-choices so far analyzed (by age, gender and class, but also by education and mobility, not reported here), indicates that the philosophical assumption of the 'cognitive autonomy of subjects', as postulated by dominant variants of radical constructivism (cf. Schmidt 1987), seems to rest on an imaginary anthropology.

16

Product Perception and Preference in Consumer Decision-making

H. M. J. J. (Dirk) Snelders and Mia J. W. Stokmans

16.1 THE USE OF CORRESPONDENCE ANALYSIS IN CONSUMER RESEARCH

For the most part, economic theory about consumer behaviour assumes that consumers try to maximize their utility in deciding what product to buy (for an overview, see Savage 1954, Chapter 5). By this is meant that when buying a product, consumers must first assess what product best satisfies their needs. This assumption has long been thought to mean that consumers derive utility from the product itself. Lancaster (1966) has put forward the view that consumers evaluate a product on the basis of its attributes. This suggests that consumers' evaluations of products are not only based on an overall comparison of product alternatives, but also on the relative importance of the attributes of the product.

Consumer research is interested in how consumers decide what product to buy, and they distinguish two research questions.

(1) What is the position of a product in the market? That is, how do consumers compare product alternatives in the market?

(2) What attributes are important in the evaluation of the product?

The first question, of product positioning, is usually answered by delineating the consumers' perception of the market. This 'perception analysis' is discussed in section 16.1.1. Techniques that are often used to determine the importance of product attributes are laddering and conjoint analysis. These will be discussed in section 16.1.2.

16.1.1 Perception analysis

Perception analysis deals with how consumers understand products, and it is an attempt to find general underlying dimensions of the image that consumers have of products. Typically, this is done by having consumers compare different product alternatives and see what the main distinctions are that consumers make between products. Perception analysis is usually carried out with the help of factor analysis or multidimensional scaling (MDS). Both analyses result in a 'perceptual map' (Urban and Hauser 1980, Green *et al.* 1988). This is a multidimensional space in which the perception of a product relative to other products in the market is reflected. The dimensions which make up the perceptual map are regarded as the critical perceptual qualities of the products. The position of a product on a dimension can therefore be interpreted as the extent to which the product possesses such a perceptual quality.

The interpretation of the perceptual dimensions is often implicit and often left to the imagination of researchers. However, for both factor analysis and MDS it is possible to have explicit attribute ratings helping in the interpretation of the dimensions. In the case of factor analysis respondents rate the products on a given list of attributes. In practice, these attribute ratings are often used to develop a perceptual map. When applying MDS, respondents are asked how similar products are to each other. Here the perceptual map is created on the basis of these inter-product similarities. Additional attribute ratings can be fitted in this perceptual map. A few comments can be made regarding the interpretation of perceptual maps that are created by factor analysis or MDS.

(1) If a large number of products are included in the research, the data collection for both techniques is so repetitive that it is boring for the respondents.

(2) All determinant attributes which make up the product should be known in advance. But the perception of the product by the consumers could well differ from the perception by the researchers. In addition, different groups of consumers may find different attributes relevant.

(3) The attributes should be expressed in terms that are meaningful to the respondents. But consumers do not use the same words to express the same attributes.

These three problems can all be thought of as problems in the data collection procedure, and they can be overcome by using ASSPAT or natural grouping instead of the techniques described above. ASSPAT, short for 'associative patterns' (Verhallen 1988), is a technique where respondents are presented with a number of products simultaneously and asked to pick out those products that are associated with a particular attribute (for a similar approach see Shahim and Greenacre 1989). The result of this is a two-way frequency table with products in the columns and attributes in the rows, with each cell showing how often one particular product is associated with one particular attribute. From this table, a qualitative profile of products can be computed with the help of CA, showing an association pattern of a set of products and their attributes.

An alternative to ASSPAT is a technique named natural grouping. This technique was first introduced by Kuylen and Verhallen (1988), and can be regarded as a special case of 'subjective clustering' (Green *et al.* 1969). The technique starts when respondents are presented with a set of products, and they are asked to split these products into two separate groups. In addition, respondents have to label each group with an attribute name. Next, people are asked to make further splits into yet smaller groups, just as long as respondents feel that they can create meaningful groups. Again every group that is formed has to be labelled by the respondents with attribute names.

While ASSPAT has the disadvantage that the attributes should be specified in advance, with natural grouping the attributes of a set of products are generated by the respondents themselves. Consequently, all relevant attributes are included and they are all stated in terms that are meaningful to the respondent. Surprising perhaps to those researchers who feel that the more standard survey methods would be better understood by respondents, natural grouping is not a difficult task for respondents. Respondents find these grouping tasks easy to carry out, and can denominate the groups that they create with little effort. Natural grouping apparently asks people to do something which they are very good at. Indeed, it is undisputed among cognitive psychologists that the categorization of objects is an integral part of people's thinking (see Lakoff 1986 for an overview). The data resulting from natural grouping can also be analyzed by means of CA. In section 16.2 we shall illustrate this with research on how consumers view product design.

16.1.2 Attribute importance

Attribute importances can be explored in many different ways. Here we shall discuss two of the most popular techniques where correspondence analysis has been applied: laddering and conjoint analysis. Consider first, laddering, which is a method for uncovering means–end chains (Gutman 1982). A means–end chain is a cognitive structure which reflects the connections between product attributes, the consequences that these attributes have for consumption and the personal values that make these consequences important for consumers. Laddering consists of a series of direct probes based on mentioned distinctions (attributes, consequences or values) the individual has with respect to the competitive products in order to trace the means–end chain.

Laddering data can be handled in various ways. One way is the construction of a 'hierarchical value map' which summarizes the most frequently mentioned sequences of the categories that have been named in respondents' ladders. This can be done by simply mapping the most frequent sequences (Reynolds and Gutman, 1988) or, in a more formalized way, by using graph theory (Valette-Florence and Rappacchi 1989). The advantage of the latter technique is that one can specify which formal characteristics the hierarchical value map should have, and hence change its structure faster and more reliably.

An alternative way of analyzing laddering data has been proposed by Valette-Florence and Rappacchi (1989), and this involves CA. Here, however, only the frequencies of the attributes, consequences or values serve as input; no information is taken from the sequence in which these answer categories were put by the respondents. The frequency table to which CA is applied has answer categories in the columns, respondents in the rows. Analysis of this table results in a reduced Euclidean space that shows how different groups of respondents stress different product attributes, consequences and values when assessing the importance of a product.

While laddering is used to show which attributes are important and for what reasons, conjoint analysis is more suitable for estimating the extent to which these attributes are important. Essentially conjoint analysis can be summarized as follows (Green and Srinivasan 1978, 1990, Wittink and Cattin 1989). The respondent is asked to state his/her preference (the dependent variable) for a set of products, which are systematically varied in their attributes (the independent variables). The variation in these attributes is created by specifying different levels for them. For example, the attribute price can be offered to respondents at different levels ($10, $15 and $20). The part-worth, i.e. the effect of an attribute level on the preference for the product, is estimated on the basis of the stated preference scores.

The results of conjoint analysis consist of part-worths estimated for each attribute level per respondent. The relative importance of a particular attribute on the overall preference for a product, is determined by calculating the largest absolute difference in the part-worths of the product attribute: the sensitivity. These sensitivities have to be aggregated in order to identify those attributes that best fulfil the wishes and demands of homogeneous groups of consumers (Montgomery and Wittink 1980). In section 3 it will be illustrated that CA can provide a suitable aggregation procedure for conjoint analysis results.

Above we have given a brief introduction to ways in which CA can be used, both for research on product positioning as well as research on the importance of product attributes. In the following sections we shall give two examples of the use of CA in consumer research.

16.2   THE POSITIONING OF PRODUCTS ON THE BASIS OF THEIR DESIGN

The more products within a product category are converging in terms of their technical functioning, the more the competitive edge of the product is provided by its non-technical aspects. The most prominent non-technical aspect of a product is its design. Although product design is expressed in the product's concrete form, we should acknowledge that the designer's intentions go beyond creating a mere physical structure. This is because the design of a product also communicates in what context we should view it. In this sense design can be considered as an intangible aspect of the product, capable of conveying the product's symbolic aspects to the consumer.

Product design is mostly considered to be a designer's affair, but it may as well be regarded as a consumer's problem. Consumers are often bemused by product design and can wonder what exactly is meant by it. The complete range of physical and symbolic aspects of a design is hardly ever conveyed to the consumer. To make things worse, consumers can add functions to products that were never even intended by the designer (Jencks and Silver 1972). So, there need not be an inherent connection between how a product is conceived by the designer and how it is viewed by the consumer. Although products may often be referred to by the consumer as 'designed'; to begin with, they are nothing but 'extended in space' (Descartes 1977), and it is up to the consumers themselves to define and handle this space. Thus it may be worth looking into the interaction between the consumer and the product's design and see what the consumer's understanding of the product is.

Consumer interaction with product design starts with the perception of the product. Perception may influence the consumer's understanding of the product's design by determining how the product's form is being reconstructed in the mind of the consumer (e.g. Smets 1986). However, the consumer's image of product design reflects on more than what is being perceived. Product design is not only understood by consumers on the basis of its looks, but also for what it is being associated with (Reynolds and Gutman 1984). A sports-car does not have a sporty image simply because it is red, but the colour red may be associated with speed and it is through this association that the consumer regards red sports-cars in a different manner. So the consumer's image of products is based both on product perception and its associative context.

Here we want to illustrate the positioning of product design on the basis of the image the consumer has of it. Earlier we have explained how a product's image is typically measured. For a product aspect that is as intangible as design, we feel that one cannot determine in advance what attributes it is associated with (unlike Holbrook 1986). If, however, we let consumers themselves come up with the relevant attributes, what do we get? Will each consumer's responses be completely idiosyncratic and therefore incomparable to the responses of another consumer? Or, alternatively, can we position the designs of different products on the basis of these subjective qualifiers that consumers supply us with? Is there such a general thing as 'tacky' and 'cool' cars, 'stunning' and 'dull' coffee machines, 'old fashioned' and 'hyper modern' stereos?

As suggested before, the best way of positioning a product while being explicit about its attributes and leaving the choice of attributes to the consumer is natural grouping. The procedure of this technique results in a structure which shows what criteria (i.e. the product attributes) the respondents use to discriminate one set of products from another. Figure 16.1 gives an example of such a structure.

There has been some debate on how to interpret this structure. Although these structures by themselves express only how people categorize groups of products, some have argued that the number of steps down the structure during which products remain in the same group can be regarded as a measure of similarity (Green *et al.* 1969). In the example of banks in Figure 16.1: A would be regarded as very similar to B, somewhat less similar to C and D, and different from E, F and G. This way, an individual tree-structure can be transformed into measures of similarity between products. This transformation provides the opportunity to aggregate the individual structures and to create one perceptual space for all respondents, with the help of MDS.

The suggestion of converting grouping structures into similarity data goes back before the introduction of natural grouping (Poiesz 1982). Yet, when natural grouping was introduced, the most appropriate method for it was thought to be correspondence analysis. Here, the individual structures are converted into one frequency table, and this table serves as the input for a CA.

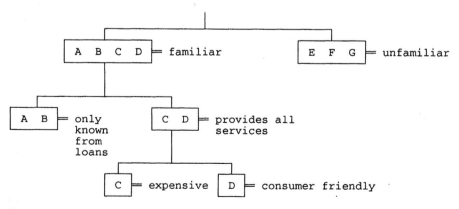

FIGURE 16.1   Example of the tree-like structure from one respondent's natural grouping of banks (after Kuylen and Verhallen 1988).

The transformation of the individual structures into an aggregate frequency table proceeds as follows. First, the attribute names that denominate the different groups are content-analyzed, resulting in a limited set of attribute categories. Next, the frequency that each product falls under one of the attribute categories is counted over all respondents, resulting in one aggregate frequency table, with the products in the columns and the attribute categories in the rows.

A somewhat similar analysis has been suggested by Steenkamp (1989) for analyzing natural grouping data, namely procrustes analysis (Gower 1975). However, there are several reasons for deciding against the use of procrustes analysis for natural grouping data, both for practical and more theoretical reasons. First, procrustes analysis needs complementary data in the form of attribute importances or saliencies. Unlike MDS, where each product has to be rated on each attribute, here one needs information only about the importance or saliency of the attributes themselves. Although this is relatively little work for respondents, it still implies that the natural grouping task has to be complemented with another task in order to conduct the analyses properly. Secondly, it is impossible to ask respondents how salient an attribute is without referring to a specific product. This implies that one can only ask consumers to rate general importances of attributes. However, there are severe difficulties with the notion of 'general importance'. Attribute importances do not say anything about the consumer's general understanding of attributes, but only about their function in specific situations of use (Myers and Shocker 1981). Thus procrustes analysis undermines the reason for giving respondents a natural grouping task, namely to provide perceptual, rather than preferential data.

Results from CA on natural grouping data have been compared to MDS solutions (Verhoeven 1991) and have been shown to be remarkably similar. However, there are two reasons for using CA instead of MDS. First, with CA, the dimensions on which products are portrayed can be interpreted with the help of the attribute names that respondents have given. Secondly, we feel that the information in the tree-like structures that natural grouping provides is not strictly ordinal. The number of times that two objects share the same group only expresses the similarity between these objects to the extent that every shared group is of the same relevance to the meaning of these objects. Since this is not necessarily the case, the measures of similarity on which MDS is based are error-prone.

So here we shall look at the positioning of products on the basis of what consumers think of their design. This shall be done with the help of a CA of natural grouping data. These data describe how consumers group products, and how they do this on the basis of their own impressions of these groups. The products under investigation are telephones. The question we shall try to answer is whether it is possible to interpret the results of a CA of natural grouping data, considering that these have been derived from the impressions of a very subjective aspect of these telephones, namely, their design.

Method

Forty-three members of the consumer panel of the Delft University of Technology were invited to our laboratory to take part in the natural grouping task. The panel is a random sample of the inhabitants of Delft and the surrounding area. Seven telephones were used in the task, four of these were selected on the basis of the diversity in their appearance, and three more were added randomly. The telephones that were used are all regular telephones, on sale at the telephone shop in The Netherlands with the largest market share. They are shown in Figure 16.2.

On average, the procedure of natural grouping lasted about 10 minutes[1]. The procedure started with instructions to the respondents. First, they were shown how the natural grouping task worked, and they were asked only to make distinctions between sets of telephones on the basis of their design. They should think of the telephones as being 'all of the same price and quality.' Next, they were shown the seven telephones, which they had to look at carefully. Then the actual task started, and the respondents had to split the group of telephones and denominate each group. In total one can make six splits of seven telephones, and thus create 12 denominated groups of telephones. After no more splits could be made by the respondents, the respondents were thanked and informed of our intentions with their data.

The 43 respondents gave 516 denominated groups of telephones which were categorized into 50 distinct attribute categories. The 50 answer categories were created by three judges who first had independently looked at these distinctions and after that had discussed what answer categories to choose.

Results

Every time a product was put in a group by a respondent, it was recorded as the event that this product fell under a specific attribute category. This resulted in a two-way table of frequencies with 7 columns (the products) and 50 rows (the attributes), listing the frequencies of products falling under the answer categories (Table 16.1). The total frequency of this table added up to 882, from which it can be derived that the respondents had made groups which contained, on average, $882/516 = 1.709$ products.

---

[1] Because of its short duration, this research was carried out in combination with some other enquiries of other researchers.

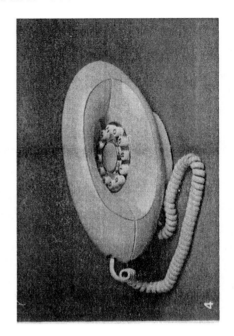

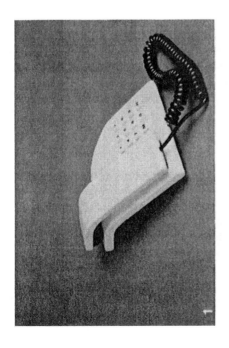

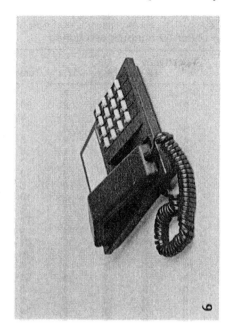

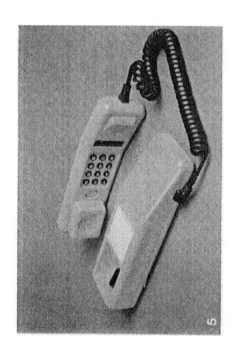

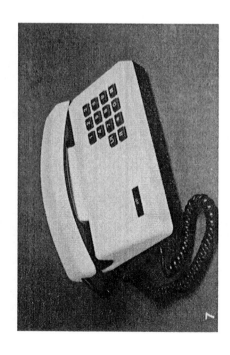

FIGURE 16.2 The seven telephones, as used in the natural grouping task. Colours: T1 white, T2 beige, T3 dark grey, T4 apricot, T5 yellow, T6 grey, T7 black and white.

TABLE 16.1
Frequency table of 7 telephones, falling under 50 attribute categories.

Attributes	T1	T2	T3	Telephones T4	T5	T6	T7	Total
Coarse	0	2	1	1	1	1	2	8
Elegant	8	1	4	7	5	3	2	30
Smooth	0	1	2	1	1	2	1	8
Big	0	8	2	3	0	13	5	31
Kitsch	0	0	0	7	2	0	0	9
Ugly colour	1	0	0	2	4	0	1	8
Round receiver	0	3	1	1	1	1	0	7
Rectangular receiver	1	1	2	0	0	2	3	9
Awkward receiver	3	0	1	0	0	1	0	5
Convenient receiver	0	2	1	2	2	0	3	10
Lightweight	3	1	1	0	0	2	1	8
Light colour	2	2	1	0	0	0	6	11
Dark grey	1	0	4	0	0	1	1	7
Practical	5	11	11	2	7	10	15	61
Impractical	5	5	2	8	6	7	2	35
Eye-catching	7	5	1	13	6	3	0	35
Ordinary	5	8	13	0	6	9	15	56
Keys in receiver	0	1	1	0	5	0	1	8
Keys next to receiver	3	2	0	0	0	4	0	9
Small keys	4	0	0	2	2	2	2	12
Large keys	0	1	7	0	0	2	3	13
Colourful	0	1	3	5	9	1	1	20
Square	2	0	0	0	1	3	0	6
Round	1	1	1	2	2	1	3	11
Chic	1	0	3	5	1	1	2	13
Modern	15	3	5	5	10	13	7	58
Classic	0	6	4	6	0	1	4	21
Playful	2	0	3	2	1	0	0	8
Sombre	0	0	3	0	0	0	0	3
Flat	2	1	3	0	2	1	0	9
Cheap	0	0	4	0	2	0	0	6
Attractive	7	3	5	6	5	2	1	29
Ugly	3	3	4	12	5	1	6	34
Businesslike	3	12	4	0	2	10	8	39
Living room	8	0	8	7	8	3	4	38
Complicated	1	22	2	2	1	11	3	42
Uncomplicated	7	1	7	1	4	4	15	39
Big receiver	0	3	0	0	0	1	1	5
Easily soiled	3	1	0	0	1	0	1	6
Easily cleaned	0	0	1	0	2	0	0	3
Unreliable	2	1	0	1	0	3	0	7
Reliable	3	7	3	0	0	4	6	23
Oblique angles	2	0	0	0	0	0	0	2
Stylized	4	1	0	1	1	2	0	9
Unusual	2	1	2	4	1	1	0	11
Horizontal receiver	0	0	4	3	0	0	4	11
Vertical receiver	3	4	0	0	3	4	0	14
Fits in everywhere	2	0	1	2	1	0	0	6
Rich in contrast	0	0	0	0	0	0	2	2
Small	10	2	2	2	11	0	0	27
Total	131	127	127	115	121	130	131	882

The content of this table was clarified by means of CA[2]. A three-dimensional solution was chosen, explaining 74% of the total inertia. A graphical display of the results, in the form of a symmetric plot, is shown in Figure 16.3. The numerical results of this analysis are shown in Table 16.2.

Dimension 1 (34.2%)

On the positive side of this dimension we find three telephones: telephones T1, T4 and T5 of Figure 16.2. On the negative side of this dimension we find telephones T2, T6 and T7. Telephone T3 is positioned right at the centre of this dimension, with a coordinate close to zero. The contributions of the telephones show which telephones have had most impact in determining the orientation of this dimension: telephones T4 and T2 (accounting for 27.7% and 24.8% of the inertia that is explained by dimension 1), and to a lesser extent telephones T5 and T6 (19.0% and 12.8%).

For an interpretation of this dimension we turn to the coordinates and contributions of the attribute categories. Those attribute categories are of interest that have a relatively high positive or negative position on this dimension and, at the same time, contribute disproportionally to the inertia on which this dimension is based (explained inertia > 3%). On the positive side these are 'living room', 'colourful', 'small', 'ugly', 'kitsch', 'eye-catching' and 'elegant'. On the negative side these are 'complicated', 'big', 'businesslike', 'reliable', 'practical' and 'ordinary'. So, the first dimension makes a distinction between large complicated business telephones and small 'expressive' telephones for use in the living room.

Dimension 2 (21.8%)

Telephones T2 and T6 have high positive scores on the second dimension, and at the same time account for some inertia: 14.3% and 11.3%. The most extreme scores, however, are found on the negative side of this dimension for telephones T3 and T7 (accounting for 41.9% and 23.9% of total inertia). Here, attribute categories with high positive scores that contribute to the inertia are 'keys next to the receiver', 'eye-catching' and 'complicated', and 'vertical receiver'. On the negative side, these are 'sombre', 'dark grey', 'cheap', 'ordinary', 'uncomplicated', 'large keys' and 'horizontal receiver'. It is less clear how to interpret this dimension. On the one hand there is a distinction between the plain and simple, cheap telephones that are rather dull and the more complicated and eye-catching telephones, but at the same time

---

[2] The correspondence analysis was carried out with the help of CORAN (Netherlands' Central Bureau of Statistics 1988).

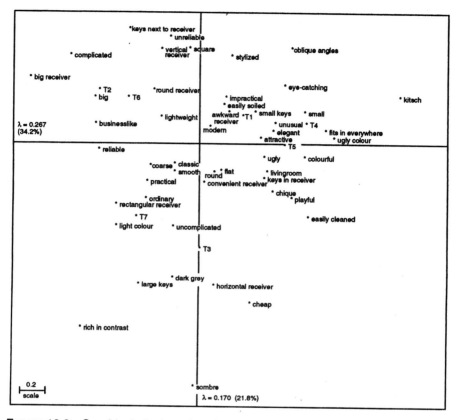

FIGURE 16.3 Graphical display of the coordinates of dimensions 1, 2 and 3. (a) Dimension 1 (horizontal) by dimension 2 (vertical).

this dimension seems to be about the position of the receiver on the telephone: horizontal or vertical.

Dimension 3 (18%)

Telephone T1 has a relatively high positive score on the third dimension, while telephones T2 and T4 have relatively high negative scores on this dimension. The remainder of the telephones are positioned close to zero. The contributions of the telephones show that T1 and T4 are of most importance to this dimension (accounting for 44.7% and 41.6% of the inertia that is explained by dimension 3). The attribute categories that both have high scores and substantial contributions are, on the positive side of dimension 3, 'oblique angles', 'awkward receiver', 'square', 'small' and 'modern'. On the negative

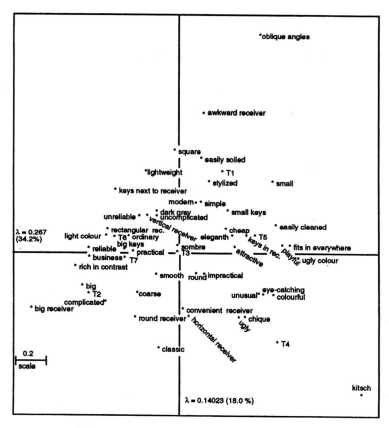

FIGURE 16.3 Graphical display of the coordinates of dimensions 1, 2 and 3. (b) Dimension 1 (horizontal) by dimension 3 (vertical).

side, these are 'kitsch', 'classic', 'ugly', 'complicated' and 'eye catching'. This dimension is fully accounted for by the two telephones with the most deviant designs, T1 and T4. It positions the avant garde telephone T1 with its oblique angles against its counterpart, T4, with its unashamedly classic appearance (even kitsch to some).

One last result should be mentioned, and this concerns the position of the attribute categories that are highly evaluative in nature. Note that for the first and second dimension 'attractive' and 'ugly' are positioned very close to each other. For the third dimension, 'ugly' loads somewhat on the negative side. However, it is not opposed to 'attractive' in this dimension, but to 'modern.' These findings clearly indicate that natural grouping data indeed reflect on product positioning, and not on product preference.

TABLE 16.2
Coordinates and contributions of dimensions 1, 2 and 3.

	Coordinates			Contributions		
	Dimensions					
	1	2	3	1	2	3
T1	0.332	0.273	0.650	6.134	6.512	44.701
T2	− 0.678	0.408	− 0.315	24.818	14.095	10.162
T3	− 0.006	− 0.703	0.028	0.002	41.888	0.080
T4	0.753	0.172	− 0.669	27.717	2.278	41.579
T5	0.608	0.029	0.125	19.010	0.067	1.527
T6	− 0.482	0.361	0.132	12.841	11.288	1.845
T7	− 0.412	− 0.522	− 0.032	9.477	23.872	0.105
Coarse	− 0.317	− 0.112	− 0.359	0.342	0.067	0.834
Elegant	0.516	0.094	0.113	3.397	0.178	0.311
Smooth	− 0.171	− 0.181	− 0.190	0.100	0.176	0.233
Big	− 0.719	0.348	− 0.250	6.816	2.510	1.567
Kitsch	1.396	0.341	−1.315	7.461	0.697	12.579
Ugly colour	0.934	0.064	− 0.073	2.968	0.022	0.035
Round receiver	− 0.321	0.375	− 0.506	0.308	0.657	1.451
Rectangular receiver	− 0.551	− 0.424	0.167	1.163	1.078	0.202
Awkward receiver	0.196	0.231	1.127	0.082	0.178	5.131
Convenient receiver	0.024	− 0.255	− 0.476	0.002	0.436	1.834
Lightweight	− 0.258	0.219	0.633	0.226	0.256	2.590
Light colour	− 0.559	− 0.546	0.124	1.461	2.191	0.136
Dark grey	− 0.163	− 0.936	0.329	0.079	4.095	0.612
Practical	− 0.353	− 0.221	0.021	3.233	1.997	0.022
Impractical	0.206	0.349	− 0.153	0.634	2.839	0.662
Eye catching	0.604	0.467	− 0.347	5.434	5.102	3.401
Ordinary	− 0.371	− 0.387	0.122	3.279	5.597	0.676
Keys in receiver	0.470	− 0.204	0.102	0.753	0.223	0.068
Keys next to receiver	− 0.492	0.830	0.549	0.929	4.134	2.192
Small keys	0.365	0.237	0.381	0.679	0.448	1.410
Large keys	− 0.436	−1.000	0.010	1.050	8.682	0.001
Colourful	0.740	− 0.090	− 0.314	4.663	0.108	1.591
Square	− 0.056	0.670	0.811	0.008	1.798	3.189
Round	0.114	− 0.182	− 0.167	0.061	0.244	0.247
Chic	0.503	− 0.304	− 0.496	1.401	0.803	2.589
Modern	0.120	0.167	0.384	0.357	1.075	6.927
Classic	− 0.157	− 0.123	− 0.735	0.222	0.211	9.179
Playful	0.668	− 0.361	0.057	1.518	0.695	0.021
Sombre	− 0.012	−1.706	0.074	0.000	5.826	0.013
Flat	0.151	− 0.199	0.430	0.087	0.237	1.348
Cheap	0.384	−1.114	0.161	0.377	4.968	0.126
Attractive	0.430	0.083	0.054	2.281	0.134	0.069
Ugly	0.459	− 0.095	− 0.498	3.045	0.205	6.816
Businesslike	− 0.699	0.148	− 0.027	8.101	0.573	0.023
Living room	0.491	− 0.192	0.141	3.904	0.938	0.613
Complicated	− 0.877	0.613	− 0.386	13.750	10.536	5.050

TABLE 16.2
(*Continued*).

	Coordinates			Contributions		
	Dimensions					
	1	2	3	1	2	3
Uncomplicated	−0.165	−0.542	0.296	0.454	7.644	2.754
Big receiver	−1.134	0.515	−0.450	2.737	0.886	0.819
Easily soiled	0.166	0.296	0.769	0.070	0.352	2.869
Easily cleaned	0.781	−0.522	0.247	0.778	0.546	0.148
Unreliable	−0.196	0.765	0.272	0.114	2.737	0.419
Reliable	−0.688	−0.014	0.020	4.635	0.003	0.007
Oblique angles	0.643	0.662	1.735	0.352	0.585	4.867
Stylized	0.225	0.653	0.595	0.194	2.561	2.576
Unusual	0.548	0.138	−0.334	1.404	0.140	0.994
Horizontal receiver	0.103	−0.967	−0.491	0.049	6.869	2.140
Vertical receiver	−0.252	0.690	0.304	0.378	4.444	1.048
Fits in everywhere	0.894	0.087	0.051	2.043	0.031	0.013
Rich in contrast	−0.799	−1.268	−0.084	0.543	2.146	0.011
Small	0.727	0.252	0.589	6.081	1.141	7.586

Discussion

Correspondence analysis has made the natural grouping data intelligible on an aggregate level. It has come up with interpretable, general dimensions of how consumers understand the design of products. In our example of the design of telephones three interpretable dimensions have arisen. The first of these has clearly to do with the most likely environment of the telephones: the living room or the office. On the positive side of this dimension we find attribute categories that stress the expressive value of the design of telephones, and that explicitly place it in the living room. On the negative side of this dimension we find attribute categories that stress the practical side of telephones and that place it in a business setting. When we look at the position of the telephones we find that the telephones on the positive side (telephones T4, T5 and T1) are more colourful and playful than the telephones on the negative side (telephones T2, T6 and T7) which are bigger, blander and more businesslike. A validation of this distinction can be found in various typologies of product attributes (for an overview, see Finn 1985), that state that products in general can be understood on the basis of their 'instrumental performance,' but also on the basis of their 'expressive performance.'

The second dimension makes a distinction between the plain and cheap, and the more complicated and eye-catching qualities of telephone design. However, at the same time this dimension is important in its differentiation between telephones which have receivers that are positioned horizontally or vertically on the telephone. Rather than assume that consumers in general think that simple telephones have receivers in a horizontal position and the more conspicuous have receivers in a vertical position, we must admit to a flaw in our choice of the seven telephones. The telephones in our set that had vertical receivers, all looked more conspicuous than the telephones with horizontal receivers. So respondents could not make a distinction between conspicuousness and position of receiver, simply because they lacked the opportunity to do so. Note that this is not caused by our selection procedure, since three of the telephones were selected randomly; the problem here is more that the group of telephones was too small. A larger group would have resulted in a smaller chance that certain attribute categories consistently go together. Adding more products to a natural grouping task would not create problems for respondents. In fact, seven is a rather small number compared to most other reports of natural grouping (e.g. Green *et al.* 1969, Kuylen and Verhallen 1988).

The third dimension distinguishes telephones that are avant garde from ones that are strikingly classic. Apparently, consumers consider deviations from more normal designs as varying on this classicism-modernism dimension. Note that for our group of respondents the classic end of this dimension is regarded as being more vulgar than the modern end. Although a classic telephone can also be considered 'chic', here it is mainly associated with the word 'kitsch'.

To conclude, natural grouping is an appropriate technique for finding out how consumers in general have an understanding of products. Its main advantages over other techniques is that it leaves respondents free to name whatever attribute they want and that it is a very time-efficient mode of enquiry. Within the whole field of perception analysis, CA seems to be most fruitful at showing generalities in data that may seem purely idiosyncratic. In the case of telephone design, it has shown that people's subjective understanding of design is not as idiosyncratic as is often suggested. Hyper-individual statements on the design of telephones did show up frequently in this research, but these statements turned out to coincide with other statements which were tapping from the same way of understanding things. From this it can be concluded that the combination of natural grouping with CA has a genuine advantage over other ways of perceptual mapping.

16.3 AGGREGATING CONJOINT ANALYSIS RESULTS

As mentioned earlier, in conjoint analysis the relative importance of an attribute is given by its sensitivity. The aggregation of these sensitivities is usually accomplished by averaging them within a homogeneous group of consumers. In a commercial setting, however, one is primarily involved in selecting attributes based on the importance ranking and not the average importance. Moreover, ranking of average sensitivities can give a different order compared to the ranking resulting from counting the number of subjects for whom a particular attribute has a specific importance ranking (Wittink *et al.* 1989). This means that the aggregation of the sensitivities should follow the line of aggregating importance rankings, rather than simply ranking the averages.

Therefore, the data which contain the most crucial information for market research are the rankings of the sensitivities per respondent. The aggregation procedure should indicate which attribute is ranked most frequently as most important, second most important, etc. CA can help provide this kind of information. The use of correspondence analysis for aggregating the sensitivities is illustrated by an example regarding a set of 27 telephones. Here, the following research questions are posed:

(1) What is the relative importance of the attributes for a homogeneous group of consumers?
(2) To what extent do homogeneous groups of consumers differ with respect to the importance ranking of the attributes?

Method

Sixty members of the consumer panel of the Delft University of Technology were invited to our laboratory to participate in the experiment. In order to estimate the relative importance of attributes of a telephone nine attributes were systematically varied. These were: a memory key (present, absent), a redial key (present, absent), a display of the number which is dialled (present, absent), a display of the number which is calling (present, absent), the form of the telephone (seven different forms), the position of the dialling mechanism (left, middle, right), the dialling mechanism (push-buttons, dial), a wireless receiver (present, absent) and price (Dfl 75, Dfl 100, Dfl 150). The full profile approach was used, meaning that the total description of each telephone is presented on separate cards (Green and Srinivasan 1978, Cattin and Wittink 1982). The full profile descriptions of the telephones used in the conjoint analysis task were chosen on the basis of an orthogonal design. Each subject was asked to rank the full profile descriptions of 27 telephones. The part-worths of the systematically varied attribute levels were estimated for each individual separately[3].

Before aggregating conjoint analysis results, homogeneous groups of subjects are constructed for which the aggregated sensitivities reflect the relative

---

[3] The estimation of the part-worths was accomplished by means of the computer package CONJOINT ANALYZER (Bretton-Clark 1986).

effect of an attribute to the group. In order to find homogeneous groups, a Q-factor analysis was done on the Spearman rank correlations of the preference ranking stated by the subjects. The scree-test showed that three factors represented the variation in the subjects quite well. The three factor solution accounted for 54.9% of the variance. In this solution 15 subjects had a low communality (< 0.40). These 15 subjects constituted group A. The other subjects were divided over three groups on the basis of a high factor loading on one of the (varimax) rotated factors. Thus the factor analysis resulted in four groups of subjects, with $n_A = 15$, $n_B = 25$, $n_C = 9$, $n_D = 11$.

For each group of subjects a contingency table (attributes by importance ranking) was constructed by counting, for each attribute varied, the number of subjects for whom the sensitivity of the attribute had a particular importance ranking. Ties were handled by taking the highest successive rank. This results in a $4 \times 9 \times 9$ three-way contingency table (groups by importance ranking by attributes). There are two ways to analyze three-way contingency tables with correspondence analysis (van der Heijden 1987). The first way would be to perform a separate analysis for each group. In this approach it is difficult to compare the multidimensional maps which result from the CA for each group, because for each group one probably needs a different number of dimensions to get the same explained inertia. Furthermore, the successive dimensions of the different mappings do not explain the same amount of inertia.

Here, however, we have used a CA of the supermatrix groups x attributes by importance ranking. The supermatrix is constructed by stacking the attributes by importance ranking tables for the four respective groups. This resulted in a 36×9 matrix. The analysis of this matrix results in one multidimensional map, in which both the attributes for each group and the importance rankings are represented. Consequently, this analysis can answer both of the research questions that were stated above.

Results

In this section the CA is described according to the research questions stated. All the numerical diagnostics presented are calculated in a full space (eight dimensions), although a three-dimensional space is a good approximation of the full space (80% of the inertia explained and almost all categories (except eight) have an explained inertia which is larger than 40%). In comparing the numerical diagnostics calculated in the three-dimensional space with those calculated in a full space some differences arose which affected the interpretation of the diagnostics. This was particularly true for the relationship between the attributes for each group and the importance rankings.

The first research question regarded the relative importance of the attributes for a homogeneous group. This research question is answered in two steps: the differences between the importance rankings are considered first, this is followed by determining the aggregated ranking of the attributes.

The differences between importance rankings

In CA, this difference is given by the Euclidean distance between the importance rankings in the full (eight dimensions) symmetric map (in the SPSS procedure ANACOR a 'PRINCIPAL' normalization). The distances between the importance rankings are given in Table 16.3. This table shows that the importance rankings which are close to each other in the arithmetical chain, are close to each other in the map. If the distances between successive importance rankings are considered, it turns out that they are almost the same (approximately 1), except for the differences between the rankings 7–8 and 8–9 which are approximately 2. This means that the difference in relative importance of the attributes is more pronounced for the three most important attributes compared to the less important attributes.

Aggregated importance of the attributes

The aggregated importance ranking of an attribute for a homogeneous group of consumers is based on the relationship between an attribute for a particular group (which is a row category) and importance rankings (which are column categories). The interpretation of the relationship between a row and a column category is an object of discussion (Carroll *et al.* 1986, 1987, 1989, Greenacre 1989). An interpretation of the relationship between a row- and column-point can be obtained by realizing that the eigenvalues are derived in such a way that the scale values $\lambda(r^k, c^k)$ of respectively the row and column variable are

TABLE 16.3
Euclidean distances between the importance rankings.

Importance ranking	Importance ranking							
	1	2	3	4	5	6	7	8
2	0.75							
3	0.99	1.01						
4	0.90	1.04	0.92					
5	1.18	1.03	1.23	0.97				
6	1.17	1.07	0.78	0.94	1.10			
7	1.45	1.48	1.29	1.33	1.31	1.02		
8	2.03	2.07	1.96	1.99	2.05	1.89	1.72	
9	2.49	2.46	2.42	2.47	2.52	2.40	2.44	2.10

Note, 1 indicates the least important attribute and 9 the most important attribute.

maximally correlated under the condition that these scale values are uncorrelated with other scale values (Bethlehem and Sikkel 1985). Because a correlation coefficient is invariant under addition of a constant to the scale values or multiplication by a constant of the scale values, the scale values can be standardized in such a way that they have expectation (E) zero and variance (V) one (Bethlehem and Sikkel 1985). This correlation is expressed in the eigenvalue of the dimension (λ) on the basis of which the standard coordinates ($r_{ik}; c_{jk}, k = 1, 2, ..., m - 1; j = 1, 2, ..., n$) are calculated (Sikkel 1981b). Consequently:

$$\Sigma_i \Sigma_j p_{ij} r_{ik} c_{jl} = 0, \quad \text{if } k \neq l$$
$$= \lambda_k, \quad \text{if } k = l \tag{16.1}$$

If the number of rows (m) is larger than the number of columns (n), $m - 1$ scales are sufficient to describe the contingency table. This can be expressed as follows:

$$p_{ij} = p_{i+} p_{+j} (1 + \Sigma_k \lambda_k r_{ik} c_{jk}) \tag{16.2}$$

This equation can be rewritten as:

$$\frac{p_{ij}}{p_{i+} p_{+j}} - 1 = \Sigma_k \lambda_k^{-1} (\lambda_k r_{ik})(\lambda_k c_{jk}) \tag{16.3}$$

The right part of this equation represents an inner product of two vectors (Sikkel and Bethlehem 1981: 20). This inner product is linearly related to the cosine of the angle which is enclosed by these two vectors. These two vectors are the standard coordinate and the principal coordinate. (In the equation there are two principal coordinates: $\lambda_k r_{(ik)}, \lambda_k c_{(jk)}$, but these are preceded by a constant λ_k^{-1}. Consequently, one of the principal coordinates changes into a standard coordinate.)

Therefore, the relationship between a row and column category can be derived from the cosine of the angle: attribute–origin–importance ranking in the asymmetric map (Sikkel and Bethlehem 1981). (In the SPSS procedure ANACOR an asymmetric map is obtained by either the 'RPRINCIPAL' or 'CPRINCIPAL' normalization.) Thus when one of the categorical variables is represented in standard coordinates and the other one in principal coordinates the interpretation of the cosine of the angle between a row-point and a column-point is as follows:

(1) If the left part of (16.3) is zero ($p_{(ij)} = p_{(i.)} p_{(.j)}$) the two categories are independent. For the right part of (16.3) this means that the angle α is ninety degrees.

(2) If the left part of (16.3) is larger than zero, then the corresponding row and column are positively related: category i of variable A is over-represented in category j of variable B and vice versa. For the right part of (16.3) this means that the angle α is smaller than ninety degrees.

(3) If the left part of (16.3) is smaller than zero, then the corresponding row and column are negatively related: category *i* of variable A is under-represented in category *j* of variable B and vice versa. For the right part of (16.3) this means that the angle α is larger than ninety degrees.

Table 16.4 shows the cosines of the angle between attributes and the importance rankings.

The aggregated importance ranking is derived from the pattern of the cosines for the angles between the attribute and the importance rankings. For the interpretation of this pattern values larger than 0.55 are considered because a large positive value means that the importance ranking is often assigned to that attribute (Sikkel and Bethlehem 1981). Given this rule of thumb, some attributes are assigned a similar aggregated importance ranking. In that case the cosine of importance rankings preceding and following the tied attributes are considered. If one of those values is large (larger than 0.55) the tie is resolved by assigning the attribute the importance ranking which goes with this (second) large value. For three attributes it was not possible to resolve the tie according to this rule. Consequently these attributes are still tied. Some attributes, on the other hand, have a pattern without high values of the cosine (in absolute terms) of the angle between the attribute and any of the importance rankings. In that case a 0 is assigned to the attribute, to indicate that the attribute is viewed in a heterogeneous way with respect to its importance. Table 16.5 displays the aggregated importance rankings of the attributes for each group.

Now a comparison can be made between the homogeneous groups of consumers on the basis of the aggregated importance ranking of the attributes. In doing so we turn to the second research question. This research question concerned the extent that homogeneous groups of consumers differ with respect to the importance ranking of the attributes. This question is answered by first describing the groups in terms of importance rankings of the attributes and then considering the differences between the groups.

Description of the groups

The homogeneous groups of consumers can now be described on the basis of the aggregated importance ranking of the attributes. These aggregated importance rankings are given in Table 16.5. For group A, 'form' is the most important attribute. For this group no attribute is unanimously assigned as second most important. The third most important attribute is the 'position of the dialling mechanism'. The 'display number dialled' is considered to be the fourth most important attribute. Two attributes are regarded as fifth most important, namely 'redial key' and the 'dialling mechanism'. The sixth most important attribute is 'wireless receiver'. The least important attribute is

TABLE 16.4
Cosine of the angle attribute–origin–importance ranking.

| Attribute | Importance ranking | | | | | | | | |
	1	2	3	4	5	6	7	8	9
a_{11}	−0.30	−0.34	−0.35	−0.36	−0.31	−0.40	−0.12	0.47	0.92
a_{12}	−0.05	−0.49	−0.18	−0.24	−0.15	0.26	0.86	0.26	−0.29
a_{13}	−0.08	−0.76	−0.28	0.49	−0.24	−0.02	0.14	0.09	0.37
a_{14}	−0.06	0.48	0.19	−0.74	0.19	0.37	0.12	−0.21	−0.17
a_{15}	−0.05	0.06	−0.17	−0.22	0.93	−0.27	0.08	−0.13	−0.11
a_{16}	−0.47	0.36	0.48	−0.26	−0.16	0.58	0.09	−0.16	−0.32
a_{17}	0.59	0.73	0.37	0.03	−0.13	−0.47	−0.36	−0.12	−0.25
a_{18}	−0.08	0.51	−0.27	0.06	0.64	0.39	−0.26	−0.22	−0.45
a_{19}	−0.05	0.05	−0.16	0.75	0.62	−0.25	−0.37	−0.30	−0.10
a_{21}	−0.23	−0.26	−0.27	−0.28	−0.24	−0.30	−0.15	0.99	0.18
a_{22}	0.03	0.01	0.08	0.39	0.31	−0.29	0.62	−0.43	−0.37
a_{23}	−0.20	−0.49	−0.12	0.15	0.42	0.42	0.66	−0.35	−0.29
a_{24}	0.59	0.21	−0.15	0.39	−0.33	0.32	0.23	−0.43	−0.36
a_{25}	0.02	0.01	0.87	−0.02	0.08	0.45	−0.30	−0.37	−0.32
a_{26}	0.03	0.40	0.08	0.58	0.09	0.51	−0.52	−0.42	−0.36
a_{27}	0.77	0.72	0.07	0.16	0.09	−0.43	−0.13	−0.38	−0.32
a_{28}	−0.26	−0.30	−0.30	−0.31	−0.27	−0.34	−0.17	0.23	0.99
a_{29}	−0.39	0.32	−0.22	0.28	0.15	0.49	0.36	−0.43	−0.36
a_{31}	−0.18	−0.20	−0.20	−0.21	−0.18	−0.23	−0.18	−0.14	0.97
a_{32}	−0.35	−0.02	−0.40	−0.42	0.03	−0.09	0.70	0.54	−0.24
a_{33}	−0.51	−0.04	0.03	0.48	0.62	−0.13	0.00	−0.01	−0.35
a_{34}	0.88	−0.03	0.03	0.43	0.04	−0.11	−0.00	−0.37	−0.31
a_{35}	0.20	−0.48	0.54	−0.05	−0.43	0.34	−0.00	0.32	−0.29
a_{36}	0.32	0.69	0.04	−0.09	0.06	0.56	0.00	−0.56	−0.48
a_{37}	0.64	0.68	0.02	−0.04	−0.34	0.27	−0.33	−0.27	−0.23
a_{38}	−0.36	−0.41	0.45	−0.05	−0.37	0.29	0.72	−0.00	−0.25
a_{39}	0.19	−0.03	−0.47	0.39	0.50	−0.10	−0.42	0.31	−0.28
a_{41}	−0.33	−0.37	−0.37	−0.39	−0.33	0.14	0.74	0.57	−0.04
a_{42}	0.07	−0.09	0.60	0.77	−0.04	−0.15	−0.34	−0.27	−0.23
a_{43}	0.06	0.45	−0.35	−0.10	0.81	0.13	−0.31	−0.25	−0.21
a_{44}	−0.39	−0.10	0.32	−0.47	−0.40	0.17	0.57	0.43	−0.27
a_{45}	−0.29	−0.33	0.53	−0.09	−0.30	0.63	−0.29	−0.23	0.29
a_{46}	0.55	−0.10	0.32	0.56	−0.40	−0.17	−0.39	0.18	−0.27
a_{47}	0.06	0.48	−0.05	−0.39	−0.33	−0.14	−0.32	−0.26	0.69
a_{48}	0.11	−0.66	−0.09	0.32	−0.06	−0.25	−0.11	0.64	−0.07
a_{49}	0.52	0.23	0.31	−0.45	0.30	−0.49	−0.07	−0.06	−0.05

Note a_{ij} indicates attribute j (1 = form, 2 = position dial, 3 = price, 4 = memory, 5 = redial, 6 = display number dialled, 7 = display number calling, 8 = dialling mechanism, 9 = wireless receiver) for group i (i = A, B, C, D).

TABLE 16.5
Aggregated importance ranking.

Attribute	Group A	Group B	Group C	Group D
Form	9	8	9	7
Position dial	7	7	8	3
Price	0	6	5	5
Memory	0	1	1	7
Redial	5	0	0	6
Display no. dialled	6	4	2	4
Display no. calling	2	2	1	9
Dialling mechanism	5	9	7	8
Wireless receiver	4	0	0	0

1 indicates the least important, 9 the most important attribute and 0 indicates that the attribute has no high correlations to any of the importance rankings.

'display number calling'. Two attributes are perceived heterogeneously with respect to their importance ranking ('price' and 'memory').

For group B the two most important attributes are 'dialling mechanism' (most important) and 'form' (second most important). The 'position of the dial' and 'price' are, respectively, considered to be the third and fourth most important attributes. The attribute 'display number dialled' is regarded as less important. The least important attribute for group B is 'memory'. Two attributes are perceived heterogeneously with respect to their importance ranking ('redial' and 'wireless receiver').

The three most important attributes for group C are respectively 'form', 'position dialling mechanism' and 'dialling mechanism'. The 'price' is regarded as the fifth most important attribute. The least important attributes for this group are 'display number calling', 'display number dialled' and 'memory'. For group C two attributes are perceived heterogeneously with respect to their importance ranking ('redial' and 'wireless receiver').

The two most important attributes for group D are respectively 'display number calling' and 'dialling mechanism'. Two attributes are considered to be third most important, namely 'form' and 'memory'. 'Redial' is considered to be the fourth most important attribute and 'price' and 'display number dialled' as respectively fifth and sixth in the importance ranking. The least important attribute is 'position dial'. One attribute is perceived heterogeneously with respect to its importance ranking: namely 'wireless receiver'.

The differences between the groups based on the attributes

In CA, the difference between the groups based on the attributes is given by the Euclidean distance between representations of a particular attribute for different groups in the full (eight dimensions) symmetric map (in the SPSS procedure ANACOR a 'PRINCIPAL' normalization). Table 16.6 displays these distances.

Group A and group C are the most similar to one another. The average distance between the corresponding attributes is 0.99. The groups A and C differ mostly regarding form (1.56), redial (1.39) and dialling mechanism (1.30).

The two groups which are the least similar are Group D and group C (average distance between corresponding attributes is 1.52). They differ mostly regarding the attributes form (3.17), position dial (1.96) and display number calling (1.53).

Discussion

Correspondence analysis can aggregate conjoint analysis results following the line of aggregating importance rankings. If the supermatrix groups of subjects x attributes by importance rankings is analyzed, the map resulting from CA provides answers regarding three important research questions concerning attribute importance. First: what is the difference between successive importance rankings? Or phrased in another way: is the difference between the most and second most important attributes larger than the difference between the fourth and fifth most important attributes? These differences are given by the Euclidean distance between the importance rankings in a symmetric map.

Second: what is the aggregated importance ranking of an attribute? The aggregated ranking of an attribute for a group of consumers is derived from the pattern of the cosines for the angles between the attribute and the importance rankings in an asymmetric map. Sometimes this pattern of cosines

TABLE 16.6
Euclidean distances between representations of a particular attribute for the groups distinguished.

| Attributes | Groups | | | | | |
	A:B	A:C	A:D	B:C	B:D	C:D
Shape	1.57	1.56	1.84	3.12	1.58	3.17
Position dial	0.95	0.69	1.62	1.27	1.01	1.96
Price	0.99	0.89	1.53	0.81	1.32	1.02
Memory	0.81	0.97	0.97	0.46	1.26	1.46
Redial	1.13	1.39	1.71	0.89	1.07	1.24
Display no. dialled	0.79	0.58	1.19	0.52	0.93	1.06
Display no. calling	0.48	0.73	1.56	0.70	1.56	1.53
Dialling mechanism	2.49	1.30	1.01	2.55	2.05	1.16
Wireless receiver	0.92	0.82	1.21	0.93	1.00	1.11
Average distance	1.13	0.99	1.40	1.25	1.31	1.52

shows no high values (in absolute terms). In that case the attribute has no extreme high or low frequency in any of the importance rankings. In practice this means that the subjects in the group perceive this attribute in a heterogeneous way regarding its importance. In the average approach it is impossible to identify such attributes.

Third: to what extent do groups of consumers differ regarding the importance assigned to the attributes? These differences can be derived from the Euclidean distance in a symmetric map between the representations of a particular attribute for the groups distinguished.

As a general practical implication one can state that CA provides a way to aggregate conjoint analysis results which is in line with commercial use of these results, namely selecting attributes based on the importance ranking instead of the average importance.

References

Agresti, A. 1984. *Analysis of Ordinal Categorical Data*. New York: Wiley.

Agresti, A. 1990. *Categorical Data Analysis*. New York: Wiley.

Amaturo, E., Costagliola, S. and G. Ragone. 1987. Furnishing and status attributes. *Environment and Behavior*, **19**, 228–249.

Andersen, E. B. 1980. *Discrete Statistical Models with Social Science Applications*. Amsterdam: North-Holland.

Andersen, E. B. 1990. *The Statistical Analysis of Categorical Data*. Berlin: Springer.

Andrews, D. F. 1972. Plots of high dimensional data. *Biometrics*, **28**, 125–136.

Babeau A. and L. Lebart. 1984. Les Conditions de vie et Aspirations des Français. *Futuribles*, **1**, 37–53.

Badawy, M. K. 1982. *Developing Managerial Skills in Engineers and Scientists*. New York: Van Nostrand.

Barlow, R. E., Bartholomew, D. J., Bremner J. M. and H. D. Brunk. 1972. *Statistical Inference under Order Restrictions*. New York: Wiley.

Baudrillard, J. 1972. *Pour une critique de l'économie politique du signe*. Paris: Gallimard.

Beck, U. 1983. Jenseits von Stand und Klasse?, in R. Kreckel (ed.), *Soziale Ungleichheiten, Soziale Welt, Sonderband 2*, pp. 35–74.

Beck, U. 1986. *Risikogesellschaft. Auf dem Weg in eine andere Moderne*. Frankfurt/Main: Suhrkamp.

Becker, M. P. 1989. Square contingency tables having ordered categories and GLIM. *Glim Newsletter* No. 19, 22–31.

Bekker, P. and J. De Leeuw. 1988. Relations between variants of nonlinear principal component analysis, in J. L. A. van Rijckevorsel and J. De Leeuw (eds), *Component and Correspondence Analysis*, pp. 1–31. New York: Wiley.

Bentler, P. M. and D. G. Weeks. 1978. Restricted multidimensional scaling models. *Journal of Mathematical Psychology*, **17**, pp. 138–151.

Benzécri, J.-P. 1963. *Cours de Linguistique Mathématique*. Rennes: Université de Rennes.

Benzécri, J.-P. 1969. Philosophie thomiste et connaissance mathématique de la nature. *La Pensée Catholique*, n. **118**, 11–24.

Benzécri, J.-P. 1970. Sur l'analyse des matrices de confusion. *Revue de statistique appliquée*, **XVIII**, 5–62.

Benzécri, J.-P. 1973. La place de l'*a priori*. *Encyclopédia universalis*, volume *17*, 11–24. Paris: Organum.

Benzécri, J.-P. 1976. Histoire et préhistoire de l'analyse des données. *Les cahiers d'analyse des données*, N. 1–4.

Benzécri, J.-P. 1977. Histoire et préhistoire de l'analyse des données. *Les cahiers d'analyse des données*, N. 1.

Benzécri, J.-P. 1979. Sur le calcul des taux d'inertie dans l'analyse d'un questionnaire. Addendum et erratum à [BIN.MULT]. *Cahiers de L'analyse des Données*, 4, 377–378.

Benzécri, J.-P. 1982. Construction d'une classification ascendante hiérarchique par la recherche en chaîne de voisins réciproques. *Cahiers d'Analyse des Données*, 7, 209–218.

Benzécri, J.-P. 1992. *Correspondence Analysis Handbook*. New York: Dekker.

Benzécri, J.-P. *et collaborateurs*. 1973. *L'Analyse des Données. L'Analyse de Correspondence*. Paris: Dunod.

Benzécri, J.-P. *et collaborateurs*. 1980. *Pratique de l'analyse des données* (Vols 1–3). Paris: Dunod.

Bernard, J. M., Le Roux, B., Rouanet, H. and M. A. Schiltz. 1989. L'analyse des données multidimensionnelles par le language d'interrogation des données (LID). *Bulletin de Méthodologie Sociologique*, N. 23, 3–46.

Bertier, R. and J. M. Bouroche. 1975. *Analyse des données multidimensionnelles*. Paris: Presses universitaires de France.

Best, H. 1990. *Die Männer von Bildung und Besitz*. Düsseldorf: Econ.

Bethlehem, J. G. and D. Sikkel. 1985. *Coran user manual*. Voorburg: Netherlands Central Bureau of Statistics. Voorburg.

Bishop, Y., Fienberg, S. and P. W. Holland. 1975. *Discrete Multivariate Analysis: Theory and Practice*. Cambridge USA: MIT Press.

Blasius, J. 1993. *Gentrification und Lebensstile. Eine empirische Untersuchung*. Wiesbaden: Deutscher-Universitäts-Verlag.

Blasius, J. and H. Rohlinger. 1988. KORRES – a program for analysing qualitative data. *Psychometrika*, 53, 425–426.

Blasius, J. and H. Rohlinger. 1989. KORRES – A program for multivariate analysis of categorical data from contingency tables, in SAS Institute Inc. (eds), *SEUGI'89, Proceedings of the SAS Users Group International Conference*, pp. 98–117. Gary, N. C.: SAS Institute Inc.

Blasius, J. and J. Winkler. 1989. Gibt es die 'feinen Unterschiede'? Eine empirische Überprüfung der Bourdieuschen Theorie. *Kölner Zeitschrift für Soziologie und Sozialpsychologie*, 41, 72–94.

Blau, P. M. and O. D. Duncan. 1967. *The American Occupational Structure*. New York: Wiley.

Blossfeld, H.-P. and A. Hamerle. 1989. Using Cox models to study multiepisode processes. *Sociological Methods & Research*, 17, 432–448.

Bock, R. D. 1975. *Multivariate Statistical Methods in Behavioral Research*. New York: McGraw Hill.

Böckenholt, U. and I. Böckenholt. 1990. Canonical analysis of contingency tables with linear constraints. *Psychometrika*, 55, 633–639.

Böckenholt, U. and I. Böckenholt. 1991. Constrained latent class analysis: simultaneous classification and scaling of discrete choice data. *Psychometrika*, **56**, 699–716.

Borgatta, E. F. and M. L. Borgatta. 1992. *Encyclopedia of Sociology*, vol. 1, pp. 621–638. New York: Macmillan.

Bourdieu, P. 1979. *La distinction. Critique sociale du jugement*. Paris: Les éditions de minuit.

Bourdieu, P. 1982. *Die feinen Unterschiede*. Frankfurt: Suhrkamp (translation of Bourdieu, P., 1979. *La distinction. Critique sociale du jugement*. Paris: Les éditions de minuit).

Bourdieu, P. 1984. *Homo Academicus*. Paris: Les éditions de minuit.

Bourdieu, P. 1985a. The social space and the genesis of groups. *Theory and Society*, **14**, 723–744.

Bourdieu, P. 1985b. The forms of capital, in J. G. Richardson (ed.), *Handbook of Theory and Research for the Sociology of Education*, pp. 241–260. New York: Greenwood Press.

Bourdieu, P. 1985c. The market of symbolic goods. *Poetics*, **14**, 13–44.

Bourdieu, P. 1988. Vive la crise! for heterodoxy in social science. *Theory and Society*, **17**, 773–787.

Bourdieu, P. 1989a. Social space and symbolic power. *Sociological Theory*, **7**, 14–25.

Bourdieu, P. 1989b. *Satz und Gegensatz. Über die Verantwortung des Intellektuellen*. Berlin: Wagenbach.

Bourdieu, P. 1989c. *La noblesse d'état. Grandes écoles et ésprit de corps*. Paris: Les éditions de minuit.

Bourdieu, P. 1991. Inzwischen kenne ich alle Krankheiten der soziologischen Vernunft. Pierre Bourdieu im Gespräch mit Beate Krais, in P. Bourdieu, J.-C. Chamboredon and J.-C. Passeron, *Soziologie als Beruf. Wissenschaftstheoretische Voraussetzungen soziologischer Erkenntnis*, pp. 269–283. Berlin: Walter de Gruyter.

Bourdieu, P., L. Boltanski, R. Castel, J.-C. Chamboredon, G. Lagneau and D. Schnapper. 1981. *Eine illegitime Kunst. Die sozialen Gebrauchsweisen der Photographie*. Frankfurt: Europäische Verlagsanstalt.

Bouroche, J. M. and G. Saporta. 1980. L'analyse des données. *Collection Que sais-je?*, n. 1854. Paris: Presses universitaires de France.

Bretton-Clark. 1986. *Conjoint Analyzer*. New York: Bretton-Clark.

Cailliez, F. and J.-P. Pagès. 1976. *Introduction à l'analyse des données*. Paris: SMASH.

Carroll, J. D., Green, P. E. and C. M. Schaffer. 1986. Interpoint distance comparisons in correspondence analysis. *Journal of Marketing Research*, **23**, 271–280.

Carroll, J. D., Green, P. E. and C. M. Schaffer. 1987. Comparing interpoint distances in correspondence analysis: a clarification. *Journal of Marketing Research*, **24**, 445–450.

Carroll, J. D., Green, P. E. and C. M. Schaffer. 1989. Reply to Greenacre's commentary on the Carroll–Green–Schaffer scaling of two-way correspondence analysis solution. *Journal of Marketing Research*, **26**, 366–368.

Carroll, J. D., Pruzansky, S. and J. B. Kruskal. 1980. CANDELINC: A general approach to multidimensional analysis of many-way arrays with linear constraints on parameters. *Psychometrika*, **45**, 3–24.

Castel, R. 1981. Bilder und Phantasiebilder, in P. Bourdieu, L. Boltanski, R. Castel, J.-C. Chamboredon, G. Lagneau and D. Schnapper, *Eine illegitime Kunst*, pp. 235–267. Frankfurt: Europäische Verlagsanstalt.

Cattin, P. & Wittink, D.R. 1982. Commercial use of conjoint analysis: A survey. *Journal of Marketing*, **46**, 44–53.

Cazes, P. 1986. Correspondance entre deux ensembles et partition de ces deux ensembles. *Les Cahiers de l'Analyse des Données*, **11**, 335–340.

Central Bureau for Statistics 1987. *Statistiek der verkiezingen 1986. Tweede Kamer der Staten-Generaal 21 mei.* Den Haag: Staatsuitgeverij.

Chambers, J. M., Cleveland, W. S., Kleiner, B. and P. A. Tukey. 1983. *Graphical Methods for Data Analysis.* Pacific Grove: Wadsworth.

Chodorow, N. 1989. *Psychoanalytic Theory and Feminism.* Cambridge: Polity Press.

Cibois, Ph. 1981. Analyse des données et sociologie. *L'Année Sociologique*, N. 31, 333–348.

Cibois, Ph. 1983. Méthodes post-factorielles pour le dépouillement d'enquete. BMS, N. 1, 41–78.

Clogg, C. C. 1981. Latent structure models of mobility. *American Journal of Sociology*, **86**, 836–868.

Connolly, Th. 1983. *Scientists, Engineers, and Organizations.* Monterey: Brooks/Cole Engineering Division.

Converse, P. E., Dotson, J. D., Hoag, W. J. and W. H. McGee, 1980. *American Social Attitudes Data Sourcebook, 1947–1978.* Cambridge, Massachusetts: Harvard University Press.

Coombs, C. H. 1964. *A Theory of Data.* New York: Wiley.

Coombs, C. H., McClelland, G. H. and L. Coombs, 1973. The measurement and analysis of family composition preferences. *Michigan Mathematical Psychology Program*, 73–75.

Cordier-Escofier, B. 1965. *L'analyse des correspondances.* Doctoral thesis. Rennes: Université de Rennes.

Cronbach, L. J. 1951. Coefficient alpha and the internal structure of tests. *Psychometrika*, **16**, 297–334.

Csikszentmihalyi, M. and E. Rochberg-Halton. 1981. *The Meaning of Things. Domestic Symbols and the Self.* Cambridge: Cambridge University Press.

Daudin, J. J. and P. Trécourt. 1980. Analyse factorielle des correspondances et modèle log-linéaire: comparaison des deux méthodes sur un exemple. *Revue des Statistique Appliquée*, **28**(1), 5–24.

De Falguerolles, A. and B. Francis (in press). Algorithmic approaches for fitting bilinear models, in Y. Dodge and J. Whittaker (eds), *Computational Statistics, Proceedings of the Tenth Symposium of Computational Statistics.* Volume 1, 77–82.

Delbeke, L. 1978. *Enkele analyses op voorkeuroordelen voor gezinssamenstellingen* [Some analyses of preference judgments for family compositions]. Centrum voor Mathematische Psychologie en Psychologische Methodologie, Katholieke Universiteit Leuven.

De Leeuw, J. 1973. *Canonical Analysis of Categorical Data.* Doctoral dissertation, Leiden: University of Leiden.

De Leeuw, J. 1984. *Statistical Properties of Multiple Correspondence Analysis.* Internal Report RR-84-06. Leiden: Department of Data Theory.

De Leeuw, J. 1988. Models and techniques. *Statistica Neerlandica*, **42**, 1988, 91–98.

De Leeuw, J. and J. L. A. Van Rijckevorsel. 1980. HOMALS and PRINCALS: some generalizations of principal components analysis, in E. Diday *et al.* (eds), *Data Analysis and Informatics.* Amsterdam: North-Holland.

De Leeuw, J. and E. van der Burg. 1986. The permutational limit distribution of generalized canonical correlations, in E. Diday *et al.* (eds), *Data Analysis and Informatics*, Vol. IV, pp. 509–521. Amsterdam: North-Holland.

De Leeuw, J. and P. G. M. van der Heijden. 1988. Correspondence analysis of incomplete tables. *Psychometrika*, **53**, 223–233.

De Leeuw, J., Heiser, W., Meulman, J. and F. Critchley. 1989. *Multidimensional Data Analysis.* Leiden: DSWO Press.

De Leeuw, J. and P. G. M. van der Heijden. 1991. Reduced rank models for contingency tables. *Biometrika*, **78**, 229–232.

De Leeuw, J., van der Heijden, P. G. M. and P. Verboon. 1990. A latent time budget model. *Statistica Neerlandica*, **44**, 1–22.

Descartes, R. 1977. *Over de methode.* Amsterdam: Boom (Discours de la methode, 1637).

Dessens, J., Jansen, W, and R. Luijkx. 1985. Fitting log-multiplicative association models. *Glim Newsletter*, no. **11**.

Deville, J. C. and G. Malinvaud. 1983. Data Analysis in Official Socio-Economic Statistics (with discussion). *Journal of the Royal Statistical Society, Series A*, **146**, 335–361.

Diday, E. 1972. Optimisation en classification automatique et reconnaissance des formes. *Revue Française de Recherche Opérationnelle*, **3**, 61–96.

Du Toit S. H. C., Steyn, A. G. W. and R. H. Stumpf. 1986. *Graphical Exploratory Data Analysis.* New York: Springer.

Eckart, R., Hahn, A. and M. Wolf. 1989. *Die ersten Jahre junger Ehen. Verständigung durch Illusionen?* Frankfurt/Main: Campus.

Eckart, C. and G. Young. 1936. The Approximation of one matrix by another of lower rank. *Psychometroika*, **1**, 211–218.

Edgington, E. S. 1987. *Randomization Tests*, 2nd edn. New York: Dekker.

Efron, B. 1982. *The Jackknife, the Bootstrap and other Resampling Plans.* Philadelphia: SIAM.

Ekman, G. 1954. Dimensions of color vision. *Journal of Psychology*, **38**, 467–474.

Engel, J. F. and R. D. Blackwell. 1982. *Consumer Behavior.* Chicago: The Druyden Press.

Escofier, B. 1983. Analyse de la différence entre deux mesures sur le produit de deux mêmes ensembles. *Cahiers de l'analyse des données*, **8**, 325–329.

Escofier, B. 1984. Analyse factorielle en reference a un modele: application a l'analyse de tableaux d'échanges. *Revue de Statistique Appliquee*, **32**, 25–36.

Escoufier, Y. 1982. L'analyse des tableaux de contingence simples et multiples, in R. Coppi (ed.): *Proc. International Meeting on the Analysis of Multidimensional Contingency Tables* (Rome 1981) *Metron*, **40**, 53–77.

Escoufier, Y. 1988. Beyond correspondence analysis, in H. H. Bock (ed.), *Classification and Related Methods of Data Analysis*, pp. 505–514. Amsterdam: North-Holland.

Escoufier, Y. and S. Junca. 1986. Least-squares approximation of frequencies and their logarithms. *International Statistical Review*, **54**, 279–283.

Feininger, A. 1965. *Die neue Fotolehre*. Düsseldorf: Econ.

Finn, A. 1985. A theory of the consumer evaluation process for new product concepts, in J. N. Sheth (ed.), *Research in Consumer Behavior*, vol. 1, pp. 35–65. Greenwich, CT: JAI Press.

Fisher, R. A. 1940. The Precision of Discriminant Functions. *Annals of Eugenics*, **10**, 422–429.

Früh, W. 1981. *Inhaltsanalyse*. München.

Früh, W. 1984. Konventionelle und maschinelle Inhaltsanalyse im Vergleich: Zur Evaluierung computerunterstützter Bewertungsanalysen, in H.-D. Klingemann (ed.), *Computerunterstützte Inhaltsanalyse in der empirischen Sozialforschung*, pp. 35–53. Frankfurt: Campus.

Gans, H. J. 1985. American popular culture and high culture in a changing class structure, in H. H. Balfe and M. J. Wyszomirski (eds.), *Art, Ideology and Politics*, pp. 40–57. New York: Praeger.

Gartman, D. 1991. Culture as class symbolization or mass reification? A critique of Bourdieu's *Distinction*. *American Journal of Sociology*, **97**, 421–447.

Gerpott, T., Domsch, M. and R. T. Keller. 1988. Career orientations in different countries and companies: an empirical investigation of West German, British and US industrial R & D professionals. *Journal of Management Studies*, **25**, 440–462.

Giegler, H. 1992. Zur computergestützten Analyse sozialwissenschaftlicher Textdaten. Quantitative und qualitative Strategien, in J. Hoffmeyer-Zlotnik (ed.), *Analyse qualitativer sozialwissenschaftlicher Daten*, pp. 335–388. Opladen: Westdeutscher Verlag.

Giele, J. Z. 1989. Gender and sex roles, in N. J. Smelser (ed.), *Handbook of Sociology*, pp. 291–326. Newbury Park: Sage.

Gifi, A. 1981. *Nonlinear Multivariate Analysis*. Leiden: Department of Data Theory.

Gifi, A. 1985. *PRINCALS*. Research Report UG-85-03. Leiden: Department of Data Theory.

Gifi, A. 1990. *Nonlinear Multivariate Analysis*. Chichester: Wiley.

Gilula, Z. 1986. Grouping and association in contingency tables: an exploratory canonical correlation approach. *Journal of American Statistical Association*, **81**, 773–779.

Gilula, Z. and S. J. Haberman. 1986. Canonical analysis of contingency tables by maximum likelihood. *Journal of the American Statistical Association*, **81**, 780–788.

Gilula, Z. and S. J. Haberman. 1988. The analysis of multivariate contingency tables by restricted canonical and restricted association models. *Journal of the American Statistical Association*, **83**, 760–771.

Gnanadesikan, R., Kettenring, J. R. and J. M. Landwehr. 1977. Interpreting and assessing the results of cluster analyses. *Bulletin of the International Statistical Institute*, **47**, 451–463.

Goldner, F. H. and R. R. Ritti. 1967. Professionalization as career immobility. *American Journal of Sociology*, **72**, 489–502.

Golub, G. H. and H. Underwood. 1970. Stationary values of the ratio of quadratic forms subject to linear constraints. *Zeitschrift für Angewandte Mathematik und Physik*, **21**, 318–326.

Goodman, L. A. 1969. How to ransack social mobility tables and other kinds of cross-classification tables. *American Journal of Sociology*, **75**, 1–40.

Goodman, L. A. 1974. Exploratory latent structure analysis using both identifiable and unidentifiable models. *Biometrika*, **61**, 215–231.

Goodman, L. A. 1979. Simple models for the analysis of association in cross-classifications having ordered categories. *Journal of the American Statistical Association*, **74**, 755–768.

Goodman, L. A. 1981a. Criteria for determining whether certain categories in a cross-classification table should be combined with special reference to occupational categories in an occupational mobility table. *American Journal of Sociology*, **87**, 612–650.

Goodman, L. A. 1981b. Association models and canonical correlation in the analysis of cross-classifications having ordered categories. *Journal of the American Statistical Association*, **76**, 320–334.

Goodman, L. A. 1985. The analysis of cross-classified data having ordered and/or unordered categories: association models, correlation models and asymmetry models for contingency tables with or without missing entries. *The Annals of Statistics*, **13**, 10–69.

Goodman, L. A. 1986. Some useful extensions to the usual correspondence analysis approach and the usual loglinear approach in the analysis of contingency tables (with comments). *International Statistical Review*, **54**, 243–309.

Goodman, L. A. 1987. New methods for analyzing the intrinsic character of qualitative variables using cross-classified data. *American Journal of Sociology*, **93**, 529–583.

Goodman, L. A. 1991. Measures, models and graphical displays in the analysis of cross-classified data (with discussion). *Journal of the American Statistical Association*, **86**, 1085–1137.

Goodman, L. A. and W. H. Kruskal. 1979. *Measures of Association for Cross Classifications*. Heidelberg: Springer.

Govaert, G. 1984. Classification simultanée de tableaux binaires, in E. Diday, M. Jambu, L. Lebart, J. Pagès and R. Tomassone (eds.), *Data Analysis and Informatics, III*, pp. 223–236. Amsterdam: North-Holland.

Gower, J. C. 1975. Generalized procrustes analysis. *Psychometrika*, **40**, 33–51.

Gower, C. and P. G. N. Digby. 1981. Expressing complex relationships in two dimensions, in V. Barnett (ed.), *Interpreting Multivariate Data*, pp. 83–118. Chichester: Wiley.

Green, M. 1989. In discussion of van der Heijden, P.G.M., de Falguerolles, A., and de Leeuw, J.: A combined approach to contingency table analysis using correspondence analysis and loglinear analysis. *Applied Statistics*, **38**, 249–292.

Green, P. E., F. J. Carmone, and L. B. Fox. 1969. Television programme similarities: an application of subjective clustering. *Journal of the Marketing Research Society*, **11**, 70–90.

Green, P. E. and V. Srinivasan. 1978. Conjoint analysis in consumer research: Issues and outlook. *Journal of Consumer Research*, **5**, 103–123.

Green, P. E. and V. Srinivasan. 1990. Conjoint analysis in marketing: New developments with implications for research and practice. *Journal of Marketing*, **54**, 3–19.

Green, P. E., D. S. Tull, and G. Albaum. 1988. *Research for Marketing Decisions*, fifth edition. Englewood Cliffs: Prentice-Hall International Editions.

Greenacre, M. J. 1981. Practical correspondence analysis, in V. Barnett (ed.), *Interpreting Multivariate Data*, pp. 119–146. Chichester: Wiley.

Greenacre, M. J. 1984. *Theory and Applications of Correspondence Analysis*. London: Academic Press.

Greenacre, M. J. 1986. SimCA: A Program to Perform Simple Correspondence Analysis. *American Statistician*, **51**, 230–231.

Greenacre, M. J. 1988. Clustering the rows and columns of a contingency table. *Journal of Classification*, **5**, 39–51.

Greenacre, M. J. 1989. The Carroll–Green–Schaffer scaling in correspondence analysis: a theoretical and empirical appraisal. *Journal of Marketing Research*, **26**, 358–365.

Greenacre, M. J. 1990a. Some limitations of multiple correspondence analysis. *Computational Statistics Quarterly*, **3**, 249–256.

Greenacre, M. J. 1990b. *SimCA Version 2*. Irene, South Africa.

Greenacre, M. J. 1991. Interpreting multiple correspondence analysis. *Applied Stochastic Models and Data Analysis*, **7**, 195–210.

Greenacre, M. J. 1993. *Correspondence Analysis in Practice*. London: Academic Press.

Greenacre, M. J. and T. Hastie. 1987. The geometric interpretation of correspondence analysis. *Journal of the American Statistical Association*, **82**, 437–447.

Gutman, J. 1982. A means-end chain model based on consumer categorisation processes. *Journal of Marketing*, **46**, 60–72.

Guttman, L. 1941. The quantification of a class of attributes: a theory and a method of scale construction, in P. Horst (ed.), *The Prediction of Personal Adjustment*. New York: SSRC.

Guttman, L. 1946. An approach for quantifying paired comparisons and rank order. *Annals of Mathematical Statistics*, **17**, 144–163.

Guttman, L. 1959. Metricizing rank-ordered or unordered data for a linear factor analysis. *Sankhya, A*, **21**, 257–268.

Guttman, L. 1984. What is not what in statistics. Statistical inference revisited – 1984. The illogic of statistical inference for cumulative science. *BMS*, **4**, 3–35.

Haberman, S. J. 1978. *Analysis of Qualitative Data*. (Vol. 1). Chicago: Academic Press.

Haberman, S. J. (1979). *Analysis of Qualitative Data* (Vol. 2). New York: Academic Press.

Hack, L. and I. Hack. 1985. *Die Wirklichkeit, die Wissen schafft*. Frankfurt/ Main: Campus.

Hayashi C., Hayashi, F., Kuroda, Y., Lebart, L. and T. Suzuki. 1987. Comparative study of quality of life and multidimensional data analysis: Japan, France and Hawaii, in E. Diday, L. Lebart, J. Pagès and R. Tomassone (eds), *Data Analysis and Informatics*, **4**, pp. 549–562. Amsterdam: North-Holland.

Hayashi, C., Sasaki, M. and T. Suzuki. 1992. *Data Analysis for Comparative Social Research, International Perspectives*. Amsterdam: North-Holland.

Heiser, W. J. 1981. *Unfolding Analysis of Proximity Data*. Doctoral dissertation, Leiden: University of Leiden.

Heiser, W. J. 1986. Undesired nonlinearities in nonlinear multivariate analysis, in E. Diday *et al.* (eds), *Data Analysis and Informatics*, Vol. IV, pp. 455–469. Amsterdam: North-Holland.

Hill, M. O. 1974. Correspondence analysis: a neglected multivariate method. *Applied Statistics*, **23**(3), 340–354.

Hirschfeld, H. O. 1935. A connection between correlation and contingency. *Proceedings of the Cambridge Philosophical Society*, **31**, 520–524.

Holbrook, M. B. 1986. Aims, concepts, and methods for the representation of individual differences in esthetic responses to design features. *Journal of Consumer Research*, **13**, 337–47.

Hradil, St. 1987. *Sozialstrukturanalyse in einer fortgeschrittenen Gesellschaft. Von Klassen und Schichten zu Lagen und Milieus*. Opladen: Leske und Budrich.

Hubert, L. and Ph. Arabie. 1985. Comparing partitions. *Journal of Classification*, **2**, 193–218.

Hutton, St., and P. Lawrence. 1981. *German Engineers: The Anatomy of a Profession*. Oxford: Clarendon Press.

Ihm, P. and H. van Groenewoud. 1984. Correspondence analysis and Gaussian ordination. *Compstat Lectures*, **3**, 5–60.

Israëls, A. Z. 1984. Redundancy analysis for qualitative variables. *Psychometrika*, **49**, 331–346.

Israëls, A. Z. 1987. *Eigenvalue Techniques for Qualitative Data*. Doctoral dissertation. Leiden: DSWO Press.

Jambu, M. 1978. *Classification Automatique pour l'Analyse des Données, I-Méthodes et algorithmes*. Paris: Dunod.

Jencks, C. and N. Silver. 1972. *Adhocism*. London: Secker and Warburg.

Kaufmann, F.-X. 1988. Familie und Modernität, in K. Lüscher, F. Schultheis, and M. Wehrspann (eds), *Die 'postmoderne' Familie. Familiale Strategien und Familienpolitik in einer Übergangszeit*. München.

Klein, H. 1991. *INTEXT/ PC 2.2 Manual*. Lengerich, Germany.

Klein, H. 1992. Validity problems and their solutions in computer-aided content. Analysis with INTEXT/PC and other new features, in F. Faulbaum, R. Harx and K.-H. Jöckel (eds), *Advances in Statistical Software 3*, pp. 483–488. Stuttgart: Fischer.

Kossbiel, H., Bamme, A. and B. Martens. 1987. *Ingenieure und Naturwissenschaftler in der industriellen Forschung und Entwicklung*. Frankfurt: Campus.

Krais, B. 1993. Geschlechterverhältnis und symbolische Gewalt, in G. Gebauer and C. Wulf (eds), *Praxis und Ästhetik. Neue Perspektiven im Denken Pierre Bourdieus*, pp. 208–250. Frankfurt: Suhrkamp.

Kruskal, J. B. 1964. Nonmetric multidimensional scaling: a numerical method. *Psychometrika*, **29**, 115–129.

Kuylen, T. and Th. M. M. Verhallen. 1988. *The natural grouping of banks: A new methodology for positioning research*. Research International Nederland: Rotterdam.

Lakoff, G. 1986. *Women, Fire, and Dangerous Thinks: What Categories Reveal about the Mind*. Chicago: The University of Chicago Press.

Lancaster, H. O. 1958. The structure of bivariate distributions. *Annals of Mathematical Statistics*, **29**, 719–736.

Lancaster, K. J. 1966. A new approach to consumer theory. *Journal of Political Economy*, **74**, 132–157.

Lash, S. 1988. Discourse or figure. Postmodernism as a 'regime of signification'. *Theory, Culture and Society*, **7**, 311–336.

Lash, S. 1990. *Sociology of Postmodernism*. London: Allen & Unwin.

Laumann, E. O. and J. S. House. 1970. Living room styles and social attributes: the patterning of material artefacts in a modern urban community, in E. O. Laumann, P. M. Siegel and R. W. Hodge (eds) *The Logic of Social Hierarchies*, pp. 189–204. Chicago: Markham Publishing Company.

Lebart, L. 1976. The significance of eigenvalues issued from correspondence analysis of contingency tables, in J. Gordesch and P. Naeve (eds), *Proceedings COMPSTAT 1976*. Vienna: Physika Verlag.

Lebart, L. 1986. Qui pense quoi? Evolution et structure des opinions en France de 1978 à 1984. *Consommation Revue de Socio-Economie*, **4**, 3–22.

Lebart, L. 1987. Conditions de vie et aspirations des Français. Evolution et structure des opinions de 1978 à 1984. *Futuribles*, **1**, 25–56.

Lebart, L. 1988. Contribution of classification to the processing of longitudinal socio-economic surveys, in H. Bock (ed.), *Classification and Related Methods of Data Analysis*, pp. 113–120. Amsterdam: North Holland.

Lebart, L. and J.-P. Fénélon. 1971. *Statistique et Informatique Appliquées*. Paris: Dunod.

Lebart, L., Morineau A. and N. Tabard. 1977. *Techniques de la description statistique*. Paris: Dunod.

Lebart L. and Y. Houzel. 1980. Le système d'enquête sur les aspirations des Français. *Consommation Revue de Socio-Economie*, **1**, 3–25.

Lebart, L., Morineau, A. and K. M. Warwick. 1984. *Multivariate Descriptive Statistical Analysis: Correspondence Analysis and Related Techniques for Large Matrices*. New York: Wiley.

Lebart L., Morineau A. and T. Lambert. 1987. *SPAD.N, Système Portable pour l'Analyse des Données*, CISIA: Paris. (Updated version of this software available at CISIA).

Lebart L., A. Salem and E. Berry. 1991. Recent development in the statistical processing of textual data, in *Applied Stochastic Models and Data Analysis*, **7**, pp. 47–62. New York: Wiley.

Le Guen, M. and C. Jaffre. 1988. *La conjonction analyse de données et statistique inférentielle pour conduire à une meilleure perception visuelle*. Orléans: University of Orléans.

Le Roux, B. and H. Rouanet. 1984. L'Analyse multidimensionnelle des données structurées. *Mathématique et Sciences humaines*, **85**.

Lord, F. M. 1958. Some relations between Guttman's principal components of scale analysis and other psychometric theory. *Psychometrika*, **23**, 291–296.

Lüdtke, H. 1989. *Expressive Ungleichheit*. Opladen: Leske und Budrich.

Madsen, T. (ed.). 1988. *Multivariate Archaeology. Numerical Approaches in Scandinavian Archaeology*. Aarhus: Aarhus University Press.

Markus, M. Th. and R. A. Visser. 1992. Applying the bootstrap to generate confidence regions in multiple correspondence analysis; a Monte Carlo Study, in K.-H. Jöckel *et al.* (eds), *Bootstrapping and Related Techniques*. Berlin: Springer.

Martens, B. 1991. *Explorative Analysen zeitlicher Verläufe – Berufliche Entwicklungen von Akademikergruppen*. Berlin: Duncker und Humblot.

McQuitty, L. L. 1966. Single and multiple classification by reciprocal pairs and rank order type. *Educational Psychology Measurements*, **26**, 253–265.

Media Daten 1991. 5/91, Mainz.

Meulman, J. J. 1982. *Homogeneity Analysis of Incomplete Data*. Leiden: DSWO Press.

Meulman, J. J. 1986. *A Distance Approach to Nonlinear Multivariate Analysis*. Doctoral dissertation. Leiden: DSWO Press.

Meulman, J. J. 1992. The integration of multidimensional scaling and multivariate analysis with optimal transformations. *Psychometrika*, **57**, 539–565.

Mirkin, B. G. 1990. A sequential fitting procedure for linear data analysis models. *Journal of Classification*, **7**, 167–195.

Mirkin, B. G. 1992. *Correspondence-wise clustering for contingency tables* (submitted for publication).

Montgomery, D. B. and D. R. Wittink. 1980. The predictive validity of conjoint analysis for alternative aggregation schemes, in D. B. Montgomery and D. R. Wittink (eds), *Market Measurement and Analysis*, pp. 298–309. Cambridge, Mass: Marketing Science Institute.

Mooijaart, A. 1982. Latent structure analysis for categorical variables, in K. G. Jöoeskog and H. Wold (eds), *Systems Under Indirect Observation*, 1–18. Amsterdam: North Holland.

Moore, W. E. 1968. Social Change, in D. E. Sills (ed.), *International Encyclopedia of the Social Sciences, vol. 14.*, pp. 365–375. New York: Macmillan/Free Press.

Morineau A., and L. Lebart. 1986. Specific clustering algorithms for large data sets and implementation in SPAD software, in W. Gaul and M. Schader (eds), *Classification as a Tool of Research*, pp. 321–330. Amsterdam: North-Holland.

Müller-Schneider, Th. 1993. *Wandel der Milieustruktur. Eine Untersuchung der Bundesrepublik Deutschland anhand von Umfragedaten 1953–1987*. Wiesbaden: Deutscher Universitäts-Verlag.

Müller-Schneider, Th. 1994. *Schichten und Erlebnismilieus. Der Wandel der Milieustruktur in der Bundesrepublik Deutschland*. Wiesbaden: Deutscher Universitäts-Verlag.

Myers, J. H. and A. D. Shocker. 1981. The nature of product-related attributes, in J. N. Sheth (ed.), *Research in Marketing*, **5**, pp. 211–236. Greenwich, CONN: JAI Press.

Netherlands Central Bureau of Statistics. 1981. *CORAN, version 2.0*, Voorburg: Department for Automation/Statistical Informatics Unit.

Nishisato, S. 1980. *Analysis of Categorical Data: Dual Scaling and its Applications*. Toronto: University of Toronto Press.

Novak, T. P. and D. L. Hoffman. 1990. Residual scaling: an alternative to correspondence analysis for the graphical representation of residuals from log-linear models. *Multivariate Behavioral Research*, **25**, 351–370.

Parsons, T. 1971. *The System of Modern Societies*. Englewood Cliffs, New Jersey.

Parsons, T. and R. F. Bales (eds). 1956. *Family, Socialization, and Interaction Process*. London: Routledge and Kegan Paul.

Pearson, K. 1901. On lines and planes of closest fit to systems of points in space. *Philosophical Magazine*, **2**, 559–572.

Poiesz, Th. B. C. 1982. *Deriving similarity-scores for large stimulus sets by the method of progressive categorization*. Working paper, Department of Psychology, Tilburg University, Tilburg.

Ramsay, J. O. 1982. Some statistical approaches to multidimensional scaling. *Journal of the Royal Statistical Society-A*, **145**, 285–312.

Rao, C. R. 1964. The use and interpretation of principal component analysis in applied research. *Sankhya, Series A*, **26**, 329–358.

Realin, J. A. 1986. *The Clash of Cultures*. Boston, Mass.: Harvard Business School Press.

Reynolds, T. J. and Gutman, J. 1984. Advertising is image management. *Journal of Advertising Research*, 24, 27–36.

Riley, M. W., Foner, A. and J. Waring. 1988. Sociology of age, in N. J. Smelser (ed.), *Handbook of Sociology*, pp. 243–290. Newbury Park: Sage.

Rosenbaum, J. E. 1984. *Career Mobility in a Corporate Hierarchy*. Orlando: Academic Press.

Rouanet, H. (in collaboration with B. Everitt). 1988. *Comparative Study of Statistical Methods Applied to Social Science Data*. Paris: ESRC-CNRS.

Rouanet, H. and B. Le Roux. 1993. *Analyse des données multi-dimensionnelles – Statistique en Sciences Humaines*. Paris: Dunod.

Saporta, G. and G. Hatabian. 1986. Régions de confiance en analyse factorielle, in E. Diday *et al.* (eds), *Data Analysis and Informatics*, vol. IV, pp. 499–508. Amsterdam: North-Holland.

SAS Institute Inc. (eds), 1985. *SAS/IML User's Guide*. Cary, NC: SAS Institute Inc.

Savage, L. J. 1954. *The Foundations of Statistics*. New York: Wiley.

Scheuch, Erwin K. and J. Blasius. 1991. Nation as a frame. An Application of Correspondence Analysis to Trend Data. Paper presented at the 'German-Japanese-Symposium on Quantitative Social Research', 6–8 May, 1991, Cologne.

Schiltz, M.-A. 1990. A French reanalysis of a British survey: comparative study of statistical methods applied to social science data. *CAMS*, report P. 055, Paris.

Schiltz, M.-A. 1991. Social sciences and statistics, in *Humanities and Computing*, European Science Foundation.

Schmidt, S. (ed.). 1987. *Der Diskurs des radikalen Konstruktivismus*. Frankfurt.

Schriever, B. F. 1983. Scaling of order dependent categorical variables with correspondence analysis. *International Statistical Review*, 51, 225–238.

Schriever, B. F. 1986. *Order Dependence. CWI Tract 20*. Amsterdam: Mathematisch Centrum.

Schulz, M. and K. P. Strohmeier. 1985. Familienkarriere und Berufskarriere, in H.-W. Franz (ed.), *22. Deutscher Soziologentag 1984*, pp. 167–171. Opladen: Westdeutscher Verlag.

Schulze, G. 1992. *Die Erlebnisgesellschaft. Kultursoziologie der Gegenwart*. Frankfurt/New York: Campus.

Shahim, V. and Greenacre, M. J. 1989. A correspondence analysis approach to perceptual maps and ideal points. Proceedings of Sawtooth Software Conference, Sun Valley, Idaho, June 1989.

Siciliano, R., Mooijaart, A. and P. G. M. van der Heijden. 1990. Non-symmetric correspondence analysis by maximum likelihood. Leiden: Department of Psychometrics (PRM 05-90).

Sikkel, D. 1981a. The relationship between canonical correlations and correspondence analysis. Voorburg: Netherlands Central Bureau of Statistics.

Sikkel, D. 1981b. Grafische weergeve van tabellen met correspondentie-analyse, in A. E. Bronner, M. T. G. Meulenberg, A. P. N. Nauta, A. J. Olivier, W. F. van Raaij and J. J. M. van Tulder (eds), *Marktonderzoek en consumentengedrag, Jaarboek van de Nederlandse Vereninging van Marktonderzoek, 1981*. Amsterdam: Nederlandse Vereninging van Marktonderzoek.

Sikkel, D. and J. Bethlehem. 1981. Interpretatie van correspondentie analyse-resultaten of wat stellen die plaatjes nu eigenlijk voor. Voorburg: Netherlands Central Bureau of Statistics.

Smets, G. 1986. *Vormleer: De paradox van de vorm*. Amsterdam: Uitgeverij Bert Bakker.

Smith, S. M. 1988. *PC-MDS*. Institute of Business MGT. Brigham Young University Provo, Utah.

Spearman, C. 1904. 'General intelligence' objectively determined and measured. *American Journal of Psychology*, **15**, 201–293.

Spearman, C. 1910. Correlation calculated from faulty data. *British Journal of Psychology*, **3**, 271–295.

SPSS. 1990. *Categories*. Chicago: SPSS Inc.

Srole, L., Langner, T. S., Michael, S. T., Opler, M. K. & Rennie, T. A. C. 1962. *Mental Health in the Metropolis: The Midtown Manhattan Study*. New York: McGraw Hill.

Steenkamp, J. E. B. M. 1989. *Product Quality: An Investigation into the Concept and How it is perceived by Consumers*. Van Gorcum: Maastricht.

Strauss, A. L. 1987. *Qualitative Analysis for Social Scientists*. Cambridge.

Takane, Y. 1978. A maximum likelihood method for nonmetric multi-dimensional scaling: I. The case in which all empirical pairwise orderings are independent – evaluations. *Japanese Psychological Research*, **20**, 105–114.

Takane, Y. 1980. Analysis of categorizing behavior by a quantification method. *Behaviormetrika*, **8**, 75–86.

Takane, Y. 1981. Multidimensional successive categories scaling: A maximum likelihood method. *Psychometrika*, **46**, 9–28.

Takane, Y. 1987. Analysis of contingency tables by ideal point discriminant analysis. *Psychometrika*, **52**, 493–513.

Takane, Y. 1989. Ideal point discriminant analysis and ordered response categories. *Behaviormetrika*, **26**, 31–46.

Takane, Y., Bozdogan, H. and T. Shibayama. 1987. Ideal point discriminant analysis. *Psychometrika*, **52**, 371–392.

Takane, Y. and T. Shibayama. 1991. Principal component analysis with external information on both subjects and variables. *Psychometrika*, **56**, 97–120.

Takane, Y., de Leeuw, J. and H. Kiers. 1991a. Component analysis with different sets of constraints on different dimensions. Unpublished manuscript. McGill University.

Takane, Y., Yanai, H. and S. Mayekawa. 1991b. Relationships among several methods of linearly constrained correspondence analysis. *Psychometrika*, **56**, 667–684.

Tenenhaus M. and F. W. Young. 1985. An analysis and synthesis of multiple correspondence analysis, optimal scaling, dual scaling, and other methods for quantifying categorical multivariate data. *Psychometrika*, **50**, 97–104.

Ter Braak, C. J. F. 1986. Canonical correspondence analysis: a new eigenvector technique for multivariate direct gradient analysis. *Ecology*, **67**, 1167–1179.

Ter Braak, C. F. 1988. Partial canonical correspondence analysis, in H. H. Bock (ed.), *Classification and related methods of data analysis*, pp. 551–558. Amsterdam: North-Holland.

Ter Braak, C. J. F. 1992. Permutation versus bootstrap significance tests in multiple regression and ANOVA, in K.-H. Jöckel *et al.* (eds), *Bootstrapping and Related Techniques*, pp. 79–86. Berlin: Springer.

Thiessen, V. and H. Rohlinger. 1988. Die Verteilung von Aufgaben und Pflichten im ehelichen Haushalt. *Kölner Zeitschrift für Soziologie und Sozialpsychologie*, **40**, 640–658.

Tuma, N. B. and M. T. Hannan. 1984. *Social Dynamics*. Orlando: Academic Press.

Urban, G. L. and J. R. Hauser. 1980. *Design and Marketing of New Products*. London: Prentice Hall International.

Valette-Florence, P. and B. Rapacchi. 1989. Value analysis by linear representation and appropriate positioning: An extension of laddering methodology, in G. J. Avlonitis and N. K. Papavasilion (eds), *Marketing Thought and Practice in the 1990s*, vol. 1, pp. 969–989. Athens: The Athens School of Economics and Business Science.

Van Buuren, S. and W. J. Heiser. 1989. Clustering N objects into k groups under optimal scaling of variables. *Psychometrika*, **54**, 699–706.

Van de Geer, J. P. 1989. *Analysis of Categorical Data* (in Dutch). Deventer: van Loghum Slaterus.

Van de Geer, J. P. and J. J. Meulman. 1985. PRIMALS. Research Report UG-85-02. Leiden: Department of Data Theory.

Van der Burg, E. and J. De Leeuw. 1988. Use of the multinomial jackknife and bootstrap in generalized nonlinear canonical correlation analysis. *Applied Stochastic Models and Data Analysis*, **4**, 159–172.

Van der Burg, E. De Leeuw J. and R. Vendegaal. 1988. Homogeneity analysis with *k* sets of variables: an alternating least squares method with optimal scaling features. *Psychometrika*, **53**, 177–197.

Van der Heijden, P. G. M. 1987. *Correspondence Analysis of Longitudinal Categorical Data*. Leiden: DSWO Press.

Van der Heijden, P. G. M. 1992, in press. Three approaches to study the departure from quasi-independence. *Statistica Applicata, Italian Journal of Statistics*.

Van der Heijden, P. G. M. and J. de Leeuw. 1985. Correspondence analysis used complementary to loglinear analysis. *Psychometrika*, **50**, 429–447.

Van der Heijden, P. G. M. and K. Worsley. 1988. Comment on correspondence analysis used complementary to loglinear analysis. *Psychometrika*, **53**, 287–291.